Extreme Ocean Waves

Efim Pelinovsky · Christian Kharif
Editors

Extreme Ocean Waves

Second Edition

Springer

Editors
Efim Pelinovsky
Institute of Applied Physics
Nizhny Novgorod State Technical University
Higher School of Economics
Nizhny Novgorod
Russia

Christian Kharif
Ecole Centrale Marseille (ECM)
Institut de Recherche sur les Phénomènes Hors Equilibre (IRPHE)
Marseille
France

ISBN 978-3-319-21574-7 ISBN 978-3-319-21575-4 (eBook)
DOI 10.1007/978-3-319-21575-4

Library of Congress Control Number: 2015944750

Springer Cham Heidelberg New York Dordrecht London

Printed on acid-free paper

Springer International Publishing AG Switzerland is part of Springer Science+Business Media (www.springer.com)

Preface

Reports of extreme wave events in seas and oceans come about almost each week. Extreme water waves investigated in this book involve mainly rogue waves, but tsunami waves and storm waves are also considered. Several catalogues of extreme events like rogue waves observed in the World Ocean have been recently published.[1] For instance, during 2006–2010, 106 events can be classified as anomalous high short-lived waves. They occurred in deep and shallow waters, and also on the coast. Perhaps, one of the last events occurred on February 17, 2014 with ship "Marco Polo" on its route from the Azores to its home port in Tilbury, England, and it was carrying 735 passengers and 349 crew. A large wave hit the side of the ship between 1 and 2 p.m. local time at adverse sea conditions and killed one passenger and injured the second one.

There are a number of physical mechanisms that focus the water wave energy into a small area and produce the occurrence of extreme waves called freak or rogue waves. These events may be due to wave instability (modulational or Benjamin-Feir instability), chaotic behavior, dispersion (frequency modulation), refraction (presence of variable currents or bottom topography), soliton interactions, crossing seas, wind–wave interaction, etc. These giant waves are a real danger to ships and platforms, causing accidents with human losses. Tsunami waves are generally due to seismic motion of the sea bottom and less frequently due to underwater or subaerial landslides. Herein, the emphasis is put on tsunami waves generated by

[1]Liu, P.C. A chronology of freaque wave encounters. Geofizica, 2007, vol. 24, 57–70.
Didenkulova, I.I., Slunyaev, A.V., Pelinovsky, E.N., and Kharif, Ch. Freak waves in 2005. Natural Hazards Earth Syst. Sci. 2006, vol. 6, 1007–1015.
Nikolkina, I. and Didenkulova, I. Rogue waves in 2006–2010. Natural Hazards Earth Syst. Sci. 2011. vol. 11, 2913–2924.
Baschek, B., and Imai, J. 2011. Rogue wave observations off the US West Coast. Oceanography. 2011, vol. 24, 158–165.
Nikolkina, I. and Didenkulova, I. Catalogue of rogue waves reported in media in 2006–2010. Natural Hazards. 2012. vol. 61, 989–1006.

subaerial collapse of granular media. These waves may cause important damages in coastal areas and loss of life.

Extreme waves have been intensively studied during the past decades, and the European Geophysical Union organizes each year a special section "Extreme Waves." This book contains invited papers written mainly on the basis of works presented during the General Assembly of the European Geosciences Union in Vienna, plus five new invited papers which concern more recent researches on the present subject. In the paper "Rogue Waves in Higher Order Nonlinear Schrödinger Models" by C.M. Schober and A. Calini, it is demonstrated that a chaotic sea state appears to be an important mechanism for both generation and increased likelihood of rogue waves. In the paper "Freak-Waves: Compact Equation Versus Fully Nonlinear One" A.I. Dyachenko, D.I. Kachulin, and V.E. Zakharov derive an approximate equation which is compared to the fully nonlinear system. The compact equation is shown to describe correctly strongly nonlinear phenomena such as rogue waves due to modulational instability. L. Fernandez, M. Onorato, J. Monbaliu, and A. Toffoli in their paper "Occurrence of Extreme Waves in Finite Water Depth" discuss laboratory experiments in a large wave basin, numerical simulations with a truncated form of the potential Euler equations, and field experiments at the Lake George experimental site (Australia) to assess the role of third-order nonlinearity, namely the modulational instability, on water wave statistics. Existence of various shapes of rogue waves in shallow water is discussed in the paper "Modeling of Rogue Wave Shapes in Shallow Water" by T. Talipova, C. Kharif, and J.P. Giovanangeli. They pointed out that variable-polarity shape of a rogue wave is more probable than only one crest or one trough. The occurrence of extreme waves in shallow water is investigated in the paper "Non-Gaussian Properties of Shallow Water Waves in Crossing Seas" by A. Toffoli, M. Onorato, A.R. Osborne, and J. Monbaliu. They show that the interaction of two crossing wave trains generates steep and high amplitude peaks, thus enhancing the deviation of the surface elevation from the Gaussian statistics. The relation between observations and freak wave theories is examined in the paper "Searching for Factors that Limit Observed Extreme Maximum Wave Height Distributions in the North Sea" by G. Burgers, F. Koek, H. de Vries, and M. Stam. Observations indicate that steepness is a limiting factor for extreme wave height and at shallow water locations, extreme waves are not more frequently observed than at deep water locations. Average wave conditions, their variations, and extreme wave storms in the Baltic Sea are studied, based on long-term time series in the paper "Extremes and Decadal Variations in the Baltic Sea Wave Conditions" by T. Soomere. Significant wave heights, H_S, more than 4 m in the Baltic Sea occur with a probability of about 1 % and extreme wave conditions with $H_S > 7$ m approximately twice in a decade. The overall recorded maximum significant wave height is 8.2 m. The possibility of appearance of freak waves on a beach is analyzed in the paper "Runup of Long Irregular Waves on Plane Beach" by I. Didenkulova, E. Pelinovsky, and A. Sergeeva. It is shown that the average runup height of waves with a wide spectrum is higher than that of waves with a narrow spectrum. In the paper "Numerical Study for Run-Up of Breaking Waves of Different Polarities on a Sloping Beach" by A. Rodin, I. Didenkulova, and

E. Pelinovsky the transformation and run-up of breaking solitary waves propagating on a sloping bottom is investigated numerically within the framework of the nonlinear shallow water equations (St-Venant equations). For high wave amplitude they studied how the wave transforms into a bore (shock wave). In the paper "Tsunami Waves Generated by Cliff Collapse: Comparison Between Experiments and Triphasic Simulations" S. Viroulet, A. Sauret, O. Kimmoun, and C. Kharif investigate tsunami waves due to subaerial or submarine landslides both experimentally and numerically. They discuss recent experimental results on granular collapse in water and the influence of the physical parameters on the amplitude of the tsunami waves; such waves of landslide origin cannot be predicted. Waves of huge amplitudes can appear in the deepest layers of the ocean and the possible shapes of such waves in two-layer fluid is described in the paper "An Analytical Model of Large Amplitude Internal Solitary Waves" by N.I. Makarenko and J.L. Maltseva. Special analysis of nonlinear resonances between water waves is given in the paper "Symbolic Computation for Nonlinear Wave Resonances" by E. Tobisch (Kartashova), C. Raab, Ch. Feurer, G. Mayrhofer, and W. Schreiner. They argue the important role of nonlinear resonances in the wave dynamics that can be used to simplify the governing equations.

The book is written for specialists in the fields of fluid mechanics, applied mathematics, nonlinear physics, physical oceanography, and geophysics, and also for students learning these subjects.

Efim Pelinovsky
Christian Kharif

Contents

Contributors

Gerrit Burgers Rijkswaterstaat Water, Verkeer en Leefomgeving, Lelystad, The Netherlands

Annalisa Calini Department of Mathematics, College of Charleston, Charleston, SC, USA

Hans de Vries Royal Netherlands Meteorological Institute, De Bilt, The Netherlands

Ira Didenkulova Nizhny Novgorod State Technical University n.a. R.E. Alekseev, Nizhny Novgorod, Russia; Institute of Applied Physics, Nizhny Novgorod, Russia; Marine Systems Institute, Tallinn University of Technology, Tallinn, Estonia

A.I. Dyachenko Landau Institute for Theoretical Physics RAS, Chernogolovka, Moscow Region, Russia; Novosibirsk State University, Novosibirsk, Russia

Leandro Fernandez Department of Civil Engineering, KU Leuven, Heverlee, Belgium

Ch. Feurer Research Institute for Symbolic Computation (RISC), Johannes Kepler University Linz, Linz, Austria

J.P. Giovanangeli Aix-Marseille Université, CNRS, Marseille, France

D.I. Kachulin Novosibirsk State University, Novosibirsk, Russia

C. Kharif Aix-Marseille Université, CNRS, Ecole Centrale Marseille, Marseille, France

O. Kimmoun Aix-Marseille Université, CNRS, Ecole Centrale Marseille, Marseille, France

Frits Koek Royal Netherlands Meteorological Institute, De Bilt, The Netherlands

Nikolay I. Makarenko Lavrentyev Institute of Hydrodynamics, Novosibirsk, Russia

Janna L. Maltseva Lavrentyev Institute of Hydrodynamics, Novosibirsk, Russia

G. Mayrhofer Research Institute for Symbolic Computation (RISC), Johannes Kepler University Linz, Linz, Austria

Jaak Monbaliu Swinburne University of Technology, Melbourne, Australia; Department of Civil Engineering, KU Leuven, Heverlee, Belgium

Miguel Onorato Universitá di Torino, Turin, Italy

A.R. Osborne Universitá di Torino, Turin, Italy

Efim Pelinovsky Nizhny Novgorod State Technical University n.a. R.E. Alekseev, Higher School of Economics, Nizhny Novgorod, Russia; Institute of Applied Physics, Nizhny Novgorod, Russia; Special Research Bureau for Automation of Marine Researches, Yuzhno-Sakhalinsk, Russia

C. Raab Radon Institute for Computational and Applied Mathematics (RICAM), Johannes Kepler University Linz, Linz, Austria

Artem Rodin Nizhny Novgorod State Technical University n.a. R.E. Alekseev, Nizhny Novgorod, Russia; Institute of Cybernetics at Tallinn University of Technology, Tallinn, Estonia

A. Sauret Surface du Verre Et Interfaces, UMR 125 CNRS/Saint-Gobain, Aubervilliers, France

Constance M. Schober Department of Mathematics, University of Central Florida, Orlando, FL, USA

W. Schreiner Research Institute for Symbolic Computation (RISC), Johannes Kepler University Linz, Linz, Austria

Anna Sergeeva Nizhny Novgorod State Technical University n.a. R.E. Alekseev, Nizhny Novgorod, Russia; Institute of Applied Physics, Nizhny Novgorod, Russia

Tarmo Soomere Institute of Cybernetics at Tallinn University of Technology, Tallinn, Estonia

Martin Stam Royal Netherlands Meteorological Institute, De Bilt, The Netherlands

T. Talipova Institute of Applied Physics RAS, Nizhny Novgorod, Russia; Nizhny Novgorod Technical State University, Nizhny Novgorod, Russia

E. Tobisch (Kartashova) Institute for Analysis, Johannes Kepler University Linz, Linz, Austria

Alessandro Toffoli Swinburne University of Technology, Swinburne, VIC, Australia; Department of Civil Engineering, KU Leuven, Heverlee, Belgium

S. Viroulet School of Mathematics and Manchester Centre for Nonlinear Dynamics, University of Manchester, Manchester, UK

V.E. Zakharov Novosibirsk State University, Novosibirsk, Russia; Department of Mathematics, University of Arizona, Tucson, AZ, USA; Physical Institute of RAS, Moscow, Russia; Landau Institute for Theoretical Physics, Moscow, Russia

Rogue Waves in Higher Order Nonlinear Schrödinger Models

Constance M. Schober and Annalisa Calini

Abstract We discuss physical and statistical properties of rogue wave generation in deep water from the perspective of the focusing Nonlinear Schrödinger equation and some of its higher order generalizations. Numerical investigations and analytical arguments based on the inverse spectral theory of the underlying integrable model, perturbation analysis, and statistical methods provide a coherent picture of rogue waves associated with nonlinear focusing events. Homoclinic orbits of unstable solutions of the underlying integrable model are certainly candidates for extreme waves, however, for more realistic models such as the modified Dysthe equation two novel features emerge: (a) a chaotic sea state appears to be an important mechanism for both generation and increased likelihood of rogue waves; (b) the extreme waves intermittently emerging from the chaotic background can be correlated with the homoclinic orbits characterized by maximal coalescence of their spatial modes.

1 Introduction

Among the various mechanisms for wave amplification under different physical conditions, the Benjamin–Feir (BF) instability and nonlinear self-focusing are often proposed in relation to rogue wave generation in deep water. In particular, the work of Henderson et al. (1999) suggests that excitation of certain breather-like solutions of the focusing nonlinear Schrödinger (NLS) equation

$$iu_t + u_{xx} + 2|u|^2 u = 0, \tag{1}$$

triggers the formation of rogue waves. Trulsen and Dysthe's analysis (1997b) of the sea state during the famous New Year Wave event recorded at the North Sea Draupner

C.M. Schober (✉)
Department of Mathematics, University of Central Florida, Orlando, FL 32816, USA
e-mail: cschober@mail.ucf.edu; drschober@gmail.com

A. Calini
Department of Mathematics, College of Charleston, Charleston, SC 29424, USA
e-mail: calinia@cofc.edu

E. Pelinovsky and C. Kharif (eds.), *Extreme Ocean Waves*,
DOI 10.1007/978-3-319-21575-4_1

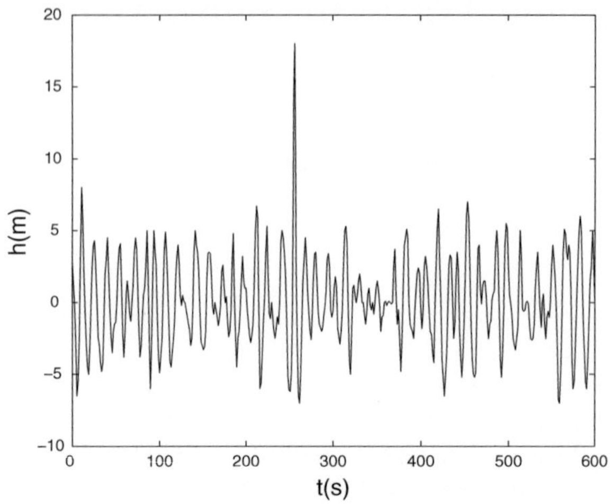

Fig. 1 Time series of the surface elevation of the 1995 New Year Wave Event

Platform in 1995 (see Fig. 1) shows a weakly nonlinear wave train with a relatively narrow frequency bandwidth, thus supporting the use of the focusing NLS equation as a basic model for studying rogue waves in deep water.

A stability analysis of solutions of the NLS equation shows that low frequency modes may become unstable and that the number of unstable modes increases with the amplitude of the carrier wave. Homoclinic orbits of unstable NLS solutions, including those of an unstable Stokes wave, exhibit many of the properties observed in rogue waves (Osborne et al. 2000; Calini and Schober 2002; Karjanto 2006). However, generic homoclinic orbits of unstable solutions of the NLS equation are unlikely to be physically observable in more realistic models of deep water wave dynamics (for example, those described by higher order generalizations of the NLS equation). It is thus important to develop criteria for rogue wave formation for general sea states, to investigate whether proximity to unstable solutions of the integrable model can be correlated to rogue wave generation, and to determine the robustness of the homoclinic orbits under physically meaningful perturbations.

A more realistic description of deep water wave dynamics is provided by the Modified Dysthe (MD) equation,

$$iu_t + u_{xx} + 2|u|^2u + i\epsilon^{1/2}\left(\frac{1}{2}u_{xxx} - 6|u|^2u_x + u^2u_x^* - 2ui\left[H\left(|u|^2\right)\right]_x\right) + \epsilon\,\frac{5}{16}u_{4x} + i\epsilon^{3/2}\,\frac{7}{32}u_{5x} = 0,$$

introduced by Trulsen and Dysthe (1996, 1997a) by retaining higher order terms in the asymptotic expansion of the surface wave displacement. (Here $H(f)$ denotes the Hilbert transform of the function f.)

Laboratory experiments conducted in conjunction with numerical simulations by the first author and her collaborators (Ablowitz et al. 2000, 2001), established that, for higher order generalizations of the NLS equations, the generic long-time dynamics of initial data near an unstable Stokes wave with two unstable modes is chaotic. Further numerical investigations of the MD equation revealed that for a general class of such initial data, high amplitude coherent structures arise intermittently above the chaotic background (Calini and Schober 2002). Remarkably, these emerging coherent structures are structurally similar to maximal homoclinic orbits of the unperturbed Stokes wave with coalesced spatial phases (which are also the homoclinic orbits of maximal amplitude). Such optimally phase modulated homoclinic solutions of the NLS equation appear to be the only homoclinic solutions which persist under perturbations, their persistence being independent of phase selection of the initial conditions (Calini and Schober 2002; Schober 2006). In other words, a chaotic sea state due to proximity to unstable NLS solutions appears to increase the occurrence of rogue waves, enhance focusing effects, and select, among homoclinic solutions, those that are good candidates for modeling physically observable rogue waves.

In order to analyze this phenomenon in more detail, we regard the MD equation as a perturbation of the NLS equation. A combination of tools from the integrable theory of the NLS equation, and a formal extension of the Melnikov method for perturbations of Hamiltonian systems with homoclinic structures are used to address both structural and statistical properties of the observed rogue waves.

In Sect. 2 we review elements of the periodic theory for the integrable NLS equation, the analytical construction of homoclinic solutions (from low-dimensional to maximal homoclinic manifolds) of the unstable Stokes wave, and discuss wave amplification due to phase coalescence, as well as the relation between phase singularities, wave compression, and wave amplification.

In Sect. 3 we study the effects of homoclinic chaos on rogue wave generation, and discuss numerical simulations of the MD equation (8) and its restriction to spatially symmetric wave trains (See e.g., Fig. 5b.). "Noisy" rogue waves emerge intermittently above a chaotic background: we discuss how the likelihood of rogue wave occurrence as well as wave focusing are found to increase in the chaotic regime.

In Sect. 4 we present a formal Melnikov-type calculation aimed at explaining the persistence of optimally phase modulated homoclinic orbits during the perturbed chaotic dynamics. These persisting coherent structures are thus natural candidates for the physically observable rogue waves.

The remaining sections use a statistical approach (in combination with the periodic theory of the integrable NLS), to develop a criterion for rogue wave prediction, and a statistical description of rogue waves associated with homoclinic chaos in both the NLS and MD models.

In Sect. 5, we discuss rogue wave generation for random sea states characterized by the Joint North Sea Wave Project (JONSWAP) power spectrum. The JONSWAP spectrum was introduced to describe developing sea states with ongoing nonlinear wave-wave interactions (Ochi 1998; Bridges and Derks 1999). A spectral quantity, the "splitting distance" between simple periodic points of the Floquet spectrum of an initial condition in a neighborhood of an unstable NLS solution, is proposed as a measurement of the proximity in spectral space to unstable waves and homoclinic data. For regimes in which few (two or three) unstable modes are present, hundreds of realizations of JONSWAP type initial data show that, in both the pure NLS and the MD models, rogue waves develop for small splitting distance, and do not when the splitting distance is large (Islas and Schober 2005).

In the final section, a statistical interpretation of rogue waves in both the NLS and MD equations is provided. Using the third and fourth statistical moments of the wave elevation for sea states characterized by JONSWAP spectra with random phases, we examine dependence of skewness, kurtosis, and likelihood of rogue waves on the proximity to unstable waves and homoclinic data. Extensive numerical studies reveal that wave strength, skeweness, and kurtosis all increase as the spectral splitting distance decreases, thus supporting the claim that modulational instability is not only an important mechanism for rogue wave generation, but also a significant source of non-Gaussianity in the water wave statistics. Finally, consistent with the numerical and analytical studies described in the first part of this article, statistically, the NLS equation appears to under predict, as compared to the MD equation, both the strength and likelihood of rogue waves.

2 Background

As is well known, the nonlinear Schrödinger (NLS) equation is equivalent to the solvability condition of the AKNS system, the pair of first-order linear systems (Zakharov and Shabat 1972):

$$\mathcal{L}^{(x)}\phi = 0, \qquad \mathcal{L}^{(t)}\phi = 0, \tag{2}$$

for a vector-valued function ϕ. The linear operators on the left-hand sides of (2) are

$$\mathcal{L}^{(x)} = \begin{pmatrix} \partial_x + i\lambda & -u \\ u^* & \partial_x - i\lambda \end{pmatrix}, \mathcal{L}^{(t)} = \begin{pmatrix} \partial_t + i(2\lambda^2 - |u|^2) & -2\lambda u - iu_x \\ 2\lambda u^* - iu_x^* & \partial_t - i(2\lambda^2 - |u|^2) \end{pmatrix},$$

and depend on x and t through the NLS potential u and on the *spectral parameter* λ.

The nonlinear spectral decomposition of an NLS initial condition (or in general of an ensemble of JONSWAP initial data) is based on the inverse spectral theory of the NLS equation. For periodic boundary conditions $u(x + L, t) = u(x, t)$, the Floquet spectrum associated with an NLS potential u (i.e., the spectrum of the linear operator $\mathcal{L}^{(x)}$ at u) can be described in terms of the *Floquet discriminant* of u, defined as the trace of the transfer matrix of a fundamental matrix solution Φ of (2) over the interval $[0, L]$ (Ablowitz and Segur 1981):

$$\Delta(u; \lambda) = \text{Trace}(\Phi(x, t; \lambda)^{-1}\Phi(x + L, t; \lambda)).$$

Then, the Floquet spectrum is defined as the region

$$\sigma(u) = \{\lambda \in \mathbb{C} \mid \Delta(u;\lambda) \in \mathbb{R}, -2 \le \Delta \le 2\}.$$

Points of the *continuous spectrum* of u are those for which the eigenvalues of the transfer matrix have unit modulus, and therefore $\Delta(u;\lambda)$ is real and between -2 and 2; in particular, the real line is part of the continuous spectrum. Points of the L-periodic/antiperiodic *discrete spectrum* of u are those for which the eigenvalues of the transfer matrix are ± 1, equivalently $\Delta(u;\lambda) = \pm 2$. Points of the discrete spectrum which are embedded in a continuous band of spectrum have to be critical points for the Floquet discriminant (i.e., $\mathrm{d}\Delta/\mathrm{d}\lambda$ must vanish at such points).

Because the transfer matrix only changes by conjugation when we shift in x or t, Δ is independent of those variables. An important consequence of this observation is that the Floquet discriminant is invariant under the NLS flow, and thus encodes an infinite family of constants of motion (parametrized by λ).

The continuous part of Floquet spectrum of a generic NLS potential consists of the real axis and of complex bands terminating in *simple points* λ_j^s (at which $\Delta = \pm 2$, $\Delta' \neq 0$). The N*-phase potentials* are those characterized by a finite number of bands of continuous spectrum (or a finite number of simple points). Figure 2 shows the spectrum of a typical N-phase potential: complex critical points (usually *double points* of the discrete spectrum for which $\Delta' = 0$ and $\Delta'' \neq 0$), such as the one appearing in the figure, are in general associated with linear instabilities of u and label its homoclinic orbits (Ercolani et al. 1990). Figure 8a shows spectrum of an N-phase potential near the one shown in Fig. 2: the complex double point has split into a pair of simple points; such a potential possesses no linear unstable modes (simple points and real double points are in general associated with neutrally stable modes).

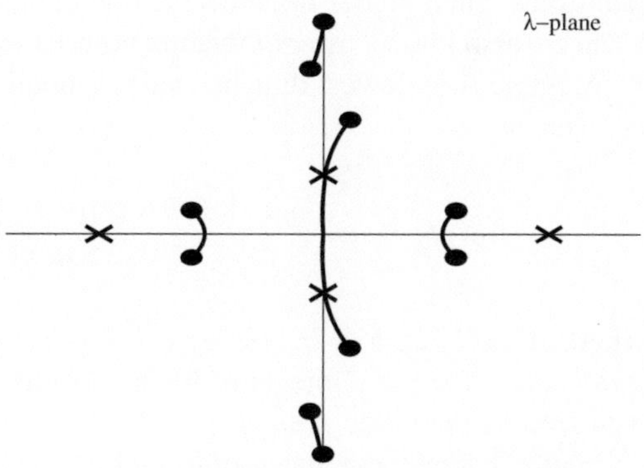

Fig. 2 Spectrum of an unstable N-phase solution. The simple periodic eigenvalues are labeled by *circles* and the double points by *crosses*

2.1 Homoclinic Solutions of the NLS Equation as Candidates for Rogue Waves

Modulationally unstable solutions of the NLS equation (e.g. N-phase solutions whose Floquet spectra have complex double points) have homoclinic orbits that can undergo large amplitude excursions away from their target solution. Such homoclinic orbits can be used as models of rogue waves.

An important (and the simplest) example of unstable NLS solution is the *plane* (or *Stokes*) *wave* potential

$$u_a(x,t) = ae^{2ia^2t}. \tag{3}$$

Elementary Fourier analysis shows that the plane wave is unstable when its amplitude a is sufficiently large: in fact, for $0 < \pi n/L < a$, the solution of the linearized NLS equation about u_a has M linearly unstable modes (UMs) $e^{\sigma_n t + 2\pi n x/L}$ with growth rates σ_n given by

$$\sigma_n^2 = \mu_n^2(\mu_n^2 - 4a^2), \quad \mu_n = 2\pi n/L,$$

where M is the largest integer satisfying $0 < M < aL/\pi$.

One can also check (see e.g. Calini and Schober 2002) that for $0 < \pi n/L < 0$ the Floquet spectrum of the plane wave potential u_a has exactly M complex double points, each "labeling" an associated unstable mode.

Using Bäcklund transformations (McLaughlin and Schober 1992; Matveev and Salle 1991) one can in principle construct the family of homoclinic orbits of an unstable NLS potential. In fact, this method gives explicit formulas for homoclinic orbits of N-phase solutions, although their expressions become rather complicated for $N > 2$. For NLS potentials with several unstable modes, iterated Bäcklund transformations will generate their entire stable and unstable manifolds, comprised of homoclinic orbits of increasing dimension up to the dimension of the invariant manifolds. Such higher-dimensional homoclinic orbits associated with two or more UMs are also known as *combination* homoclinic orbits.

A single (i.e., lowest dimensional) homoclinic orbit of the plane wave potential is given by

$$u(x,t) = ae^{-2ia^2t}\,\frac{1 + 2\cos(px)e^{\sigma_n t + 2i\phi + \rho} + Ae^{2\sigma_n t + 4i\phi + 2\rho}}{1 + 2\cos(px)e^{\sigma_1 t + \rho} + Ae^{2\sigma_n t + 2\rho}}, \tag{4}$$

where $A = 1/\cos^2\phi$, $\sigma_n = \pm p\sqrt{4a^2 - p^2}$, $\phi = \sin^{-1}(p/2a)$, and $p = \mu_n = 2\pi n/L < a$ for some integer n. Each UM has an associated homoclinic orbit characterized by the mode $p = \mu_n$.

Figure 3 shows the space-time plot of the amplitude $|u(x,t)|$ of a homoclinic orbit with one UM, for $a = 0.5$, $L = 2\sqrt{2}$ and $p = 2\pi/L$. As $t \to \pm\infty$, solution (4) limits to the plane wave potential; in fact, the plane wave behavior dominates the dynamics of the homoclinic solution for most of its lifetime. As t approaches $t_0 = 0$, nonlinear focusing occurs due to the BF instability and the solution rises

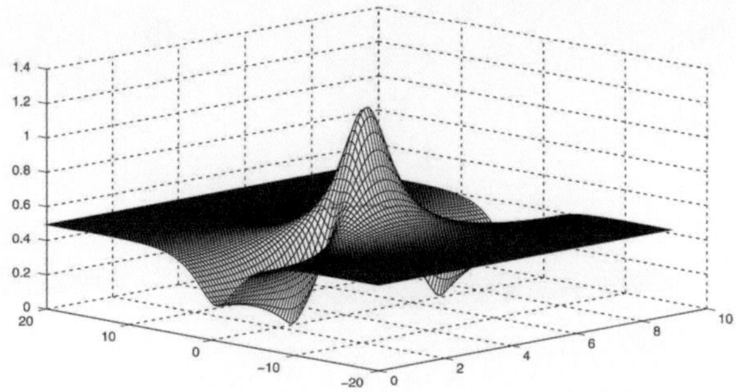

Fig. 3 Analytical rogue wave solution of the NLS corresponding to one UM

to a maximum height of $2.4a$. Thus, the homoclinic solution with one UM can be regarded as the simplest model of rogue wave.

An almost equally dramatic wave trough occurs close to the crest of the rogue wave as a result of wave compression due to wave dislocation. The amplitude amplification factor is given by

$$f = \frac{\max_{x\in[0,L],\, t\in\mathbb{R}} |u(x,t)|}{\lim_{t\to\pm\infty} |u(x,t)|} \approx 2.4. \tag{5}$$

In general, f depends upon the wavenumber of the modulation. As the wave number decreases, the amplification factor increases to the limiting value

$$f_{max} = \lim_{\kappa\to 0^+} f(\kappa) = 3, \tag{6}$$

although the waves take longer to reach their maximum height since their growth rate is smaller (Akhmediev et al. 1988).

2.2 Phase Modulated Rogue Waves

As the number of UMs increases, the space-time structure of the homoclinic solutions becomes more complex. When two or more UMs are present the initial wave train can be phase modulated to produce additional focusing.

The family of homoclinic orbits of the plane wave potential with two UMs is given by an expression of the form

$$u(x,t) = a e^{2ia^2 t} \frac{g(x,t)}{f(x,t)}, \tag{7}$$

where $f(x,t)$ and $g(x,t)$ depend on the two spatial modes $\cos(2n\pi x/L)$, $\cos(2m\pi x/L)$, and on temporal exponential factors $\exp(\sigma_n t + \rho_n)$, $\exp(\sigma_m t + \rho_m)$, with growth rates

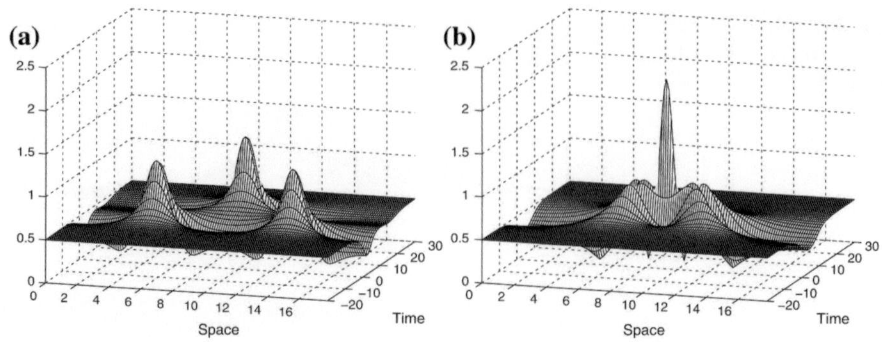

Fig. 4 Rogue wave solutions of the NLS corresponding to two unstable modes **a** without phase modulation ($\rho = -1$), and **b** with phase modulation ($\rho = -0.65$)

$\sigma_l = \mu_l \sqrt{\mu_l^2 - 4a^2}$, $\mu_l = 2\pi l/L$. (The complete formulas can be found in Calini et al. 1996, Calini and Schober 2002.)

As in the one UM case, this combination homoclinic orbit decays to the plane wave potential as $t \to \pm\infty$, and the associated rogue wave remains hidden beneath the background plane wave for most of its lifetime. The temporal separation of the two spatial modes depends upon a parameter ρ related to the difference $\rho_n - \rho_m$ in the temporal phases (Calini et al. 1996; Calini and Schober 2002).

In turn, ρ affects the amplitude amplification factor. Figures 4a, b show the combination homoclinic orbit (7) obtained with all parameters set equal except for ρ. In Fig. 4a, $\rho = 0.1$, the modes are well separated, and the amplitude amplification factor is roughly three. In Fig. 4b, the value of ρ is approximately -0.65, corresponding to the two UMs being simultaneously excited or coalesced. At such ρ-value the amplitude amplification factor is maximal and the rogue wave rises to a height of 4.1 times the height of the carrier wave (whose maximum height is 2.1).

Note that Fig. 4a shows focusing due to only weak amplitude modulation of the initial wave train; the growth in amplitude beginning at $t \approx -5$ and at $t \approx 10$ is due to the BF instability. However, in Fig. 4b focusing due to both amplitude and phase modulation occurs. The amplitude growth at $t \approx -5$ is due to the BF instability, while the additional very rapid focusing at $t \approx 3.4$ is due to the phase modulation. In general, it is possible to select the phases in a combination homoclinic orbit with N spatial modes so that any number n ($2 \leq n \leq N$) of modes coalesce at some fixed time.

3 Noisy Rogue Waves

The Broad Bandwidth Modified NLS equation was introduced by Trulsen and Dysthe (1996) as a higher order asymptotic approximation of slowly modulated periodic wave trains in deep water, assuming that the wave slope ka (where k is the wave number, and a the size of the initial displacement) is $\mathcal{O}(\epsilon)$, while the bandwidth $|\Delta k|/k$ and the quantity $(kh)^{-1}$ (where h is the water depth) are $\mathcal{O}(\epsilon^{1/2})$. The resulting Modified Dysthe (MD) equation

$$iu_t + u_{xx} + 2|u|^2 u + i\epsilon^{1/2}\left(\frac{1}{2}u_{xxx} - 6|u|^2 u_x + u^2 u_x^* - 2ui\left[H\left(|u|^2\right)\right]_x\right) + \\ + \epsilon\,\frac{5}{16}u_{4x} + i\epsilon^{3/2}\,\frac{7}{32}u_{5x} = 0, \qquad (8)$$

is the starting point of our numerical experiments, aimed at investigating the robustness of homoclinic solutions of the NLS equation, as well as the likelihood of rogue wave generation, when higher order terms are introduced in the wave dynamics.

We choose initial data for solutions with two and three UMs; for example, in the two-UM regime, the initial condition has the form

$$u(x,0) = a\left[1 + 4i\left(\epsilon_1 \sin\phi_1 e^{i\phi_1}\cos(\frac{2\pi x}{L}) + \epsilon_2 \sin\phi_2 e^{i\phi_2}\cos(\frac{4\pi x}{L})\right)\right],$$

where the parameters ϕ_i's are varied to explore a neighborhood of the unstable plane wave potential.

Figure 5a illustrates a striking rogue wave solution of Eq. (8) for $\epsilon_1 = 10^{-4}$ and $\epsilon_2 = 10^{-5}$. The solution rapidly becomes chaotic (around $t = 31$) and exhibits an irregular dynamics for a long time afterwards. At $t \approx 471.2$ a rogue wave rises from the plane wave state, developing a crest of amplitude approximately equal to four times the background wave height. The structure of this rogue wave is remarkably similar to that of the combination homoclinic solution (7) with coalesced spatial modes obtained when $\rho = -0.65$. (Compare Fig. 5a with Fig. 4b.)

Numerical simulations of the MD equation in the three-UM regime show a similar phenomenon: after the onset of chaotic dynamics, rogue waves rise intermittently

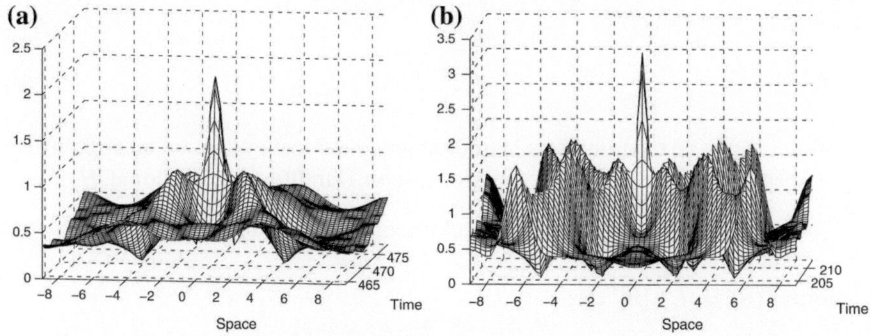

Fig. 5 Rogue waves solutions for the even MD equation when **a** two and **b** three unstable modes are present

above the chaotic background (see Fig. 5b). At $t \approx 208$ a rogue wave develops with a wave amplitude amplification factor of almost five. Again, the emerging rogue wave is close to the optimally phase modulated NLS homoclinic solution in the three-UM regime.

Extensive numerical experiments were performed for both the full MD model and its restriction to spatially even potentials (see Sect. 4) in the two- and three-UM regime, varying both perturbation strength ϵ and the values of the parameters ϕ_i's in the initial data. In all cases, the coalesced homoclinic NLS solution emerges generically as a structurally stable feature of the perturbed dynamics.

We observe how the chaotic regime produces additional focusing by effectively selecting optimal phase modulation, and how the chaotic dynamics singles out the maximally coalesced homoclinic solutions of the unperturbed NLS equation as physically observable rogue waves. Moreover, (see e.g. Fig. 5) larger amplitude waves, and more of them, are obtained for the MD equation, as also supported by the diagnostics developed in Sect. 5, correlating wave strengths in the NLS and MD models to proximity to homoclinic data. Thus, the underlying chaotic dynamics of the MD equation appears to increase the likelihood of rogue wave generation and to favor occurrence of large amplitude rogue waves, as compared to predictions obtained from the NLS equation.

4 Melnikov Analysis

In order to better understand the emergence of the coalesced homoclinic orbit in the chaotic dynamics described by the MD equation, we use perturbation methods to construct appropriate measurements of the splitting distance of the stable and unstable manifolds of a plane wave solution with two unstable modes. In this section, we briefly describe the main ingredients of the ensuing Melnikov analysis; the reader is referred to Calini and Schober (2002) for full details.

We consider, the following restriction of the MD equation to spatially even potentials $u(x,t) = u(-x,t)$:

$$iu_t + u_{xx} + 2|u|^2 u = \epsilon u_{xxxx}. \tag{9}$$

Equation (9) can be regarded as a Hamiltonian dynamical system on an appropriate Sobolev space of even, periodic functions, with Hamiltonian functional $H_\epsilon(u) = \int_0^L (|u_x| - |u|^4 - \epsilon|u_{xx}|^2)\, dx$, and an additional conserved functional given by the L^2-norm $I(u) = \int_0^L |u|^2\, dx$.

For $\epsilon = 0$, we consider a plane wave potential u_a with two UMs (equivalently, with two complex double points): linear analysis shows that u_a possesses two-dimensional stable and unstable eigenspaces and an infinite number of center modes (characterized by complex conjugate pairs of imaginary eigenvalues). Its center-stable and center-unstable invariant manifolds coincide and have codimension two. (In fact, they are explicitly parametrized in terms of the homoclinic solution (7).)

When $\epsilon \neq 0$, the plane wave potential persists as a solution of the perturbed equation, and its perturbed center-stable and center-unstable manifolds generically split. In finite-dimensional situations, invariant manifolds of unstable solutions can split under perturbation and intersect transversally. Such transversal intersections are often associated with chaotic behavior and with persistence of homoclinic orbits in the perturbed system. For PDEs the analogous situation is far more subtle, and rigorous analysis has been performed only in a handful of cases (Haller and Wiggins 1992; Li et al. 1996; Li 1999; Zeng 2000a, b).

If we assume that the perturbed invariant manifolds split transversally, we need two independent measurements for their splitting distance (one for each direction transversal to the unperturbed invariant manifold). However, the perturbation is Hamiltonian, so the splitting occurs within the codimension one energy surface $H_\epsilon = const$, thus reducing the number of measurements to one.

In order to define suitable measurements, we recall how the Floquet discriminant $\Delta(u;\lambda)$ of an NLS solution u, viewed as a functional on the NLS phase space, encodes an infinite family of constants of motion (Li and McLaughlin 1994). Given, a solution u_c with a purely imaginary critical point λ^c (such as, for example, an unstable plane wave potential), regarding λ^c as a functional on a neighborhood $\mathcal{U}$ of u_c, the functional $F : \mathcal{U} \to \mathbb{C}$,

$$\boldsymbol{F}(u) := \Delta(\lambda^c(u); u), \tag{10}$$

is locally smooth, provided $\dfrac{d^2\Delta}{d\lambda^2}(\lambda, u) \neq 0 \,, \forall\, u \in \mathcal{U}$. Then, the sequence

$$F_j(u) = \Delta(\lambda_j^c(u), u),$$

generated as λ_j^c varies among the critical points of the potential u, defines a natural family of constants of motion, which identify the critical level sets of u by labeling them in terms of the double points of its Floquet spectrum.

One of the main advantages of this representation of the constants of motion of the NLS equation is that the gradient of F_j can be explicitly expressed in terms of solutions of the AKNS system by means of the following remarkable formula (Li and McLaughlin 1994):

$$\frac{\delta F_j}{\delta \mathbf{u}}(u) = i\,\frac{\sqrt{\Delta^2-4}}{W[\psi^+,\psi^-]}\left[\begin{array}{c}\psi_2^+\psi_2^- \\ -\psi_1^+\psi_1^-\end{array}\right]\Bigg|_{\lambda=\lambda^c}. \tag{11}$$

In formula (11), $\mathbf{u} = (u, u^*)$, $\psi^\pm(x,\lambda)$ are the Bloch eigenfunctions (common eigenfunctions of the operator $\mathcal{L}^{(x)}$ and the shift operator $(\mathcal{S}\psi)(x) = \psi(x+L)$), and W denotes the Wronskian.

We observe that $\delta F_j/\delta \mathbf{u}$ vanishes at a critical potential u_c (such as the plane wave solution), reflecting the fact that F_j is critical along the critical level set. On the other hand, if u_h is a homoclinic orbit of u_c, then $\delta F_j/\delta \mathbf{u}(u_h) \neq 0$; therefore $\delta F_j/\delta \mathbf{u}\,(u^h)$, $j = 1, \ldots M$, (M being the number of complex double points in the spectrum of u_c) define directions transversal to the homoclinic manifold.

Returning to the even restriction of the MD equation (9), the components of the splitting distance of the perturbed stable and unstable manifolds of a plane wave potential with two UMs along directions ∇F_j, $j = 1, 2$ are expressed in terms of the following Melnikov-type integrals:

$$d_j = \epsilon M_j(\rho) + \mathcal{O}(\epsilon^2), \qquad M_j(\rho) = \int_{-\infty}^{+\infty} \langle \nabla F_j, \mathbf{f} \rangle \big|_{u=u_h} \, dt, \tag{12}$$

where $\mathbf{f}(\mathbf{u}) = (u_{xxxx}, u^*_{xxxx})$ is the vector of the perturbation, $< , >$ is the standard inner product in $L^2([0, L], \mathbb{C})$, and u_h is the homoclinic orbit (7). Both measurements depend on parameter ρ, the same parameter that governs the temporal separation of the spatial modes of the unperturbed homoclinic orbit (7) (see Sect. 2.2).

Consistent with the dimensional count for the splitting distance, numerical evaluation of the two Melnikov integrals show that M_1 and M_2 are mutually proportional functions of the parameter ρ (i.e., a single measurement is sufficient). Figure 6 shows existence of a unique nondegenerate zero of M_1, suggesting that a transversal homoclinic structure persists under perturbation.

A truly remarkable fact, is that the nondegenerate zero of $M_1(\rho)$ coincides (up to order ϵ) with the value of ρ at which the two spatial modes of the homoclinic solution (7) coalesce, producing a homoclinic orbit of maximal amplitude. The same structure is observed as the recurring structurally stable feature of the chaotic dynamics (See Fig. 5).

A study of the analytical structure of single and combination homoclinic orbits, together with the numerical experiments and the Melnikov analysis, supports the following:

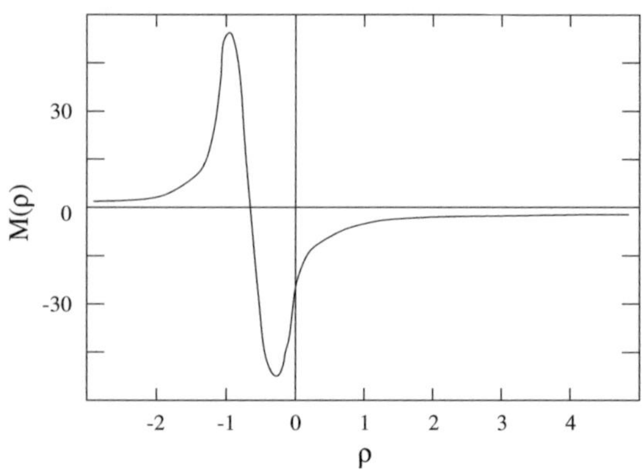

Fig. 6 Graph of the Melnikov integral M_1 as a function of parameter ρ. A transversal zero occurs at $\rho = -0.65$.

- Although homoclinic solutions of the NLS equation have many of the features of rogue waves, not all can be regarded as good candidates for modeling actual rogue waves, as not all are robust under perturbations that lead to more accurate physical models.
- For sea states characterized by a finite number of unstable modes, the homoclinic solutions that are robust under perturbation are the combination homoclinic orbits: (1) with a maximal number of spatial modes excited (this should not be a surprise, since the lower-dimensional homoclinic solutions are linear unstable); and (2) for which the spatial modes are optimally coalesced.
- A chaotic sea state enhances the occurrence of rogue waves. One should note that a homoclinic solution of the NLS will rise over the background wave only once in its life time and for a relative brief time. However, in a chaotic evolution, the maximally coalesced homoclinic orbit will occur repeatedly, although in an unpredictable fashion.

5 Random Oceanic Sea States and the Proximity to Homoclinic Data

To study the generation of rogue waves in a random sea state, we consider initial data for the surface elevation to be of the form (Onorato et al. 2001)

$$\eta(x,0) = \sum_{n=1}^{N} C_n \cos\left(k_n x - \phi_n\right), \tag{13}$$

where $k_n = 2\pi n/L$ and the random phases ϕ_n are uniformly distributed on $(0, 2\pi)$ and the spectral amplitudes, $C_n = \sqrt{2S(f_n)/L}$, are obtained from the JONSWAP spectrum (Ochi 1998):

$$S(f) = \frac{\alpha g^2}{(2\pi f)^5} \exp\left[-\frac{5}{4}\left(\frac{f_0}{f}\right)^4\right]\gamma^r, \qquad r = \exp\left[-\frac{1}{2}\left(\frac{f-f_0}{\sigma_0 f_0}\right)^2\right]. \tag{14}$$

Here f is spatial frequency, $f_n = k_n/2\pi$, f_0 is the dominant frequency, determined by the wind speed at a specified height above the sea surface, g is gravity, and $\sigma_0 = 0.07$ (0.9) for $f \le f_0$ ($f > f_0$). In contrast to physical experiments, which monitor the surface evolution at a given spatial point (probe) in time, here we take time slices and examine the features in space.

JONSWAP spectra describe developing sea states since for $\gamma > 1$ the wave spectra continues to evolve through nonlinear wave-wave interactions for very long times and distances. As γ is increased, the spectrum becomes narrower about the dominant peak (see Fig. 6). In this sense, γ is considered the "peak-shape" parameter.

The scale parameter α is related to the amplitude and energy content of the wave-field. Based on an "Ursell number", the ratio of the nonlinear and dispersive terms of the NLS equation (1) in dimensional form, the NLS equation is considered to be applicable for $2 < \gamma < 8$ (Onorato et al. 2001). Typical values of alpha are $0.008 < \alpha < 0.02$.

In the numerical experiments, the NLS and MD equations are integrated using a pseudo-spectral scheme with 256 Fourier modes in space and a fourth order Runge-Kutta discretization in time ($\Delta t = 10^{-3}$). The nonlinear mode content of the data is numerically computed using the direct spectral transform described above, i.e., the system of ODEs (2) is numerically solved to obtain the discriminant Δ. The zeros of $\Delta \pm 2$ are then determined with a root solver based on Muller's method (Ercolani et al. 1990). The spectrum is computed with an accuracy of $\mathcal{O}(10^{-6})$, whereas the spectral quantities we are interested in range from $\mathcal{O}(10^{-2})$ to $\mathcal{O}(10^{-1})$ Fig. 7.

Under perturbation complex double points typically split into two simple points, $\lambda_{\pm}$, thus opening a gap in the band of spectrum (see Fig. 2). We refer to the distance between these two simple points, $\delta(\lambda_+, \lambda_-) = |\lambda_+ - \lambda_-|$, as the splitting distance. As mentioned, homoclinic solutions arise as an appropriate degeneration of a finite gap solution (Its et al. 1988), i.e., when the resulting double point, $\delta(\lambda_+, \lambda_-) \to 0$, is complex. Consequently, we can use δ to measure the proximity in the spectral plane to homoclinic data, i.e., to complex double points and their corresponding instabilities. Since the NLS spectrum is symmetric with respect to the real axis and real double-points correspond to inactive modes, in subsequent plots only the spectrum in the upper half complex λ-plane will be displayed.

Our first step is to determine the spectrum of JONSWAP initial data given by (13) for various combinations of $\alpha = 0.008, 0.012, 0.016, 0.02$, and $\gamma = 1, 2, 4, 6, 8$. For each such pair (γ, α), we performed 50 simulations, each with a different set of randomly generated phases. As expected, the spectral configuration depends on the energy α and the enhancement coefficient γ. However, the extent of the dependence of features of the spectrum, such as the proximity to complex double points, upon the phases in the initial data is surprising.

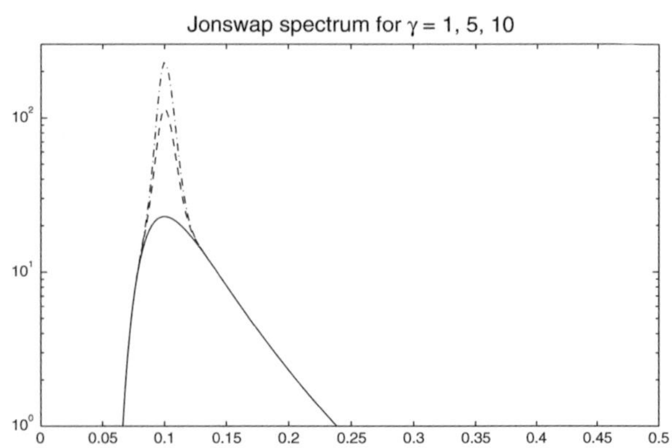

Fig. 7 The JONSWAP spectrum for $\gamma = 1$ (*solid line*), $\gamma = 5$ (*dashed line*), and $\gamma = 10$ (*dash-dot line*) with $f_0 = 0.1$ Hz and $\alpha = 0.0081$

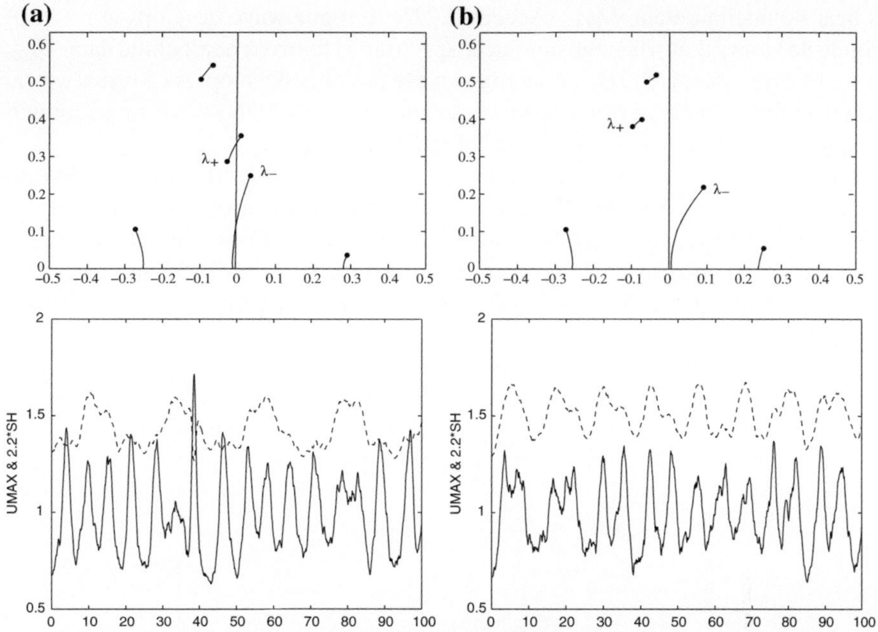

Fig. 8 Spectrum and evolution of U_{max}: **a** "near", and **b** "far" from homoclinic data

Typical examples of the results are given in Fig. 8a, b which show the numerically computed nonlinear spectrum of JONSWAP initial data when $\gamma = 4$ and $\alpha = 0.016$ for two different realizations of the random phases.

We find that JONSWAP data correspond to "semi-stable" N-phase solutions, i.e., JONSWAP data can be viewed as perturbations of N-phase solutions with one or more unstable modes (compare Fig. 2 with the spectrum of an unstable N-phase solution in Fig. 8). In Fig. 8a the splitting distance $\delta(\lambda_+, \lambda_-) \approx 0.07$, while in Fig. 8b $\delta(\lambda_+, \lambda_-) \approx 0.2$. Thus, the JONSWAP data can be quite "near" homoclinic data as in Fig. 8a or "far" from homoclinic data as in Fig. 8b, depending on the values of the phases ϕ_n in the initial data. For all the examined values of α and γ we find that, when α and γ are fixed, as the phases in the JONSWAP data are varied, the spectral distance δ of typical JONSWAP data from homoclinic data varies.

Most importantly, irrespective of the values of the JONSWAP parameters α and γ, in simulations of the NLS equation (1) we find that extreme waves develop for JONSWAP initial data that is "near" NLS homoclinic data, whereas the JONSWAP data that is "far" from NLS homoclinic data typically does not generate extreme waves. Figure 8c, d show the corresponding evolution of the maximum surface elevation, U_{max}, obtained with the NLS equation. U_{max} is given by the solid curve and as a reference, $2.2 H_S$ (the threshold for a rogue wave) is given by the dashed curve. H_S is the significant wave height and is calculated as four times the standard deviation of the surface elevation. Figure 8c shows that when the nonlinear spectrum is near homoclinic data, U_{max} exceeds $2.2H_S$ (a rogue wave develops at $t \approx 40$). Figure 8d shows that when the nonlinear spectrum is far from homoclinic data, U_{max} is significantly below $2.2H_S$ and a rogue wave does not develop. As a result we can correlate the occurrence of rogue waves characterized by JONSWAP spectrum with the proximity to homoclinic solutions of the NLS equation.

The results of hundreds of simulations of the NLS and MD equations consistently show that proximity to homoclinic data is a crucial indicator of rogue wave events. Figures 9 and 10 provide a synthesis of 200 random simulations of the NLS equation and of the MD equation for two perturbation strengths ($\epsilon = 0.005$ and $\epsilon = 0.01$) for JONSWAP initial data with different (γ, α) pairs (with $\gamma = 2, 4, 6, 8,$ and

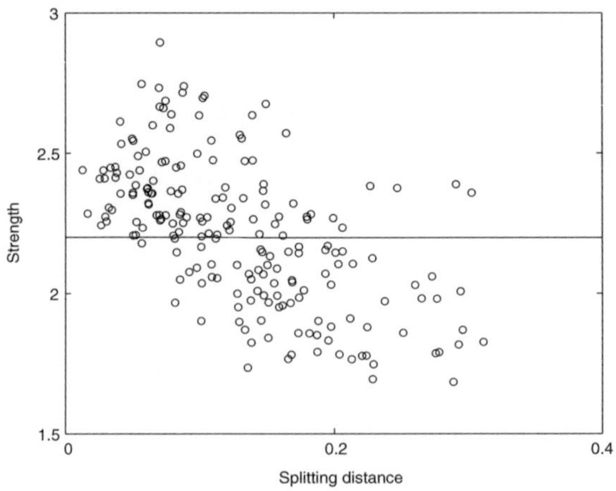

Fig. 9 Strength of U_{max}/H_s versus the splitting distance $\delta(\lambda_+, \lambda_-)$ for solutions of the NLS equation

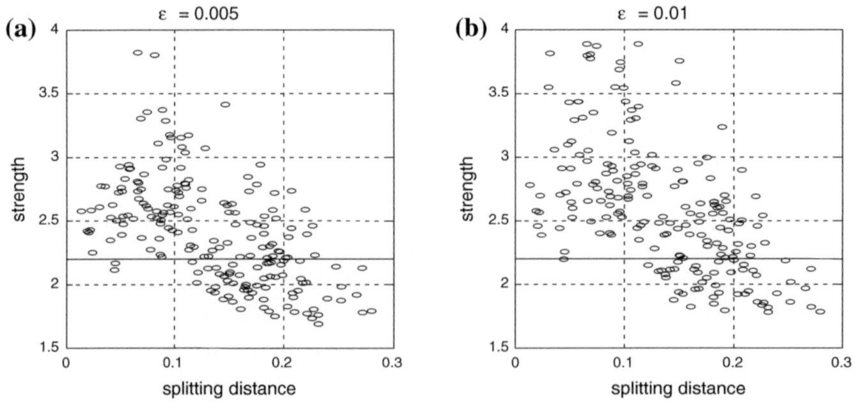

Fig. 10 Strength of U_{max}/H_s versus the splitting distance $\delta(\lambda_+, \lambda_-)$ for solutions of the MD equation when **a** $\epsilon = 0.005$, and **b** $\epsilon = 0.01$

$\alpha = 0.012, 0.016$). For each such pair (γ, α), we performed 25 simulations, each with a different set of randomly generated phases. We restrict our consideration to semi-stable N-phase solutions near unstable solutions of the NLS with one UM. Each circle represents the strength of the maximum wave (U_{max}/H_S) attained during one simulation as a function of the splitting distance $\delta(\lambda_+, \lambda_-)$. The results for the particular pair ($\gamma = 4$, $\alpha = 0.012$) is represented with an asterisk. A horizontal line at $U_{max}/H_S = 2.2$ indicates the reference strength for rogue wave formation. We identify two critical values $\delta_1(\epsilon)$ and $\delta_2(\epsilon)$ that clearly show that: (a) if $\delta < \delta_1$ (near homoclinic data) rogue waves will occur; (b) if $\delta_1 < \delta < \delta_2$, the likelihood of obtaining rogue waves decreases as δ increases and, (c) if $\delta > \delta_2$ the likelihood of a rogue wave occurring is extremely small.

This behavior is robust. As α and γ are varied, the strength of the maximum wave and the occurrence of rogue waves are well predicted by the proximity to homoclinic solutions. The individual plots of the strength versus δ for particular pairs (γ, α) are qualitatively the same regardless of the pair chosen. As noted in Sect. 4 on the MD equation, enhanced focusing occurs in the chaotic regime. Figure 10 shows that as ϵ increases the average strength and the likelihood of rogue waves increases. Clarification on the likelihood of rogue waves through an examination of the kurtosis is provided in the next section. These results give strong evidence of the relevance of homoclinic solutions of the NLS equation in investigating rogue wave phenomena for more realistic oceanic conditions and identifies the nonlinear spectral decomposition as a simple diagnostic tool for predicting the occurrence and strength of rogue waves.

6 Non-Gaussian Statistics and the Dependence of Kurtosis on the Proximity to Homoclinic Data

In Longuet-Higgins (1952) the probability distribution of crest-to-trough wave heights was formulated to be given by the Rayleigh distribution when the wave spectrum is narrow banded and the phases in the reconstruction of the surface elevation are uniformly distributed. Various studies using experimental and field wave data have shown that this can be a reasonable assumption for water waves in the linear regime.

In the nonlinear regime, the relation of the probability density function of wave heights to the nonlinear parameters describing various sea states is not generally known. Simply assuming a Gaussian distribution can be risky. If the kurtosis is in fact much greater than that for the Gaussian distribution, then the probability of an extreme event will be underpredicted.

The main questions we address in this section are whether the modulational instability and the presence of coherent structures yield non-Gaussian statistics of surface gravity waves in the nonlinear regime and whether this can be captured by the spectral parameter δ. In our earlier work with the NLS equation it appeared that homoclinic chaos increases the likelihood of rogue waves. After a short time the waves become chaotic resulting in a sea state characterized by intermittent rogue waves. To more precisely quantify rogue wave events, in our current numerical experiments we monitor the evolution of the skewness, m_3, and the kurtosis, m_4, of the wavefield which are related to the third and fourth statistical moments of the probability density function of the surface elevation by

$$m_3(\eta) = \sum_{j=1}^{N} \frac{(\eta_j - \bar{\eta})^3}{N\sigma^3}, \qquad m_4(\eta) = \sum_{j=1}^{N} \frac{(\eta_j - \bar{\eta})^4}{N\sigma^4},$$

where σ is the standard deviation of the surface elevation, $\bar{\eta}$ is the average surface elevation and N is the number of data points sampled.

Skewness is a measure of the vertical asymmetry of the wavefield. Positive values indicate the wavefield is skewed above average height, i.e., the crests are bigger than the troughs. Negative values indicate that the wavefield is skewed below average height.

The kurtosis is a measure of whether the distribution for the wavefield is peaked or flat, relative to a Gaussian distribution and defines the contribution of large waves to the wavefield. The kurtosis for a Gaussian distribution is three. Wavefields with high kurtosis (in excess of 3) tend to have a distinct peak near the mean, decline rapidly, and have heavy tails. That is, fewer observations or events cluster near the average and more observations populate the extremes either far above or far below the average compared to the bell curve of the normal distribution. For this reason, excess kurtosis much above three indicates that the contribution of large waves is significant and corresponds to a higher probability of a rogue wave event.

Figure 11a shows the plot of the kurtosis as a function of time for the analytical two unstable mode homoclinic solutions of the NLS (7) (the corresponding waveforms are given in Fig. 4) in the uncoalesced case, $\rho = -1$, and in the coalesced case, $\rho = -0.65$. Here we are using the kurtosis as a formal tool to obtain a rough estimate

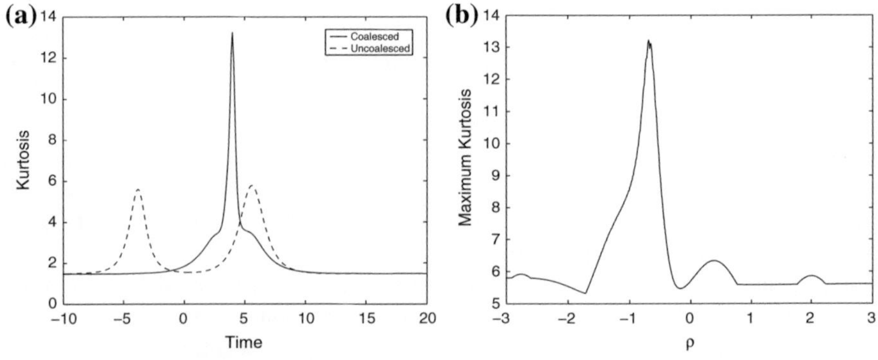

Fig. 11 For the two unstable mode homoclinic solution of NLS: **a** the evolution of the kurtosis for the coalesced and uncoalesced cases and **b** the maximum of the kurtosis as a function of ρ

of the peakedness of the waveform. In both cases the kurtosis starts to increase with the onset of the BF instability and reaches a maximum when the instability saturates. In the uncoalesced case there are two excursions in the kurtosis. In the coalesced case, the increased height achieved by the waveform is reflected in a significantly larger kurtosis. Figure 11b shows the plot of the maximum of the kurtosis of the two unstable mode homoclinic orbit, as a function of the phase parameter ρ. Interestingly, the maximum of the kurtosis is optimized by the robust coalesced homoclinic solution which also gives the zero of the Melnikov integrals and is persistent in the MD equation.

Janssen (2003) formulated the relation between the kurtosis of the surface elevation and the probability of rogue wave occurrence for $1D$ weakly non-Gaussian waves. The PDF of the envelope A of the wave train follows from an integration of the joint probability distribution over the phase ϕ. The first term gives the Rayleigh distribution while the terms involving the skewness integrate to zero. The third term gives a contribution depending on the kurtosis and the narrow-band approximation of the PDF of the envelope is computed as

$$p(A) = Ae^{-\frac{1}{2}A^2}\left[1 + \frac{1}{3}m_4\left(1 - A^2 + \frac{1}{8}A^4\right)\right].$$

The probability of occurrence of a rogue wave as a function of N (the number of waves) and the kurtosis is

$$\mathcal{P}_{rogue} = 1 - \exp\left[-e^{-8}N\left(1 + 8m_4\right)\right].$$

In this way, as the kurtosis increases, the probability that rogue waves will occur increases. We examine the evolution of the skewness and kurtosis for three ranges of δ: (i) $\delta \leq 0.1$, (ii) $0.1 < \delta < 0.2$, and (iii) $\delta \geq 0.2$. The skewness and kurtosis is computed at each time step, first as an average over space and then averaging over the ensemble. As before, we begin by determining the nonlinear spectrum of the JONSWAP initial data for various combinations of (α, γ). We used a sufficient number of realizations of the random phases to produce 250 cases for each range of values of δ. Figure 12 provides the evolution of the skewness and kurtosis for three different values of the nonlinear spectral gap size δ. Both the skewness and kurtosis grows initially and then relax to their asymptotic value. The asymptotic value of the kurtosis is approximately (i) 3.5 when $\delta \leq 0.1$, (ii) 3.2 when $0.1 < \delta < 0.2$, and (iii) 3 when $\delta \geq 0.2$. It is clear in Fig. 12 the kurtosis is strongly dependent on δ and attains larger values for JONSWAP data closer to homoclinic data. The proximity to homoclinic data changes the wave statistics and increases the likelihood of rogue waves.

Using the inverse spectral theory of the NLS equation, we have shown that the development of extreme waves in random oceanic sea states characterized by JONSWAP power spectra is well predicted by the proximity to homoclinic data of the NLS equation. We observe that the modulational instability generates a significant

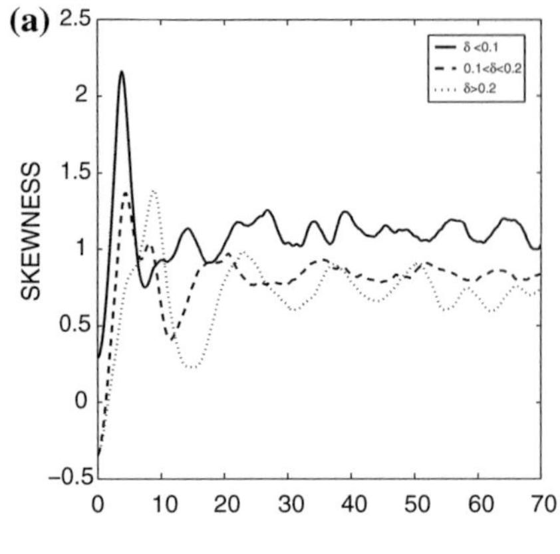

Fig. 12 Evolution of the **a** skewness and **b** kurtosis for Jonswap initial data

deviation from Gaussianity. In particular, we find: (i) the kurtosis and wave strength appear to be strongly dependent on δ, the proximity to instabilities and homoclinic structures; (ii) the likelihood of rogue waves increases for JONSWAP data near homoclinic data of the NLS; (iii) the NLS equation underpredicts, as compared to the MD equation, both the wave strength and likelihood of rogue waves.

References

Ablowitz MJ, Segur H (1981) Solitons and the inverse scattering transform. SIAM, Philadelphia

Ablowitz MJ, Hammack J, Henderson D, Schober CM (2000) Modulated periodic stokes waves in deep water. Phys Rev Lett 84:887–890

Ablowitz MJ, Hammack J, Henderson D, Schober CM (2001) Long time dynamics of the modulational instability of deep water waves. Phys D 152–153:416–433

Akhmediev NN, Korneev VI, Mitskevich NV (1988) N-modulation signals in a single-mode optical waveguide under nonlinear conditions. Sov Phys JETP 67:1

Bridges TJ, Derks G (1999) Unstable eigenvalues and the linearization about solitary waves and fronts with symmetry. Proc R Soc Lond A 455:2427

Cai D, McLaughlin DW, McLaughlin KTR (1995) The nonlinear Schrödinger equation as both a PDE and a dynamical system. Preprint

Calini A, Schober CM (2001) Chaotic dynamics for a symmetry breaking perturbation of the NLS equation. J Math Comput Simul 55:351–364

Calini A, Schober CM (2002) Homoclinic chaos increases the likelihood of rogue waves. Phys Lett A 298:335–349

Calini A, Ercolani NM, McLaughlin DW, Schober CM (1996) Mel'nikov analysis of numerically induced chaos in the nonlinear Schrödinger equation. Phys D 89:227–260

Dysthe K, Trulsen K (1999) Note on breather type solutions of the NLS as model for freak waves. Phys Scr T82:48–52

Ercolani N, Forest MG, McLaughlin DW (1990) Geometry of the modulational instability part III: homoclinic orbits for the periodic Sine-Gordon equation. Phys D 43:349–384

Haller G, Wiggins S (1992) Orbits homoclinic to resonances: the Hamiltonian case. Phys D 66:298–346

Henderson KL, Peregrine DH, Dold JW (1999) Unsteady water wave modulations: fully nonlinear solutions and comparison with the nonlinear Schrödinger equation. Wave Motion 29:341

Islas A, Schober CM (2005) Predicting rogue waves in random oceanic sea states. Phys Fluids 17:1–4

Its AR, Salle MA, Rybin AV (1988) On exact integration of nonlinear Schrödinger equation. Teor Mat Fiz 74:29–45

Janssen P (2003) Nonlinear four-wave interactions and freak waves. J Phys Oceanogr 33:863–884

Karjanto N (2006) Mathematical aspects of extreme water waves. Ph D thesis, Universiteet Twente

Kharif C, Pelinovsky E (2001) Focusing of nonlinear wave groups in deep water. JETP Lett 73: 170–175

Kharif C, Pelinovsky E (2004) Physical mechanisms of the Rogue wave phenomenon. Eur J Mech B/Fluids 22:603–634

Krichever IM (1977) Methods of algebraic geometry in the theory of nonlinear equations. Russ Math Surv 32:185–213

Li Y (1999) Homoclinic tubes in the nonlinear Schrödinger equation under Hamiltonian perturbations. Prog Theor Phys 101:559–577

Li Y, McLaughlin DW (1994) Morse and Mel'nikov functions for NLS PDE's discretized perturbed NLS systems I. Homoclinic orbits. Commun Math Phys 612:175–214

Li Y, McLaughlin DW, Shatah J, Wiggins S (1996) Persistent homoclinic orbits for a perturbed nonlinear Schrödinger equation. Commun. Pure Appl Math 49:1175–1255

Longuet-Higgins MS (1952) On the statistical distribution of the heights of sea waves. J Mar Res 11:1245

Matveev VB, Salle MA (1991) Darboux Transformations and solitons. Springer, Berlin

McLaughlin DW, Schober CM (1992) Chaotic and homoclinic behavior for numerical discretizations of the nonlinear Schrodinger equation. Phys D 57:447–465

Ochi MK (1998) Ocean waves: the stochastic approach. Cambridge University Press, Cambridge

Osborne A, Onorato M, Serio M (2000) The nonlinear dynamics of rogue waves and holes in deep-water gravity wave trains. Phys Lett A 275:386

Onorato M, Osborne A, Serio M, Bertone S (2001) Freak wave in random oceanic sea states. Phys Rev Lett 86:5831

Schober C (2006) Melnikov analysis and inverse spectral analysis of Rogue waves in deep water. Eur J Mech B-Fluids 25:602–620

Torcini A, Frauenkron H, Grassberger P (1997) Studies of phase turbulence in the one-dimensional complex Ginzburg-Landau equation. Phys Rev E 55:5073–5081

Trulsen K, Dysthe K (1996) A modified nonlinear Schrödinger equation for broader bandwidth gravity waves on deep water. Wave Motion 24:281

Trulsen K, Dysthe K (1997a) Frequency downshift in three-dimensional wave trains in a deep basin. J Fluid Mech 352:359–373

Trulsen K, Dysthe K (1997b) Freak waves—a three dimensional wave simulation. In: Rood EP (ed) Proceedings of the 21st symposium naval hydrodynamics. National Academy Press

van Groesen EWC, Karjanto N, Peterson P, Andonowati A, Wave dislocation and nonlinear amplitude amplification for extreme fluid surface waves. Preprint

White BS, Fornberg B (1998) On the chance of freak waves at sea. J Fluid Mech 355:113–138

Zakharov VE, Shabat AB (1972) Exact theory of two-dimensional self-focusing and one-dimensional self-modulation of waves in nonlinear media. Sov Phys JETP 34:62–69

Zeng C (2000a) Homoclinic orbits for a perturbed nonlinear Schrödinger equation. Commun Pure Appl Math 53:1222–1283

Zeng C (2000b) Erratum: Homoclinic orbits for a perturbed nonlinear Schrödinger equation. Commun Pure Appl Math 53:1603–1605

Freak-Waves: Compact Equation Versus Fully Nonlinear One

A.I. Dyachenko, D.I. Kachulin and V.E. Zakharov

Abstract We compare applicability of the recently derived compact equation for surface wave with the fully nonlinear equations. Strongly nonlinear phenomena, namely modulational instability and breathers with the steepness $\mu \sim 0.4$ are compared in numerical simulations using both models.

1 Introduction

A two-dimensional potential flow of an ideal incompressible fluid of infinite depth with a one-dimensional free surface (boundary) in a gravity field is described by the following well-known set of equations:

$$\begin{aligned} \phi_{xx} + \phi_{zz} &= 0 \qquad (\phi_z \to 0, z \to -\infty), \\ \eta_t + \eta_x \phi_x &= \phi_z \Big|_{z=\eta} \\ \phi_t + \frac{1}{2}(\phi_x^2 + \phi_z^2) + g\eta &= 0 \Big|_{z=\eta}; \end{aligned} \tag{1}$$

A.I. Dyachenko (✉)
Landau Institute for Theoretical Physics RAS, Ak. Semenova 1a, Chernogolovka, Moscow Region 142432, Russia
e-mail: alexd@itp.ac.ru

A.I. Dyachenko · D.I. Kachulin · V.E. Zakharov
Novosibirsk State University, Pirogova 2, Novosibirsk 630090, Russia

D.I. Kachulin
e-mail: d.kachulin@gmail.com

V.E. Zakharov
Department of Mathematics, University of Arizona, Tucson, Arizona 857201, USA
e-mail: zakharov@math.arizona.edu

V.E. Zakharov
Physical Institute of RAS, Leninskiy Prospekt 53, Moscow 119991, Russia

V.E. Zakharov
Landau Institute for Theoretical Physics, 2 Kosygin Street, Moscow 119334, Russia

E. Pelinovsky and C. Kharif (eds.), *Extreme Ocean Waves*,
DOI 10.1007/978-3-319-21575-4_2

here $\eta \leftharpoondown x^c\psi t \leftharpoondown$ is the shape of a surface, $\phi \leftharpoondown x^c\psi z^c t \leftharpoondown$ is a potential function of the flow and g is a gravitational acceleration. As was shown in Zakharov (1968), the variables $\eta(x,t)$ and $\psi(x,t) = \phi(x,z,t)\Big|_{z=\eta}$ are canonically conjugated, and satisfy the equations

$$\frac{\partial\psi}{\partial t} = -\frac{\delta H}{\delta\eta} \qquad \frac{\partial\eta}{\partial t} = \frac{\delta H}{\delta\psi}.$$

Hamiltonian can be written as infinite series of powers of ψ and η, (see Zakharov 1968). Taking into account only three- and four-wave interactions, one can cut this series after fourth order term:

$$H=\frac{1}{2}\int g\eta^2+\psi\hat{k}\psi dx - \frac{1}{2}\int\{(\hat{k}\psi)^2-(\psi_x)^2\}\eta dx + \frac{1}{2}\int\{\psi_{xx}\eta^2\hat{k}\psi+\psi\hat{k}(\eta\hat{k}(\eta\hat{k}\psi))\}dx \tag{2}$$

We will study two strongly nonlinear problems numerically:

- modulational instability of the homogeneous wave train of the Stokes waves up to the freak-wave formation
- propagation of narrow breather with the steepness $\mu \sim 0.4 - -0.5$

The goal is to justify the applicability of approximate equation based on truncated Hamiltonian (2) for strongly nonlinear flows of fluid. The reference solutions (fully nonlinear) is performed for the Eq. (1) written in conformal variables, according to Dyachenko (2001).

2 Fully Nonlinear Conformal Equations

To study two-dimensional potential flow of fluid one can perform the conformal transformation to map the domain, filled with fluid

$$-\infty < x < \infty, \qquad -\infty < y < \eta(x,t), \qquad Z = x + iy$$

in Z-plane to the lower half-plane

$$-\infty < u < -\infty, \qquad -\infty < v < 0, \qquad W = u + iv$$

in W-plane like in Figs. 1 and 2. The shape of surface $\eta(x,t)$ is given now by parametric equations

$$y = y(u,t), \;\; x = x(u,t),$$

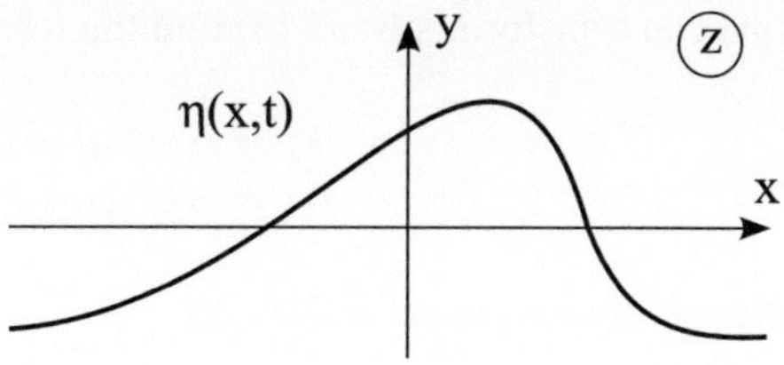

Fig. 1 Physical plane Z

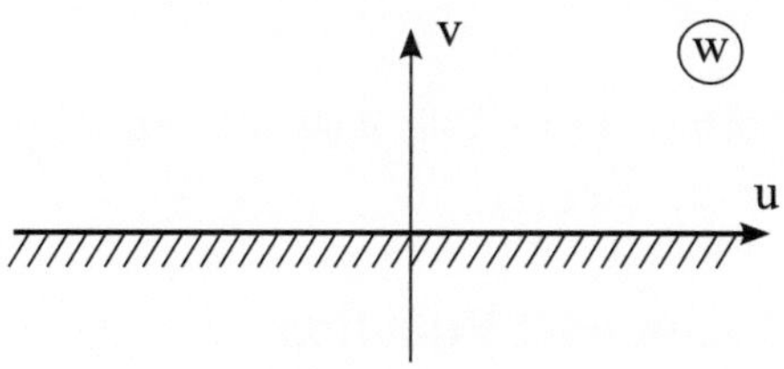

Fig. 2 Conformal plane W

As it was shown in Dyachenko (2001), Dyachenko et al. (1996a, b) the exact Eq. (1) can be written as following:

$$Z_t = iUZ_u,$$
$$\Phi_t = iU\Phi_u - B + ig(Z-u). \tag{3}$$

Here, $\Phi = \psi + i\hat{H}\psi$ is complex velocity potential, and

$$U = \hat{P}\left(\frac{-\hat{H}\psi_u}{|Z_u|^2}\right), \qquad B = \hat{P}\left(\frac{|\Phi_u|^2}{|z_u|^2}\right) = \hat{P}\left(|\Phi_z|^2\right). \tag{4}$$

In (4) $\hat{P}$ is the projector operator generating a function analytical in the lower half-plane

$$\hat{P}(f) = \frac{1}{2}\left(1 + i\hat{H}\right) f.$$

$$\hat{H}(f(u)) = P \cdot V.\frac{1}{\pi}\int_{-\infty}^{\infty}\frac{f(u')\mathrm{d}u'}{u'-u} \quad \text{is the Hilbert transformation.}$$

Functions Φ and Z can be easily analytically continued to the lower half-plane, just by changing u by w.

Introducing new variables (Dyachenko 2001)

$$R = \frac{1}{Z_w}, \qquad \text{and } V = i\Phi_z = i\frac{\Phi_w}{Z_w} \tag{5}$$

one can transform system (3) into the following one

$$\begin{aligned} R_t &= i(UR_w - RU_w), \\ V_t &= i(UV_w - RB_w) + g(R - 1). \end{aligned} \tag{6}$$

Now U and B are the following:

$$\begin{aligned} U &= \hat{P}(V\bar{R} + \bar{V}R) \\ B &= \hat{P}(V\bar{V}). \end{aligned}$$

So, these exact (fully nonlinear) Eq. (6) give us reference solutions to compare with.

3 Compact Equation

In this section, we very briefly derive compact equation based on the truncated Hamiltonian (2). All the details of the derivation can be found in Dyachenko and Zakharov (2011, 2012). It based on the following property of one-dimensional gravity surface waves: In Dyachenko and Zakharov (1994) it was shown that four-wave interaction coefficient vanishes on the following resonant manifold

$$\begin{aligned} k + k_1 &= k_2 + k_3, \\ \omega_k + \omega_{k_1} &= \omega_{k_2} + \omega_{k_3} \end{aligned}$$

with nontrivial solution:

$$\begin{aligned} k &= a(1+\zeta)^2, & k_1 &= a(1+\zeta)^2\zeta^2, \\ k_2 &= -a\zeta^2, & k_3 &= a(1+\zeta+\zeta^2)^2. \end{aligned}$$

Than only trivial resonant interaction remains in force:

$$k = k_2, k_1 = k_3 \qquad \text{or} \qquad k = k_3, k_1 = k_2.$$

Vanishing of four-waves interaction allows us:

- to consider solutions which consist of waves propagating in the same direction (all k in the initial condition and solution can be positive only)
- drastically simplify fourth order term in the truncated Hamiltonian (2).

To make this simplification one can apply appropriate canonical transformation to the Hamiltonian. But first it is convenient to introduce complex normal canonical variables in a standard way:

$$\eta_k = \sqrt{\frac{\omega_k}{2g}}(a_k + a^*_{-k}) \qquad \psi_k = -i\sqrt{\frac{g}{2\omega_k}}(a_k - a^*_{-k}) \tag{7}$$

where $\omega_k = \sqrt{g|k|}$ and the Fourier transform is defined as follow:

$$a(x,t) = \frac{1}{\sqrt{2\pi}} \int_{-\infty}^{\infty} a_k(t) e^{ikx} dx \tag{8}$$

Using normal variable $a_k(t)$ the truncated Hamiltonian can be written as follows:

$$\begin{aligned} H = & \int \omega_k |a_k|^2 dk + \\ & + \int V^{k_1}_{k_2 k_3} (a^*_{k_1} a_{k_2} a_{k_3} + a_{k_1} a^*_{k_2} a^*_{k_3}) \delta_{k_1 - k_2 - k_3} dk_1 dk_2 dk_3 + \\ & + \frac{1}{3} \int U_{k_1 k_2 k_3} (a^*_{k_1} a^*_{k_2} a^*_{k_3} + a_{k_1} a_{k_2} a_{k_3}) \delta_{k_1 + k_2 + k_3} dk_1 dk_2 dk_3 + \\ & + \frac{1}{2} \int W^{k_3 k_4}_{k_1 k_2} a^*_{k_1} a^*_{k_2} a_{k_3} a_{k_4} \delta_{k_1 + k_2 - k_3 - k_4} dk_1 dk_2 dk_3 dk_4 + \\ & + \frac{1}{3} \int G^{k_4}_{k_1 k_2 k_3} (a^*_{k_1} a^*_{k_2} a^*_{k_3} a_{k_4} + a_{k_1} a_{k_2} a_{k_3} a^*_{k_4} \delta_{k_1 + k_2 + k_3 - k_4} dk_1 dk_2 dk_3 dk_4 + \\ & + \frac{1}{12} \int R_{k_1 k_2 k_3 k_4} (a^*_{k_1} a^*_{k_2} a^*_{k_3} a^*_{k_4} + a_{k_1} a_{k_2} a_{k_3} a_{k_4}) \delta_{k_1 + k_2 + k_3 + k_4} dk_1 dk_2 dk_3 dk_4. \end{aligned} \tag{9}$$

Expressions for $V^{k_1}_{k_2 k_3}$, $U_{k_1 k_2 k_3}$, $W^{k_3 k_4}_{k_1 k_2}$, $G^{k_4}_{k_1 k_2 k_3}$, $R_{k_1 k_2 k_3 k_4}$ in Appendix (see 18, 19).

Then one applies transformation from variables a_k to b_k to exclude nonresonant cubic terms along with non resonant fourth order terms. Following Zakharov et al. (1992) canonical transformation from b_k to a_k can be written as the series:

$$\begin{aligned} a_k = b_k + & \int \Big[2\tilde{V}^{k_1}_{k k_2} b_{k_1} b^*_{k_2} \delta_{k_1 - k - k_2} - \tilde{V}^{k}_{k_1 k_2} b_{k_1} b_{k_2} \delta_{k - k_1 - k_2} - \tilde{U}_{k k_1 k_2} b^*_{k_1} b^*_{k_2} \delta_{k + k_1 + k_2} \Big] dk_1 dk_2 \\ & + \int \Big[A^{k}_{k_1 k_2 k_3} b_{k_1} b_{k_2} b_{k_3} + A^{k k_1}_{k_2 k_3} b^*_{k_1} b_{k_2} b_{k_3} + A^{k k_1 k_2}_{k_3} b^*_{k_1} b^*_{k_2} b_{k_3} + A^{k k_1 k_2 k_3} b^*_{k_1} b^*_{k_2} b^*_{k_3} \Big] dk_1 dk_2 dk_3 \end{aligned} \tag{10}$$

All coefficients in (10) are derived in Appendix (see 24, 26). After the transformation (details of it are given in Dyachenko and Zakharov (2011, 2012)) Hamiltonian takes the form in x-space:

$$\mathscr{H} = \int b^* \hat{\omega}_k b dx + \frac{1}{2} \int \left| \frac{\partial b}{\partial x} \right|^2 \left[\frac{i}{2} \left(b \frac{\partial b^*}{\partial x} - b^* \frac{\partial b}{\partial x} \right) - \hat{k} |b|^2 \right] dx. \tag{11}$$

$b(x)$ can be analytically continued to $x + iy$, $y > 0$. Motion equation for $b(x,t)$ should be understood as follow:

$$i \frac{\partial b}{\partial t} = \hat{P}^+ \frac{\delta \mathscr{H}}{\delta b^*}, \tag{12}$$

here $\hat{P}^+$ projection operator to the upper half-plane.

$$\hat{P}^+ = \frac{1}{2}(1 - i\hat{H}). \tag{13}$$

This operator is the consequence of only positive k in the system of waves. Corresponding equation of motion is the following:

$$
\begin{aligned}
i\frac{\partial b}{\partial t} = \hat{\omega}_k b + \frac{i}{4}\hat{P}^+\left[b^*\frac{\partial}{\partial x}(b'^2) - \frac{\partial}{\partial x}(b^{*\prime}\frac{\partial}{\partial x}b^2)\right] - \\
-\frac{1}{2}\hat{P}^+\left[b\cdot\hat{k}(|b'|^2) - \frac{\partial}{\partial x}(b'\hat{k}(|b|^2))\right]. \quad (14)
\end{aligned}
$$

Transformation from $b(x,t)$ to physical variables $\eta(x,t)$ and $\psi(x,t)$ can be recovered from canonical transformation. It has been derived in the Appendix. Here, we write this transformation up to the second order:

$$
\begin{aligned}
\eta(x) &= \frac{1}{\sqrt{2}g^{\frac{1}{4}}}(\hat{k}^{\frac{1}{4}}b(x) + \hat{k}^{\frac{1}{4}}b(x)^*) + \frac{\hat{k}}{4\sqrt{g}}[\hat{k}^{\frac{1}{4}}b(x) - \hat{k}^{\frac{1}{4}}b^*(x)]^2, \\
\psi(x) &= -i\frac{g^{\frac{1}{4}}}{\sqrt{2}}(\hat{k}^{-\frac{1}{4}}b(x) - \hat{k}^{-\frac{1}{4}}b(x)^*) + \frac{i}{2}[\hat{k}^{\frac{1}{4}}b^*(x)\hat{k}^{\frac{3}{4}}b^*(x) - \hat{k}^{\frac{1}{4}}b(x)\hat{k}^{\frac{3}{4}}b(x)] \\
&+ \frac{1}{2}\hat{H}[\hat{k}^{\frac{1}{4}}b(x)\hat{k}^{\frac{3}{4}}b^*(x) + \hat{k}^{\frac{1}{4}}b^*(x)\hat{k}^{\frac{3}{4}}b(x)]. \quad (15)
\end{aligned}
$$

Here $\hat{H}$ is Hilbert transformation with eigenvalue $i\,\mathbf{sign}(k)$.

4 Modulational Instability of Wave Train

In this section, we perform numerical simulation of the modulational instability of the homogeneous wave train in the framework of compact Eq. (14). Such a wave train has two parameters: wavelength and steepness, i.e., maximal slope of the surface, μ. Initial steepness of the wave train was equal to $\mu = 0.095$ and the number of waves in the periodic domain was equal to 100. These values were chosen for comparison with the earlier simulation in the framework of fully nonlinear simulation in the works (Dyachenko and Zakharov 2005; Zakharov et al. 2006, 2008). One can see in Figs. 3 and 4 that both waves coincide in details. Different time of their appearing is due to slightly different values of perturbations. Zoomed shape of the surface is shown in the inset to Fig. 3.

Couple of snapshots of development of modulational instability is shown in Fig. 5.

Dynamics of surface in fully nonlinear Eq. (6) can be found at the address http://www.itp.ac.ru/~kachulin/MInstability/Freak-0.095-end.avi and dynamics of the surface in compact equation is at the address
http://www.itp.ac.ru/~kachulin/MInstability/Surface-end.avi.

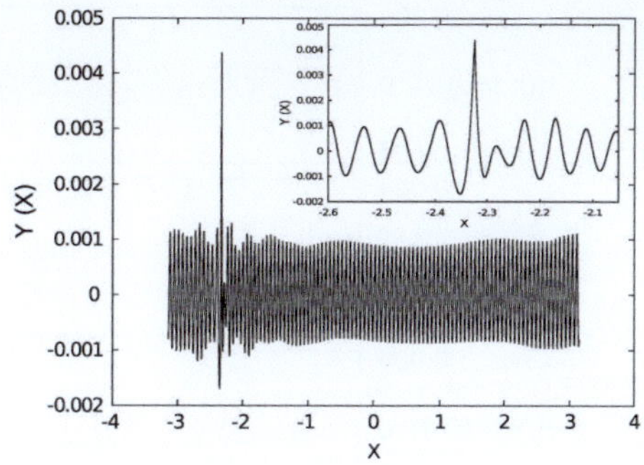

Fig. 3 Freak-wave formation after t = 802 (fully nonlinear equation)

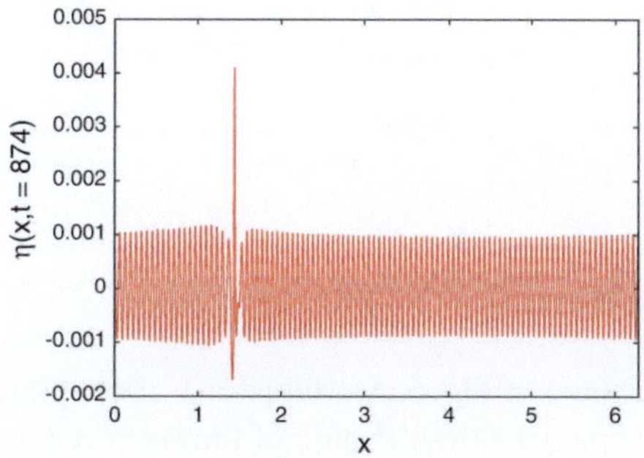

Fig. 4 Freak-wave formation after t = 874 (compact equation)

5 Breathers

Breather is the localized solution of (14) of the following type:

$$b(x,t) = B(x - Vt)e^{i(k_0 x - \omega_0 t)}, \tag{16}$$

where $B(x)$ is localized function in space, having zero asymptotic at $\pm\infty$. In K space it can be written as following:

$$b_k(t) = e^{-i(\Omega + Vk)t}\phi_k, \tag{17}$$

k_0 is the wavenumber of the carrier wave, V is the group velocity, ω_0 is the frequency close to ω_{k_0}, and Ω is close to $\frac{\omega_{k_0}}{2}$. Existence of such solution may indicate that equation is integrable one. However, in the paper (Dyachenko et al. 2013) nonintegrability of the equation was proven.

In the papers (Dyachenko et al. 2013, 2014) such solutions with different group velocities and amplitudes were found by iterative Petviashvili method. Here we tried to get numerically very narrow breather for the compact equation with carrier wavenumber $k_0 = 50$. Picture of real part of $b(x,0)$ and modulus of $b(x,0)$ is

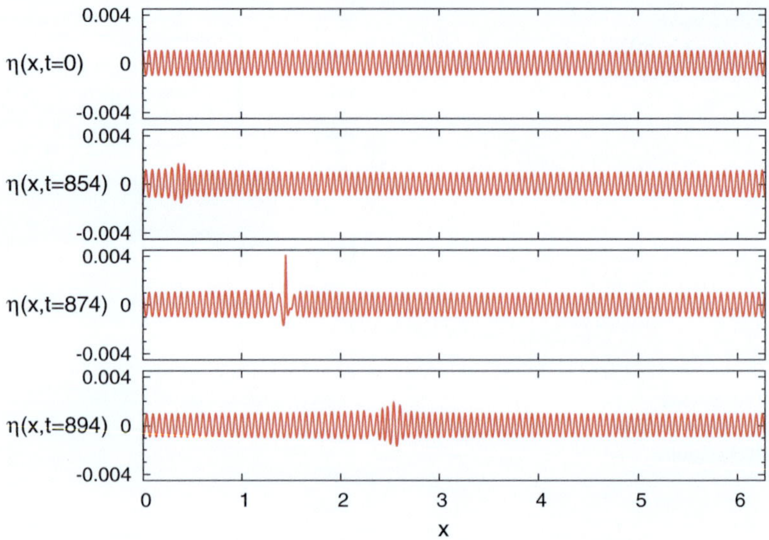

Fig. 5 Formation of the freak wave. Free surface for different times is shown (compact equation)

shown in Fig. 6. Modulus of b corresponds in some sense to the envelope in NLSE approximation. Profile of the surface calculated according to transformation (15) is shown in Fig. 7. Steepness of this solution is very high, $\mu \sim 0.45$. Profile of the

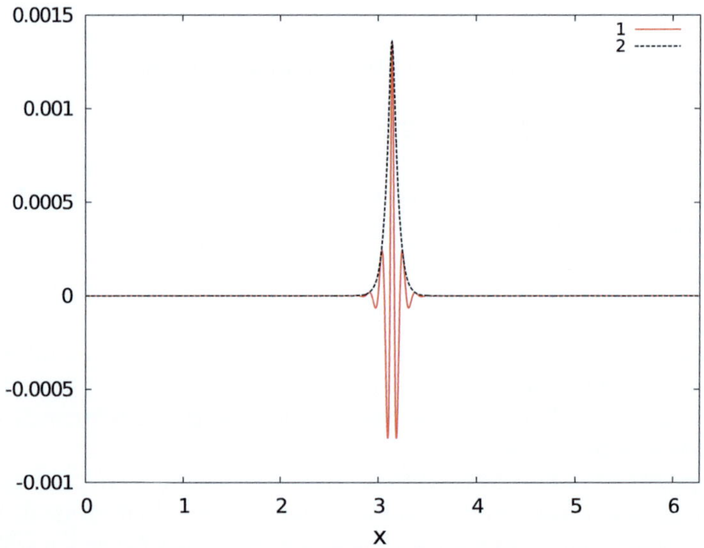

Fig. 6 Modulus of $b(x)$ and real part of $b(x)$ with $V = 1/20$ and $\Omega = 5.2$. *Solid line* (1) corresponds to the real part of $b(x)$, *dashed line* (2) corresponds to modulus of $b(x)$

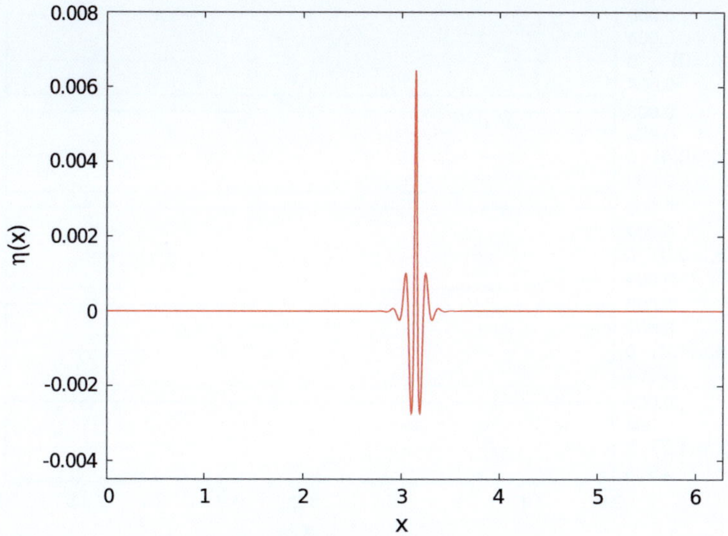

Fig. 7 Surface profile of the breather (compact equation)

steepness is shown in Fig. 8. This breather is exact solution of the compact Eq. (14) and moves on the surface without changing (see Dyachenko et al. 2013, 2014). Couple of snapshots of moving breather is given in Fig. 9. It is clearly seen that

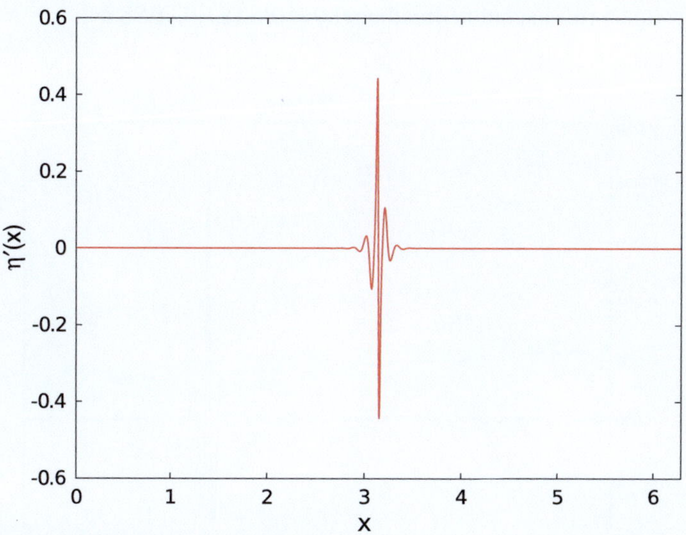

Fig. 8 Steepness of the breather (compact equation)

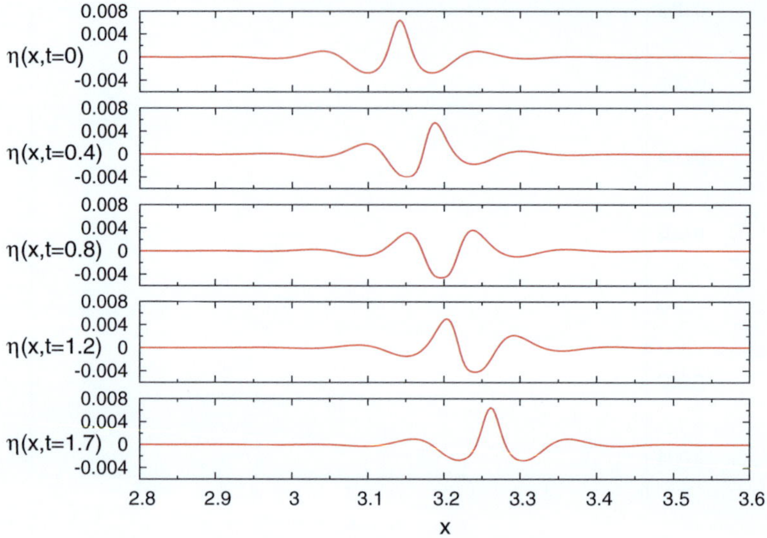

Fig. 9 Free surface corresponds to the breather solution at different times (compact equation)

breather moves with the group velocity which is $\frac{1}{2}\sqrt{\frac{g}{k}} \sim 0.0707$. Dynamics of this breather in compact equation can be found at the address http://www.itp.ac.ru/~kachulin/Breathers/k0=50/deta(x,t).avi. Breather dynamics in fully nonlinear equations can be found at
http://www.itp.ac.ru/~kachulin/Breathers/k0=50/STEEPNESS.avi.

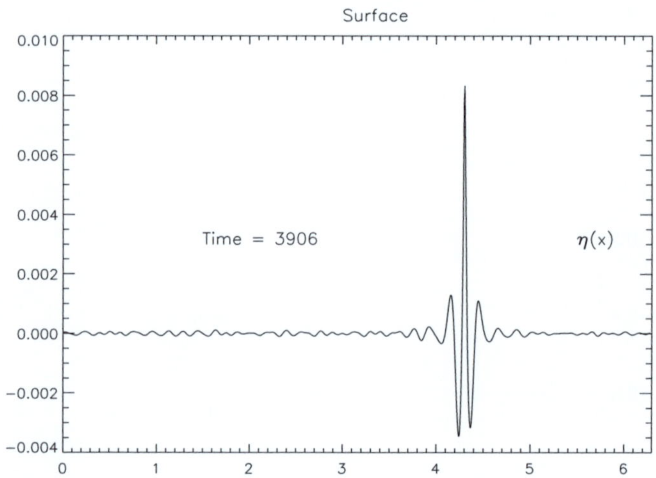

Fig. 10 Surface profile (fully nonlinear equations)

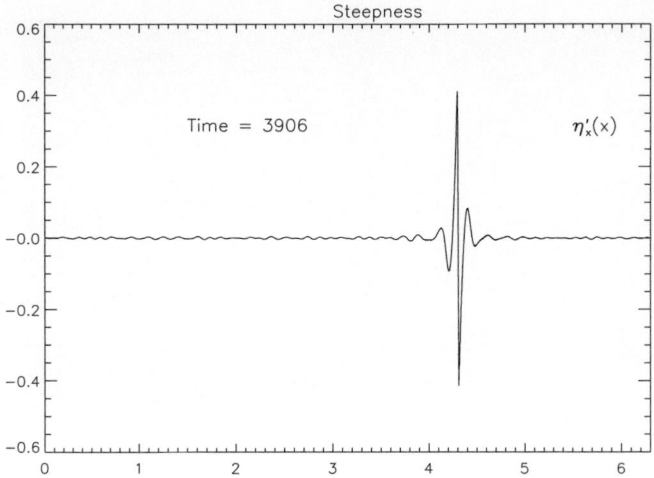

Fig. 11 Profile of steepness (fully nonlinear equations)

In the paper (Dyachenko and Zakharov 2008) we have performed similar simulations in the framework of fully nonlinear conformal equations (6). Here we present pictures from that paper, namely surface of the fluid and its steepness (Figs. 10, 11).

Figures 8 and 11 show that numerical simulation of the highly nonlinear phenomena, steep breather, in the framework of compact equation is very similar to that of fully nonlinear equation. Recent laboratory experiment (Slunyaev et al. 2013), also confirm existence of such highly narrow breathers at the surface.

6 Conclusions

We have demonstrated that compact equation although approximate, quantitatively describes strongly nonlinear phenomena at the surface of potential fluid. Especially, we have studied nonlinear stage of modulational instability up to the freak-wave formation and propagation of very steep breather. Also compact equation can be generalized for quasi one-dimensional waves propagating at the surface of 3D fluid, see Dyachenko et al. (2014). When considering waves slightly inhomogeneous in transverse direction, one can think in the spirit of Kadomtsev-Petviashvili equation for Korteveg-de-Vries equation, namely one can treat now frequency ω_k depending on both k_x and k_y as ω_{k_x,k_y}, while leaving coefficient $\tilde{T}^{kk_1}_{k_2k_3}$ not depending on y. b now depends on both x and y:

$$\mathscr{H} = \int b^* \hat{\omega}_{k_x,k_y} b dx dy + \frac{1}{2}\int |b'_x|^2 \left[\frac{i}{2}(bb'^*_x - b^* b'_x) - \hat{K}_x |b|^2\right] dx dy.$$

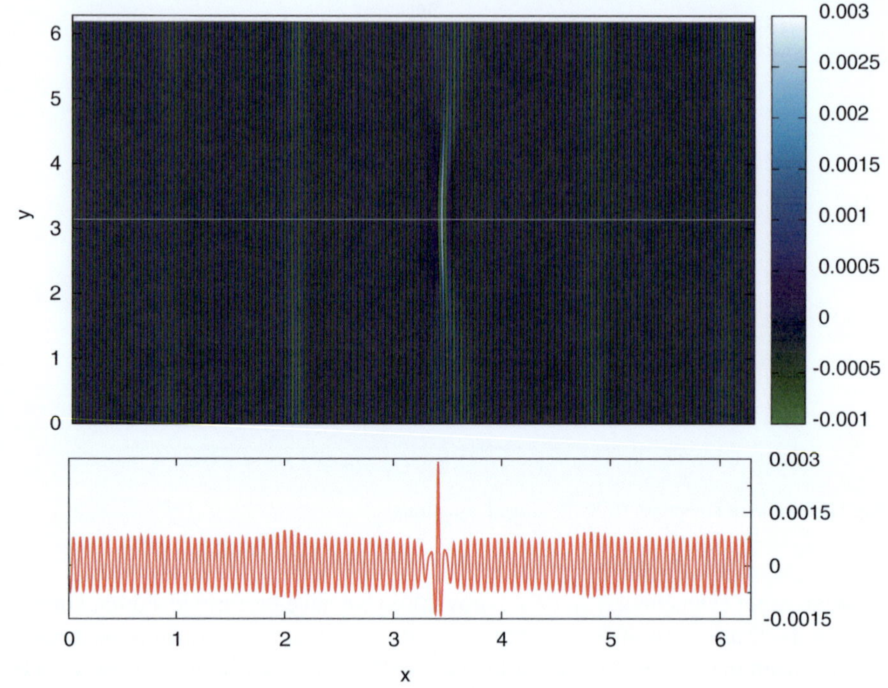

Fig. 12 2D surface with a freak-wave (2D compact equation)

Here, we can show picture of numerical simulation of quasi one-dimensional wave train. One can see in Fig. 12 top view of 100 almost 1D waves with the freak wave in some place. Profile of the surface along the white line is also shown.

Acknowledgments Main part of this work, regarding numerical simulation of modulational instability and narrow breather in the framework of compact equation and derivation of canonical transformation, was supported by Grant "Wave turbulence: theory, numerical simulation, experiment" #14-22-00174 of Russian Science Foundation.
Rest part of the work was supported by the Program "Fundamental Problems of Nonlinear Dynamics in Mathematics and Physics" from the RAS Presidium, and Grant 6170.2012.2 "Leading Scientific Schools of Russia".
Numerical simulation was performed on the Informational Computational Center of the Novosibirsk State University.

Appendix

Coefficients in the Hamiltonian (9) can be calculated plugging expressions for complex canonical variables into the (2):

$$U_{k_1k_2k_3} = \frac{1}{8}\frac{g^{\frac{1}{4}}}{\sqrt{\pi}}\left[|\frac{k_1}{k_2k_3}|^{\frac{1}{4}}L_{k_2k_3} + |\frac{k_2}{k_1k_3}|^{\frac{1}{4}}L_{k_1k_3} + |\frac{k_3}{k_1k_2}|^{\frac{1}{4}}L_{k_1k_2}\right],$$
$$V^{k_1}_{k_2k_3} = \frac{1}{8}\frac{g^{\frac{1}{4}}}{\sqrt{\pi}}\left[|\frac{k_1}{k_2k_3}|^{\frac{1}{4}}L_{k_2k_3} - |\frac{k_2}{k_1k_3}|^{\frac{1}{4}}L_{-k_1k_3} - |\frac{k_3}{k_1k_2}|^{\frac{1}{4}}L_{-k_1k_2}\right]. \quad (18)$$

$$W^{k_3k_4}_{k_1k_2} = \frac{-1}{32\pi}\left[\left|\frac{k_1k_2}{k_3k_3}\right|^{\frac{1}{4}} M^{k_1k_2}_{-k_3-k_4} + \left|\frac{k_3k_4}{k_1k_2}\right|^{\frac{1}{4}} M^{-k_3-k_4}_{k_1k_2} - \left|\frac{k_1k_3}{k_2k_4}\right|^{\frac{1}{4}} M^{k_1-k_3}_{k_2-k_4} - \left|\frac{k_2k_3}{k_1k_4}\right|^{\frac{1}{4}} M^{k_2-k_3}_{k_1-k_4} - \right.$$
$$\left. - \left|\frac{k_1k_4}{k_2k_3}\right|^{\frac{1}{4}} M^{k_1-k_4}_{k_2-k_3} - \left|\frac{k_2k_4}{k_1k_3}\right|^{\frac{1}{4}} M^{k_2-k_4}_{k_1-k_3}\right]$$
$$G^{k_4}_{k_1k_2k_3} = \frac{-1}{32\pi}\left[\left|\frac{k_3k_4}{k_1k_2}\right|^{\frac{1}{4}} M^{k_3-k_4}_{k_1k_2} + \left|\frac{k_2k_4}{k_1k_3}\right|^{\frac{1}{4}} M^{k_2-k_4}_{k_1k_3} + \left|\frac{k_1k_4}{k_2k_3}\right|^{\frac{1}{4}} M^{k_1-k_4}_{k_2k_3} - \left|\frac{k_1k_2}{k_3k_4}\right|^{\frac{1}{4}} M^{k_1k_2}_{k_3-k_4} - \right.$$
$$\left. - \left|\frac{k_1k_3}{k_2k_4}\right|^{\frac{1}{4}} M^{k_1k_3}_{k_2-k_4} - \left|\frac{k_2k_3}{k_1k_4}\right|^{\frac{1}{4}} M^{k_2k_3}_{k_1-k_4}\right]$$
$$R_{k_1k_2k_3k_4} = \frac{-1}{32\pi}\left[\left|\frac{k_3k_4}{k_1k_2}\right|^{\frac{1}{4}} M^{k_3k_4}_{k_1k_2} + \left|\frac{k_2k_4}{k_1k_3}\right|^{\frac{1}{4}} M^{k_2k_4}_{k_1k_3} + \left|\frac{k_2k_3}{k_1k_4}\right|^{\frac{1}{4}} M^{k_2k_3}_{k_1k_4} + \left|\frac{k_1k_4}{k_2k_3}\right|^{\frac{1}{4}} M^{k_1k_4}_{k_2k_3} + \right.$$
$$\left. + \left|\frac{k_1k_3}{k_2k_4}\right|^{\frac{1}{4}} M^{k_1k_3}_{k_2k_4} + \left|\frac{k_1k_2}{k_3k_4}\right|^{\frac{1}{4}} M^{k_1k_2}_{k_3k_4}\right] \quad (19)$$

Here

$$L_{k_1k_2} = |k_1k_2| + k_1k_2$$
$$M^{k_3k_4}_{k_1k_2} = |k_1k_2|(|k_1+k_3| + |k_1+k_4| + |k_2+k_3| + |k_2+k_4| - 2|k_1| - 2|k_2|). \quad (20)$$

To construct canonical transformation of general form we follow the book (Zakharov et al. 1992) and use auxiliary Hamiltonian:

$$\tilde{H} = -i\int \tilde{V}^{k_1}_{k_2k_3}(b^*_{k_1}b_{k_2}b_{k_3} - b_{k_1}b^*_{k_2}b^*_{k_3})\delta_{k_1-k_2-k_3}dk_1dk_2dk_3-$$
$$-\frac{i}{3}\int \tilde{U}_{k_1k_2k_3}(b^*_{k_1}b^*_{k_2}b^*_{k_3} - b_{k_1}b_{k_2}b_{k_3})\delta_{k_1+k_2+k_3}dk_1dk_2dk_3,$$
$$+\frac{1}{2}\int (\tilde{\tilde{W}}^{k_3k_4}_{k_1k_2} + i\tilde{W}^{k_3k_4}_{k_1k_2})b^*_{k_1}b^*_{k_2}b_{k_3}b_{k_4}\delta_{k_1+k_2-k_3-k_4}dk_1dk_2dk_3dk_4-$$
$$-\frac{i}{3}\int \tilde{G}^{k_4}_{k_1k_2k_3}(b^*_{k_1}b^*_{k_2}b^*_{k_3}b_{k_4} - b_{k_1}b_{k_2}b_{k_3}b^*_{k_4})\delta_{k_1+k_2+k_3-k_4}dk_1dk_2dk_3dk_4-$$
$$-\frac{i}{12}\int \tilde{R}_{k_1k_2k_3k_4}(b^*_{k_1}b^*_{k_2}b^*_{k_3}b^*_{k_4} - b_{k_2}b_{k_3}b_{k_4})\delta_{k_1+k_2+k_3+k_4}dk_1dk_2dk_3dk_4 \quad (21)$$

with standard symmetry conditions for coefficients. Just mention that for $\tilde{W}^{k_3k_4}_{k_1k_2}$ this condition is the following:

$$\tilde{W}^{k_3k_4}_{k_1k_2} = \tilde{W}^{k_3k_4}_{k_2k_1} = \tilde{W}^{k_4k_3}_{k_1k_2} = -\tilde{W}^{k_1k_2}_{k_3k_4}. \quad (22)$$

Again, following Zakharov et al. (1992) general canonical transformation from b_k to a_k can be written as the series:

$$a_k = b_k + \int \left[2\tilde{V}^{k_1}_{kk_2} b_{k_1} b^*_{k_2} \delta_{k_1-k-k_2} - \tilde{V}^{k}_{k_1k_2} b_{k_1} b_{k_2} \delta_{k-k_1-k_2} - \tilde{U}_{kk_1k_2} b^*_{k_1} b^*_{k_2} \delta_{k+k_1+k_2}\right] dk_1 dk_2$$
$$+ \int \left[A^{k}_{k_1k_2k_3} b_{k_1} b_{k_2} b_{k_3} + A^{kk_1}_{k_2k_3} b^*_{k_1} b_{k_2} b_{k_3} + A^{kk_1k_2}_{k_3} b^*_{k_1} b^*_{k_2} b_{k_3} + A^{kk_1k_2k_3} b^*_{k_1} b^*_{k_2} b^*_{k_3}\right] dk_1 dk_2 dk_3 \qquad (23)$$

Coefficients A with upper and lower indices are equal to:

$$\begin{aligned}
A^{k}_{k_1k_2k_3} &= \left[\frac{1}{3}\tilde{G}^{k}_{k_1k_2k_3} + \tilde{V}^{k}_{k_1k-k_1} \tilde{V}^{k_2+k_3}_{k_2k_3} - \tilde{V}^{k_1}_{kk_1-k} \tilde{U}_{-k_2-k_3k_2k_3}\right] \delta_{k-k_1-k_2-k_3},\\
A^{kk_1}_{k_2k_3} &= \left[-i\tilde{\tilde{W}}^{k_2k_3}_{kk_1} + \tilde{W}^{k_2k_3}_{kk_1} - 2\tilde{V}^{k}_{k_2k-k_2} \tilde{V}^{k_3}_{k_1k_3-k_1} - \tilde{V}^{k+k_1}_{kk_1} \tilde{V}^{k_2+k_3}_{k_2k_3} + 2\tilde{V}^{k_3}_{kk_3-k} \tilde{V}^{k_1}_{k_2k_1-k_2} +\right.\\
&\quad \left. + \tilde{U}_{-k-k_1kk_1} \tilde{U}_{-k_2-k_3k_2k_3}\right] \delta_{k+k_1-k_2-k_3},\\
A^{kk_1k_2}_{k_3} &= \left[-\tilde{G}^{k_3}_{kk_1k_2} + \tilde{V}^{k}_{k_3k-k_3} \tilde{U}_{-k_2-k_1k_2k_1} - \tilde{V}^{k_3}_{kk_3-k} \tilde{V}^{k_1+k_2}_{k_1k_2} + 2\tilde{V}^{k+k_1}_{kk_1} \tilde{V}^{k_3}_{k_2k_3-k_2} -\right.\\
&\quad \left. - 2\tilde{U}_{-k-k_1kk_1} \tilde{V}^{k_2}_{k_3k_2-k_3}\right] \delta_{k+k_1+k_2-k_3},\\
A^{kk_1k_2k_3} &= \left[-\frac{1}{3}\tilde{R}_{kk_1k_2k_3} - \tilde{V}^{k+k_1}_{kk_1} \tilde{U}_{-k_2-k_3k_2k_3} + \tilde{V}^{k_2+k_3}_{k_2k_3} \tilde{U}_{-k-k_1kk_1}\right] \delta_{k+k_1+k_2+k_3}.
\end{aligned} \qquad (24)$$

Let us now substitute transformation (23) into the Hamiltonian (9) and calculate second, third and fourth order terms.

Collecting all cubic terms after substitution and making symmetrization one can get:

$$H_3 = \int [V^{k_1}_{k_2k_3} - (\omega_{k_1} - \omega_{k_3} - \omega_{k_3})\tilde{V}^{k_1}_{k_2k_3}] b^*_{k_1} b_{k_2} b_{k_3} \delta_{k_1-k_2-k_3} dk_1 dk_2 dk_3 +$$
$$+ \frac{1}{3} \int [U_{k_1k_2k_3} - (\omega_{k_1} + \omega_{k_3} + \omega_{k_3})\tilde{U}_{k_1k_2k_3}] b^*_{k_1} b^*_{k_2} b^*_{k_3} \delta_{k_1+k_2+k_3} dk_1 dk_2 dk_3 + c.c. \qquad (25)$$

it is possible to cancel nonresonant both cubic and fourth order terms. If

$$\tilde{V}^{k}_{k_1k_2} = \frac{V^{k}_{k_1k_2}}{\omega_k - \omega_{k_1} - \omega_{k_2}}, \qquad \tilde{U}_{kk_1k_2} = \frac{U_{kk_1k_2}}{\omega_k + \omega_{k_1} + \omega_{k_2}}. \qquad (26)$$

than H_3 vanishes.

Counting all fourth terms, making symmetrization and calculating new H_4 one can get

$$\begin{aligned}
H_4 &= \frac{1}{2} \int [W^{k_3k_4}_{k_1k_2} + D^{k_3k_4}_{k_1k_2} + (\omega_{k_1} + \omega_{k_2} - \omega_{k_3} - \omega_{k_4})(\tilde{W}^{k_3k_4}_{k_1k_2} - i\tilde{\tilde{W}}^{k_3k_4}_{k_1k_2})] b^*_{k_1} b^*_{k_2} b_{k_3} b_{k_4} \delta_{k_1+}\\
&\quad {}_{k_2-k_3-k_4} dk_1 dk_2 dk_3 dk_4 +\\
&+ \frac{1}{3} \int \left[(G^{k_4}_{k_1k_2k_3} + D^{k_4}_{k_1k_2k_3} - (\omega_{k_1} + \omega_{k_2} + \omega_{k_3} - \omega_{k_4})\tilde{G}^{k_4}_{k_1k_2k_3}) b^*_{k_1} b^*_{k_2} b^*_{k_3} b_{k_4} + c.c.\right] \delta_{k_1+}\\
&\quad {}_{k_2+k_3-k_4} dk_1 dk_2 dk_3 dk_4 +
\end{aligned}$$

$$+\frac{1}{12}\int\Big[(R_{k_1k_2k_3k_4}+D_{k_1k_2k_3k_4}-(\omega_{k_1}+\omega_{k_2}+\omega_{k_3}+\omega_{k_4})\tilde{R}_{k_1k_2k_3k_4})b^*_{k_1}b^*_{k_2}b^*_{k_3}b^*_{k_4}+c.c.\Big]\delta_{k_1+k_2+k_3+k_4}dk_1dk_2dk_3dk_4. \quad (27)$$

Here

$$\begin{aligned}
D^{k_3k_4}_{k_1k_2} = {} & \tilde{V}^{k_1}_{k_3k_1-k_3}\tilde{V}^{k_4}_{k_2k_4-k_2}\left[\omega_{k_1}-\omega_{k_3}-\omega_{k_1-k_3}+\omega_{k_4}-\omega_{k_2}-\omega_{k_4-k_2}\right]+\\
&+\tilde{V}^{k_2}_{k_3k_2-k_3}\tilde{V}^{k_4}_{k_1k_4-k_1}\left[\omega_{k_2}-\omega_{k_3}-\omega_{k_2-k_3}+\omega_{k_4}-\omega_{k_1}-\omega_{k_4-k_1}\right]+\\
&+\tilde{V}^{k_1}_{k_4k_1-k_4}\tilde{V}^{k_3}_{k_2k_3-k_2}\left[\omega_{k_1}-\omega_{k_4}-\omega_{k_1-k_4}+\omega_{k_3}-\omega_{k_2}-\omega_{k_3-k_2}\right]+\\
&+\tilde{V}^{k_2}_{k_4k_2-k_4}\tilde{V}^{k_3}_{k_1k_3-k_1}\left[\omega_{k_2}-\omega_{k_4}-\omega_{k_2-k_4}+\omega_{k_3}-\omega_{k_1}-\omega_{k_3-k_1}\right]-\\
&-\tilde{V}^{k_1+k_2}_{k_1k_2}\tilde{V}^{k_3+k_4}_{k_3k_4}\left[\omega_{k_1+k_2}-\omega_{k_1}-\omega_{k_2}+\omega_{k_3+k_4}-\omega_{k_3}-\omega_{k_4}\right]-\\
&-\tilde{U}_{-k_1-k_2k_1k_2}\tilde{U}_{-k_3-k_4k_3k_4}\left[\omega_{k_1+k_2}+\omega_{k_1}+\omega_{k_2}+\omega_{k_3+k_4}+\omega_{k_3}+\omega_{k_4}\right], \quad (28)
\end{aligned}$$

$$\begin{aligned}
D^{k_4}_{k_1k_2k_3} = {} & \tilde{V}^{k_1+k_2}_{k_1k_2}\tilde{V}^{k_4}_{k_3k_4-k_3}(\omega_{k_1+k_2}-\omega_{k_1}-\omega_{k_2}-\omega_{k_4}+\omega_{k_3}+\omega_{k_3-k_4})+\\
&+\tilde{V}^{k_1+k_3}_{k_1k_3}\tilde{V}^{k_4}_{k_2k_4-k_2}(\omega_{k_1+k_3}-\omega_{k_1}-\omega_{k_3}-\omega_{k_4}+\omega_{k_2}+\omega_{k_2-k_4})+\\
&+\tilde{V}^{k_2+k_3}_{k_2k_3}\tilde{V}^{k_4}_{k_1k_4-k_1}(\omega_{k_2+k_3}-\omega_{k_2}-\omega_{k_3}-\omega_{k_4}+\omega_{k_1}+\omega_{k_1-k_4})+\\
&+\tilde{U}_{-k_1-k_2k_1k_2}\tilde{V}^{k_3}_{k_4k_3-k_4}(\omega_{k_1+k_2}+\omega_{k_1}+\omega_{k_2}-\omega_{k_3}+\omega_{k_4}+\omega_{k_3-k_4})+\\
&+\tilde{U}_{-k_1-k_3k_1k_3}\tilde{V}^{k_2}_{k_4k_2-k_4}(\omega_{k_1+k_3}+\omega_{k_1}+\omega_{k_3}-\omega_{k_2}+\omega_{k_4}+\omega_{k_2-k_4})+\\
&+\tilde{U}_{-k_2-k_3k_2k_3}\tilde{V}^{k_1}_{k_4k_1-k_4}(\omega_{k_2+k_3}+\omega_{k_2}+\omega_{k_3}-\omega_{k_1}+\omega_{k_4}+\omega_{k_1-k_4}), \quad (29)
\end{aligned}$$

$$\begin{aligned}
D_{k_1k_2k_3k_4} = {} & -\tilde{U}_{-k_1-k_2k_1k_2}\tilde{V}^{k_3+k_4}_{k_3k_4}(\omega_{k_1+k_2}+\omega_{k_1}+\omega_{k_2}+\omega_{k_3+k_4}-\omega_{k_3}-\omega_{k_4})-\\
&-\tilde{U}_{-k_1-k_3k_1k_3}\tilde{V}^{k_2+k_4}_{k_2k_4}(\omega_{k_1+k_3}+\omega_{k_1}+\omega_{k_3}+\omega_{k_2+k_4}-\omega_{k_2}-\omega_{k_4})-\\
&-\tilde{U}_{-k_1-k_4k_1k_4}\tilde{V}^{k_3+k_2}_{k_3k_2}(\omega_{k_1+k_4}+\omega_{k_1}+\omega_{k_4}+\omega_{k_3+k_2}-\omega_{k_3}-\omega_{k_2})-\\
&-\tilde{U}_{-k_2-k_3k_2k_3}\tilde{V}^{k_1+k_4}_{k_1k_4}(\omega_{k_2+k_3}+\omega_{k_2}+\omega_{k_3}+\omega_{k_1+k_4}-\omega_{k_1}-\omega_{k_4})-\\
&-\tilde{U}_{-k_2-k_4k_2k_4}\tilde{V}^{k_1+k_3}_{k_1k_3}(\omega_{k_2+k_4}+\omega_{k_2}+\omega_{k_4}+\omega_{k_1+k_3}-\omega_{k_1}-\omega_{k_3})-\\
&-\tilde{U}_{-k_3-k_4k_3k_4}\tilde{V}^{k_1+k_2}_{k_1k_2}(\omega_{k_3+k_4}+\omega_{k_3}+\omega_{k_4}+\omega_{k_1+k_2}-\omega_{k_1}-\omega_{k_2}). \quad (30)
\end{aligned}$$

To cancel nonresonant fourth order terms in (27) relations given below must be valid:

$$\begin{aligned}
\tilde{G}^{k_4}_{k_1k_2k_3} &= \frac{1}{\omega_{k_1}+\omega_{k_2}+\omega_{k_3}-\omega_{k_4}}(G^{k_4}_{k_1k_2k_3}+D^{k_4}_{k_1k_2k_3}),\\
\tilde{R}_{k_1k_2k_3k_4} &= \frac{1}{\omega_{k_1}+\omega_{k_2}+\omega_{k_3}+\omega_{k_4}}(R_{k_1k_2k_3k_4}+D_{k_1k_2k_3k_4}). \quad (31)
\end{aligned}$$

Now the Hamiltonian has only resonant four-wave interaction term ($2 \Leftrightarrow 2$):

$$
\begin{aligned}
H = & \int \omega_k |b_k|^2 dk + \\
& + \frac{1}{2} \int [W_{k_1 k_2}^{k_3 k_4} + D_{k_1 k_2}^{k_3 k_4} + (\omega_{k_1} + \omega_{k_2} - \omega_{k_3} - \omega_{k_4})(\tilde{W}_{k_1 k_2}^{k_3 k_4} \\
& - i \tilde{\tilde{W}}_{k_1 k_2}^{k_3 k_4})] b_{k_1}^* b_{k_2}^* b_{k_3} b_{k_4} \delta_{k_1+k_2-k_3-k_4} dk_1 dk_2 dk_3 dk_4
\end{aligned} \tag{32}
$$

If we put

$$
\tilde{W}_{k_1 k_2}^{k_3 k_4} - i \tilde{\tilde{W}}_{k_1 k_2}^{k_3 k_4} = 0, \tag{33}
$$

we obtain so-called Zakharov equation with the following Hamiltonian:

$$
\begin{aligned}
H &= \int \omega_k |b_k|^2 dk + \frac{1}{2} \int T_{k_1 k_2}^{k_3 k_4} b_{k_1}^* b_{k_2}^* b_{k_3} b_{k_4} \delta_{k_1+k_2-k_3-k_4} dk_1 dk_2 dk_3 dk_4 \\
T_{k_1 k_2}^{k_3 k_4} &= W_{k_1 k_2}^{k_3 k_4} + D_{k_1 k_2}^{k_3 k_4}
\end{aligned} \tag{34}
$$

At this moment the key point of the transformation takes place: we explicitly use property of vanishing of $T_{k_1 k_2}^{k_3 k_4}$ on the resonant manifold and consider waves propagating in the same direction. Then we chose instead of (33) the following expression:

$$
\tilde{W}_{k_1 k_2}^{k_3 k_4} - i \tilde{\tilde{W}}_{k_1 k_2}^{k_3 k_4} = \frac{1}{\omega_{k_1} + \omega_{k_2} - \omega_{k_3} - \omega_{k_4}} (\tilde{T}_{k_1 k_2}^{k_3 k_4} - W_{k_1 k_2}^{k_3 k_4} - D_{k_1 k_2}^{k_3 k_4}), \tag{35}
$$

here

$$
\begin{aligned}
\tilde{T}_{k_2 k_3}^{k k_1} = & \frac{\theta(k)\theta(k_1)\theta(k_2)\theta(k_3)}{8\pi} [(k k_1 (k + k_1) + k_2 k_3 (k_2 + k_3)) - \\
& - (k k_2 |k - k_2| + k k_3 |k - k_3| + k_1 k_2 |k_1 - k_2| + k_1 k_3 |k_1 - k_3|)],
\end{aligned} \tag{36}
$$

This coefficient $\tilde{T}_{k_2 k_3}^{k k_1}$ gives us simple Hamiltonian (11).

Now we can calculate symmetrized coefficients A of the cubic part of the transformation:

$$
\begin{aligned}
A_{k_3 k_4}^{k_1 k_2} = & \frac{1}{\omega_{k_1} + \omega_{k_2} - \omega_{k_3} - \omega_{k_4}} \Big[\tilde{T}_{k_1 k_2}^{k_3 k_4} - W_{k_1 k_2}^{k_3 k_4} + 2(U_{-k_1 - k_2 k_1 k_2} \tilde{U}_{-k_3 - k_4 k_3 k_4} + V_{k_1 k_2}^{k_1 + k_2} \tilde{V}_{k_3 k_4}^{k_3 + k_4} \\
& - V_{k_3 k_1 - k_3}^{k_1} \tilde{V}_{k_2 k_4 - k_2}^{k_4} - \tilde{V}_{k_3 k_2 - k_3}^{k_2} V_{k_1 k_4 - k_1}^{k_4} - V_{k_4 k_1 - k_4}^{k_1} \tilde{V}_{k_2 k_3 - k_2}^{k_3} - \tilde{V}_{k_4 k_2 - k_4}^{k_2} V_{k_1 k_3 - k_1}^{k_3}) \Big]
\end{aligned} \tag{37}
$$

$$
\begin{aligned}
A^{k_1 k_2 k_3 k_4} = & \frac{1}{3(\omega_{k_1} + \omega_{k_2} + \omega_{k_3} + \omega_{k_4})} \Big[-R_{k_1 k_2 k_3 k_4} + 2(U_{-k_1 - k_2 k_1 k_2} \tilde{V}_{k_3 k_4}^{k_3 + k_4} + U_{-k_1 - k_3 k_1 k_3} \tilde{V}_{k_2 k_4}^{k_2 + k_4} \\
& + U_{-k_1 - k_4 k_1 k_4} \tilde{V}_{k_2 k_3}^{k_2 + k_3} + \tilde{U}_{-k_2 - k_3 k_2 k_3} V_{k_1 k_4}^{k_1 + k_4} + \tilde{U}_{-k_2 - k_4 k_2 k_4} V_{k_1 k_3}^{k_1 + k_3} + \tilde{U}_{-k_3 - k_4 k_3 k_4} V_{k_1 k_2}^{k_1 + k_2}) \Big]
\end{aligned} \tag{38}
$$

$$A_{k_4}^{k_1k_2k_3} = \frac{-1}{\omega_{k_1}+\omega_{k_2}+\omega_{k_3}-\omega_{k_4}}\Big[G_{k_1k_2k_3}^{k_4} + 2(V_{k_1k_2}^{k_1+k_2}\tilde{V}_{k_3k_4-k_3}^{k_4} + V_{k_1k_3}^{k_1+k_3}\tilde{V}_{k_2k_4-k_2}^{k_4}$$
$$+ U_{-k_1-k_2k_1k_2}\tilde{V}_{k_4k_3-k_4}^{k_3} + U_{-k_1-k_3k_1k_3}\tilde{V}_{k_4k_2-k_4}^{k_2} - \tilde{V}_{k_2k_3}^{k_2+k_3}V_{k_1k_4-k_1}^{k_4} - \tilde{U}_{-k_2-k_3k_2k_3}V_{k_4k_1-k_4}^{k_1})\Big] \tag{39}$$

$$A_{k_2k_3k_4}^{k_1} = \frac{-1}{3(\omega_{k_1}-\omega_{k_2}-\omega_{k_3}-\omega_{k_4})}\Big[G_{k_2k_3k_4}^{k_1} - 2(\tilde{V}_{k_2k_3}^{k_2+k_3}V_{k_4k_1-k_4}^{k_1} + \tilde{V}_{k_2k_4}^{k_2+k_4}V_{k_3k_1-k_3}^{k_1}$$
$$+\tilde{V}_{k_3k_4}^{k_3+k_4}V_{k_2k_1-k_2}^{k_1} + \tilde{U}_{-k_2-k_3k_2k_3}V_{k_1k_4-k_1}^{k_4} + \tilde{U}_{-k_2-k_4k_2k_4}V_{k_1k_3-k_1}^{k_3} + \tilde{U}_{-k_3-k_4k_3k_4}V_{k_1k_2-k_1}^{k_2})\Big]. \tag{40}$$

Below we calculate $A_{k_2k_3k_4}^{k_1}$, $A^{k_1k_2k_3k_4}$, $A_{k_4}^{k_1k_2k_3}$ and $A_{k_1k_2}^{k_3k_4}$ for the case when canonical variable b_k has harmonics with positive k only.

Let us start with $A_{k_2k_3k_4}^{k_1}$, expression (40). According to δ-function in (24) k_1 is also positive. One can finally get:

$$A_{k_2k_3k_4}^{k_1} = \frac{\omega_{k_1}+\omega_{k_2}+\omega_{k_3}+\omega_{k_4}}{48\pi g}k_1(k_1k_2k_3k_4)^{\frac{1}{4}}. \tag{41}$$

Coefficient $A^{k_1k_2k_3k_4}$ has to be calculated for negative k_1 (according to δ-function in (24), so we will calculate it as $A^{-k_1k_2k_3k_4}$.

$$A^{-k_1k_2k_3k_4} = \frac{\omega_{k_1}-\omega_{k_2}-\omega_{k_3}-\omega_{k_4}}{48\pi g}k_1(k_1k_2k_3k_4)^{\frac{1}{4}}. \tag{42}$$

Coefficient $A_{k_4}^{k_1k_2k_3}$ has to be calculated both for positive and negative k_1. For $k_i > 0$ the following is valid:

$$A_{k_4}^{k_1k_2k_3} = \frac{\omega_{k_1}+\omega_{k_2}+\omega_{k_3}+\omega_{k_4}}{16\pi g}k_1(k_1k_2k_3k_4)^{\frac{1}{4}}. \tag{43}$$

For $k_1 < 0$ we will calculate it as $A_{k_4}^{-k_1k_2k_3}$. Let us start with the case $k_4 > k_2$, $k_3 > k_1$:

$$A_{k_4}^{-k_1k_2k_3} = \frac{\omega_{k_4}+\omega_{k_3}+\omega_{k_2}-\omega_{k_1}}{16\pi g}(k_1k_2k_3k_4)^{\frac{1}{4}}k_1\frac{3\sqrt{k_1k_4}-\sqrt{k_2k_3}}{\sqrt{k_1k_4}+\sqrt{k_2k_3}} \tag{44}$$

In the case $k_2 > k_1, k_4 > k_3$:

$$A_{k_4}^{-k_1k_2k_3} = \frac{\omega_{k_4}+\omega_{k_3}+\omega_{k_2}-\omega_{k_1}}{16\pi g}(k_1k_2k_3k_4)^{\frac{1}{4}}k_1\frac{\sqrt{k_1k_4}(2k_3+k_1)-\sqrt{k_2k_3}(2k_3-k_1)}{\sqrt{k_1k_4}+\sqrt{k_2k_3}} \tag{45}$$

In the case $k_3 > k_1, k_4 > k_2$:

$$A_{k_4}^{-k_1k_2k_3} = \frac{\omega_{k_4} + \omega_{k_3} + \omega_{k_2} - \omega_{k_1}}{16\pi g}(k_1k_2k_3k_4)^{\frac{1}{4}}k_1\frac{\sqrt{k_1k_4}(2k_2+k_1) - \sqrt{k_2k_3}(2k_2-k_1)}{\sqrt{k_1k_4}+\sqrt{k_2k_3}} \tag{46}$$

In the case $k_1 > k_2, k_3 > k_4$:

$$A_{k_4}^{-k_1k_2k_3} = \frac{\omega_{k_4} + \omega_{k_3} + \omega_{k_2} - \omega_{k_1}}{16\pi g}(k_1k_2k_3k_4)^{\frac{1}{4}}k_1\frac{\sqrt{k_1k_4}(2k_4+k_1) - \sqrt{k_2k_3}(2k_4-k_1)}{\sqrt{k_1k_4}+\sqrt{k_2k_3}} \tag{47}$$

Coefficient $A_{k_3k_4}^{k_1k_2}$ has to be calculated both for positive and negative k_1. Below we calculate $A_{k_3k_4}^{k_1k_2}$ for the case $k_1, k_2, k_3, k_4 > 0$. Let us start with the case $k_2 > k_3, k_4 > k_1$:

$$A_{k_3k_4}^{k_1k_2} = \frac{1}{\omega_{k_1} + \omega_{k_2} - \omega_{k_3} - \omega_{k_4}}\left[\tilde{T}_{k_1k_2}^{k_3k_4} - \frac{(k_1k_2k_3k_4)^{\frac{1}{4}}}{8\pi}k_1\left(3\sqrt{k_1k_2} + \sqrt{k_3k_4}\right)\right] \tag{48}$$

In the case $k_1 > k_3, k_4 > k_2$:

$$A_{k_3k_4}^{k_1k_2} = \frac{1}{\omega_{k_1} + \omega_{k_2} - \omega_{k_3} - \omega_{k_4}} \times \times \left[\tilde{T}_{k_1k_2}^{k_3k_4} - \frac{(k_1k_2k_3k_4)^{\frac{1}{4}}}{8\pi}\left(\sqrt{k_1k_2}(2k_2+k_1) + \sqrt{k_3k_4}(2k_2-k_1)\right)\right] \tag{49}$$

In the case $k_4 > k_1, k_2 > k_3$:

$$A_{k_3k_4}^{k_1k_2} = \frac{1}{\omega_{k_1} + \omega_{k_2} - \omega_{k_3} - \omega_{k_4}} \times \times \left[\tilde{T}_{k_1k_2}^{k_3k_4} - \frac{(k_1k_2k_3k_4)^{\frac{1}{4}}}{8\pi}\left(\sqrt{k_1k_2}(2k_3+k_1) + \sqrt{k_3k_4}(2k_3-k_1)\right)\right] \tag{50}$$

In the case $k_3 > k_1, k_2 > k_4$:

$$A_{k_3k_4}^{k_1k_2} = \frac{1}{\omega_{k_1} + \omega_{k_2} - \omega_{k_3} - \omega_{k_4}} \times \times \left[\tilde{T}_{k_1k_2}^{k_3k_4} - \frac{(k_1k_2k_3k_4)^{\frac{1}{4}}}{8\pi}\left(\sqrt{k_1k_2}(2k_4+k_1) + \sqrt{k_3k_4}(2k_4-k_1)\right)\right] \tag{51}$$

For $k_1 < 0$ we will calculate it as $A_{k_3k_4}^{-k_1k_2}$ and $k_1, k_2, k_3, k_4 > 0$:

$$A^{-k_1k_2}_{k_3k_4} = \frac{1}{\omega_{k_1}+\omega_{k_2}-\omega_{k_3}-\omega_{k_4}}\left[\frac{(k_1k_2k_3k_4)^{\frac{1}{4}}}{8\pi}k_1\left(\sqrt{k_1k_4}+\sqrt{k_1k_3}-\sqrt{k_3k_4}\right)\right] =$$
$$= \frac{(k_1k_2k_3k_4)^{\frac{1}{4}}}{16\pi g}k_1\left(\omega_{k_2}+\omega_{k_3}+\omega_{k_4}-\omega_{k_1}\right) \quad (52)$$

It appears that if spectrum of $b(x)$ consists of harmonics with positive k only, transformation from b_k to η_k and ψ_k can be considerably simplified. To prove that, let us calculate η_k and ψ_k for positive k using transformations (23) and (7). To recover η_k and ψ_k for negative k one can use the following relations:

$$\eta_{-k} = \eta_k^*, \qquad \psi_{-k} = \psi_k^*. \quad (53)$$

But first let us write η_k and ψ_k as a power series of b_k up to the third order:

$$\eta_k = \eta_k^{(1)} + \eta_k^{(2)} + \eta_k^{(3)}, \qquad \psi_k = \psi_k^{(1)} + \psi_k^{(2)} + \psi_k^{(3)}. \quad (54)$$

Obviously

$$\eta_k^{(1)} = \sqrt{\frac{\omega_k}{2g}}[b_k + b_{-k}^*], \qquad \psi_k^{(1)} = -i\sqrt{\frac{g}{2\omega_k}}[b_k - b_{-k}^*]. \quad (55)$$

Or

$$\eta^{(1)}(x) = \frac{1}{\sqrt{2}g^{\frac{1}{4}}}(\hat{k}^{\frac{1}{4}}b(x) + \hat{k}^{\frac{1}{4}}b(x)^*), \qquad \psi^{(1)}(x) = -i\frac{g^{\frac{1}{4}}}{\sqrt{2}}(\hat{k}^{-\frac{1}{4}}b(x) - \hat{k}^{-\frac{1}{4}}b(x)^*). \quad (56)$$

Operators $\hat{k}^{\alpha}$ act in Fourier space as multiplication by $|k|^{\alpha}$.

Quadratic terms in (54) are the following:

$$\eta_k^{(2)} = \sqrt{\frac{\omega_k}{2g}}\left[2\int(\tilde{V}^{k_2}_{kk_1} + \tilde{V}^{k_1}_{-kk_2})b^*_{k_1}b_{k_2}\delta_{k+k_1-k_2}dk_1dk_2 - \right.$$
$$\left. - \int(\tilde{V}^{k}_{k_1k_2} + \tilde{U}_{-kk_1k_2})b_{k_1}b_{k_2}\delta_{k-k_1-k_2}dk_1dk_2\right],$$
$$\psi_k^{(2)} = -i\sqrt{\frac{g}{2\omega_k}}\left[2\int(\tilde{V}^{k_2}_{kk_1} - \tilde{V}^{k_1}_{-kk_2})b^*_{k_1}b_{k_2}\delta_{k+k_1-k_2}dk_1dk_2 - \right.$$
$$\left. - \int(\tilde{V}^{k}_{k_1k_2} - \tilde{U}_{-kk_1k_2})b_{k_1}b_{k_2}\delta_{k-k_1-k_2}dk_1dk_2\right]. \quad (57)$$

All coefficients in (57) can be easily calculated using expressions (18), (26), properties (53) and little algebra. The following formulae are valid for both positive and negative k:

$$\eta_k^{(2)} = \frac{|k|}{4\sqrt{2g\pi}}\left[\int k_1^{\frac{1}{4}} b_{k_1} k_2^{\frac{1}{4}} b_{k_2} \delta_{k-k_1-k_2} dk_1 dk_2 + \int k_1^{\frac{1}{4}} b_{k_1}^* k_2^{\frac{1}{4}} b_{k_2}^* \delta_{k+k_1+k_2} dk_1 dk_2 \right.$$
$$\left. - 2\int k_1^{\frac{1}{4}} b_{k_1}^* k_2^{\frac{1}{4}} b_{k_2} \delta_{k+k_1-k_2} dk_1 dk_2 \right],$$
$$\psi_k^{(2)} = -\frac{i}{4\sqrt{2\pi}}\left[\int (\sqrt{k_1}+\sqrt{k_2}) k_1^{\frac{1}{4}} b_{k_1} k_2^{\frac{1}{4}} b_{k_2} \delta_{k-k_1-k_2} dk_1 dk_2 - \right.$$
$$- \int (\sqrt{k_1}+\sqrt{k_2}) k_1^{\frac{1}{4}} b_{k_1}^* k_2^{\frac{1}{4}} b_{k_2}^* \delta_{k+k_1+k_2} dk_1 dk_2 -$$
$$\left. -2\mathbf{sign}(k)\int (\sqrt{k_1}+\sqrt{k_2}) k_1^{\frac{1}{4}} b_{k_1}^* k_2^{\frac{1}{4}} b_{k_2} \delta_{k+k_1-k_2} dk_1 dk_2 \right]. \tag{58}$$

Applying Fourier transformation to (58) one can get

$$\eta^{(2)}(x) = \frac{\hat{k}}{4\sqrt{g}}[\hat{k}^{\frac{1}{4}} b(x) - \hat{k}^{\frac{1}{4}} b^*(x)]^2,$$
$$\psi^{(2)}(x) = \frac{i}{2}[\hat{k}^{\frac{1}{4}} b^*(x)\hat{k}^{\frac{3}{4}} b^*(x) - \hat{k}^{\frac{1}{4}} b(x)\hat{k}^{\frac{3}{4}} b(x)] +$$
$$+ \frac{1}{2}\hat{H}[\hat{k}^{\frac{1}{4}} b(x)\hat{k}^{\frac{3}{4}} b^*(x) + \hat{k}^{\frac{1}{4}} b^*(x)\hat{k}^{\frac{3}{4}} b(x)]. \tag{59}$$

Here $\hat{H}$—is Hilbert transformation with eigenvalue $i\mathbf{sign}(k)$.

Cubic terms in (54) are the following (k, k_1, k_2 and k_3 are positive):

$$\eta_k^{(3)} = \sqrt{\frac{\omega_k}{2g}}\left[\int (A^k_{k_1k_2k_3} + A_{-kk_1k_2k_3}) b_{k_1} b_{k_2} b_{k_3} \delta_{k-k_1-k_2-k_3} dk_1 dk_2 dk_3 + \right.$$
$$+ \int (A^{kk_1}_{k_2k_3} + A^{-kk_2k_3}_{k_1}) b_{k_1}^* b_{k_2} b_{k_3} \delta_{k+k_1-k_2-k_3} dk_1 dk_2 dk_3 +$$
$$\left. + \int (A^{kk_1k_2}_{k_3} + A^{-kk_3}_{k_1k_2}) b_{k_1}^* b_{k_2}^* b_{k_3} \delta_{k+k_1+k_2-k_3} dk_1 dk_2 dk_3 \right],$$
$$\psi_k^{(3)} = -i\sqrt{\frac{g}{2\omega_k}}\left[\int (A^k_{k_1k_2k_3} - A_{-kk_1k_2k_3}) b_{k_1} b_{k_2} b_{k_3} \delta_{k-k_1-k_2-k_3} dk_1 dk_2 dk_3 + \right.$$
$$+ \int (A^{kk_1}_{k_2k_3} - A^{-kk_2k_3}_{k_1}) b_{k_1}^* b_{k_2} b_{k_3} \delta_{k+k_1-k_2-k_3} dk_1 dk_2 dk_3 +$$
$$\left. + \int (A^{kk_1k_2}_{k_3} - A^{-kk_3}_{k_1k_2}) b_{k_1}^* b_{k_2}^* b_{k_3} \delta_{k+k_1+k_2-k_3} dk_1 dk_2 dk_3 \right] \tag{60}$$

Some of coefficients in (60) can be easily calculated using expressions for A and little algebra :

$$A^k_{k_1k_2k_3} + A_{-kk_1k_2k_3} = \frac{\omega_k}{24\pi g} k (kk_1k_2k_3)^{\frac{1}{4}}$$
$$A^k_{k_1k_2k_3} - A_{-kk_1k_2k_3} = \frac{\omega_{k_1}+\omega_{k_2}+\omega_{k_3}}{24\pi g} k (kk_1k_2k_3)^{\frac{1}{4}} \tag{61}$$

$$
\begin{aligned}
A_{k_3}^{kk_1k_2} + A_{k_1k_2}^{-kk_3} &= \frac{\omega_{k_1} + \omega_{k_2} + \omega_{k_3}}{8\pi g} k(kk_1k_2k_3)^{\frac{1}{4}} \\
A_{k_3}^{kk_1k_2} - A_{k_1k_2}^{-kk_3} &= \frac{\omega_k}{8\pi g} k(kk_1k_2k_3)^{\frac{1}{4}}
\end{aligned} \tag{62}
$$

For $k, k_1, k_2, k_3 > 0$

$$
\begin{aligned}
&A_{k_2k_3}^{kk_1} + A_{k_1}^{-kk_2k_3} = \frac{\tilde{T}_{kk_1}^{k_2k_3}}{\omega_k + \omega_{k_1} - \omega_{k_2} - \omega_{k_3}} - \frac{\omega_k}{8\pi g}(kk_1k_2k_3)^{\frac{1}{4}} k - \\
&- \frac{(kk_1k_2k_3)^{\frac{1}{4}}}{8\pi g} min(k, k_1, k_2, k_3) \times \\
&\times \left[\frac{\sqrt{kk_1} + \sqrt{k_2k_3}}{\sqrt{kk_1} - \sqrt{k_2k_3}} \left(\omega_k + \omega_{k_1} + \omega_{k_2} + \omega_{k_3}\right) + \frac{\sqrt{kk_1} - \sqrt{k_2k_3}}{\sqrt{kk_1} + \sqrt{k_2k_3}} \left(\omega_k - \omega_{k_1} - \omega_{k_2} - \omega_{k_3}\right) \right] \\
&A_{k_2k_3}^{kk_1} - A_{k_1}^{-kk_2k_3} = \frac{\tilde{T}_{kk_1}^{k_2k_3}}{\omega_k + \omega_{k_1} - \omega_{k_2} - \omega_{k_3}} - \frac{\omega_{k_1} + \omega_{k_2} + \omega_{k_3}}{8\pi g}(kk_1k_2k_3)^{\frac{1}{4}} k - \\
&- \frac{(kk_1k_2k_3)^{\frac{1}{4}}}{8\pi g} min(k, k_1, k_2, k_3) \times \\
&\times \left[\frac{\sqrt{kk_1} + \sqrt{k_2k_3}}{\sqrt{kk_1} - \sqrt{k_2k_3}} \left(\omega_k + \omega_{k_1} + \omega_{k_2} + \omega_{k_3}\right) - \frac{\sqrt{kk_1} - \sqrt{k_2k_3}}{\sqrt{kk_1} + \sqrt{k_2k_3}} \left(\omega_k - \omega_{k_1} - \omega_{k_2} - \omega_{k_3}\right) \right]
\end{aligned} \tag{63}
$$

Using properties (53) expressions for $\eta_k^{(3)}$ and $\psi_k^{(3)}$ can be extended for negative k, so that the following formulae are valid for both positive and negative k:

$$
\begin{aligned}
\eta_k^{(3)} &= \frac{k^2}{24\pi g^{\frac{3}{4}}\sqrt{2}} \int k_1^{\frac{1}{4}} b_{k_1} k_2^{\frac{1}{4}} b_{k_2} k_3^{\frac{1}{4}} b_{k_3} \delta_{k-k_1-k_2-k_3} dk_1 dk_2 dk_3 + \\
&+ \frac{k^2}{24\pi g^{\frac{3}{4}}\sqrt{2}} \int k_1^{\frac{1}{4}} b_{k_1}^* k_2^{\frac{1}{4}} b_{k_2}^* k_3^{\frac{1}{4}} b_{k_3}^* \delta_{k+k_1+k_2+k_3} dk_1 dk_2 dk_3 + \\
&+ \int \left[\sqrt{\frac{\omega_k}{2g}} (A_{k_2k_3}^{kk_1} + A_{k_1}^{-kk_2k_3}) + \frac{k^{\frac{3}{2}} (k_1k_2k_3)^{\frac{1}{4}} \left(k_1^{\frac{1}{2}} + k_2^{\frac{1}{2}} + k_3^{\frac{1}{2}}\right)}{8\pi g^{\frac{3}{4}}\sqrt{2}} \right] \times \\
&\times b_{k_1}^* b_{k_2} b_{k_3} \delta_{k+k_1-k_2-k_3} dk_1 dk_2 dk_3 + \\
&+ \int \left[\sqrt{\frac{\omega_k}{2g}} (A_{k_2k_1}^{-kk_3} + A_{k_3}^{kk_2k_1}) + \frac{k^{\frac{3}{2}} (k_1k_2k_3)^{\frac{1}{4}} \left(k_1^{\frac{1}{2}} + k_2^{\frac{1}{2}} + k_3^{\frac{1}{2}}\right)}{8\pi g^{\frac{3}{4}}\sqrt{2}} \right] \times \\
&\times b_{k_1}^* b_{k_2}^* b_{k_3} \delta_{k+k_1+k_2-k_3} dk_1 dk_2 dk_3
\end{aligned} \tag{64}
$$

$$
\begin{aligned}
\psi_k^{(3)} = & -i\frac{|k|}{24\pi g^{\frac{1}{4}}\sqrt{2}}\int\left(k_1^{\frac{3}{4}}k_2^{\frac{1}{4}}k_3^{\frac{1}{4}}+k_1^{\frac{1}{4}}k_2^{\frac{3}{4}}k_3^{\frac{1}{4}}+k_1^{\frac{1}{4}}k_2^{\frac{1}{4}}k_3^{\frac{3}{4}}\right)b_{k_1}b_{k_2}b_{k_3}\delta_{k-k_1-k_2-k_3}dk_1dk_2dk_3+\\
& +i\frac{|k|}{24\pi g^{\frac{1}{4}}\sqrt{2}}\int\left(k_1^{\frac{3}{4}}k_2^{\frac{1}{4}}k_3^{\frac{1}{4}}+k_1^{\frac{1}{4}}k_2^{\frac{3}{4}}k_3^{\frac{1}{4}}+k_1^{\frac{1}{4}}k_2^{\frac{1}{4}}k_3^{\frac{3}{4}}\right)b^*_{k_1}b^*_{k_2}b^*_{k_3}\delta_{k+k_1+k_2+k_3}dk_1dk_2dk_3+\\
& +\int\left[-i\sqrt{\frac{g}{2\omega_k}}(A^{kk_1}_{k_2k_3}-A^{-kk_2k_3}_{k_1})+i\frac{k^{\frac{3}{2}}(k_1k_2k_3)^{\frac{1}{4}}}{8\pi g^{\frac{1}{4}}\sqrt{2}}\right]b^*_{k_1}b_{k_2}b_{k_3}\delta_{k+k_1-k_2-k_3}dk_1dk_2dk_3+\\
& +\int\left[i\sqrt{\frac{g}{2\omega_k}}(A^{-kk_3}_{k_2k_1}-A^{kk_2k_1}_{k_3})-i\frac{k^{\frac{3}{2}}(k_1k_2k_3)^{\frac{1}{4}}}{8\pi g^{\frac{1}{4}}\sqrt{2}}\right]b^*_{k_1}b^*_{k_2}b_{k_3}\delta_{k+k_1+k_2-k_3}dk_1dk_2dk_3
\end{aligned}
\tag{65}
$$

References

Dyachenko AI (2001) On the dynamics of an ideal fluid with a free surface. Dokl Math 63(1): 115–118

Dyachenko AI, Kachulin DI, Zakharov VE (2013) On the nonintegrability of the free surface hydrodynamics. JETP Lett 98(1):43–47

Dyachenko AI, Kachulin DI, Zakharov VE (2013) Collisions of two breathers at the surface of deep water. Nat Hazards Earth Syst Sci, Spec Issue: Extrem Seas Ship Oper 13:3205–3210

Dyachenko AI, Kachulin DI, Zakharov VE (2014) Freak waves at the surface of deep water. J Phys: Conf Ser 510:012050

Dyachenko AI, Kuznetsov EA, Spector MD, Zakharov VE (1996a) Analytical description of the free surface dynamics of an ideal fluid. Phys Lett A 221:73–79

Dyachenko AI, Kuznetsov EA, Zakharov VE (1996b) Nonlinear dynamics of the free surface of an ideal fluid. Plasma Phys Rep 22(10):829–840

Dyachenko AI, Zakharov VE (1994) Is the free-surface hydrodynamics an integrable system? Phys Lett A 190:144–148

Dyachenko AI, Zakharov VE (2005) Modulation instability of stokes wave → freak wave. JETP Lett 81(6):255–259

Dyachenko AI, Zakharov VE (2008) On the formation of freak waves on the surface of deep water. JETP Lett 88(5):307–311

Dyachenko AI, Zakharov VE (2011) Compact equation for gravity waves on deep water. JETP Lett 93(12):701–705

Dyachenko AI, Zakharov VE (2012) A dynamic equation for water waves in one horizontal dimension. Eur J Mech B 32:17–21

Slunyaev A, Clauss GF, Klein M, Onorato M (2013) Simulations and experiments of short intense envelope solitons of surface water waves. Phys Fluids 25:067105

Zakharov VE (1968) Stability of periodic waves of finite amplitude on the surface of a deep fluid. J Appl Mech Tech Phys 9(2):190–194

Zakharov VE, Dyachenko AI, Prokofiev AO (2006) Freak waves as nonlinear stage of Stokes wave modulation instability. Eur J Mech B/Fluids 120(5):677–692

Zakharov VE, Dyachenko AI, Prokofiev AO (2008) Freak waves: peculiarities of numerical simulations. In: Pelinovsky E, Kharif C (eds) Extreme ocean waves. Springer, Berlin, pp 1–30

Zakharov VE, Lvov VS, Falkovich G (1992) Kolmogorov spectra of turbulence I. Springer, Berlin

Occurrence of Extreme Waves in Finite Water Depth

Leandro Fernandez, Miguel Onorato, Jaak Monbaliu and Alessandro Toffoli

Abstract For random unidirectional wave fields propagating on water of infinite depths, high-order nonlinearity such as modulational instability is responsible for strong departures from Gaussian-and second-order-based statistics. For finite water depth, on the contrary, wave instability attenuates and eventually vanishes for relative water depths as low as $kh = 1.36$ (where k is the wavenumber of the dominant waves and h the water depth). Experimental results, nonetheless, indicates that oblique (directional) perturbations are capable of triggering and sustaining modulational instability and possibly lead to a wave amplitude growth even if $kh < 1.36$. Here a comparative analysis of the statistical properties of surface gravity waves in water of finite depth is discussed by analyzing laboratory experiments, field measurements, and numerical simulations to assess the role of modulational instability on wave statistics and particularly on the occurrence of extremes. Despite evidence of modulational instability in regular wave fields, results presented herein seems to indicates that wave instability has a negligible effect on wave statistics, which remains primarily affected by second-order contributions.

1 Introduction

Statistical properties of random wave fields and the probability of occurrence of extremely large waves (also known as freak or rogue waves) have received much attention during the past decade in many different fields of physics such as nonlinear optics (Solli et al. 2007; Kibler et al. 2010), plasma physics (Bailung et al. 2011)

L. Fernandez (✉) · J. Monbaliu
Department of Civil Engineering, KU Leuven, Kasteelpark Arenberg
40 Box 2448, 3001 Heverlee, Belgium
e-mail: Leandro.Fernandez@bwk.kuleuven.be

M. Onorato
Department of General Physics, University of Torino, Via Pietro Giuria 1, 10125 Torino, Italy

A. Toffoli
Center for Ocean Engineering, Science and Technology, Swinburne University of Technology, P.O. Box 218, Hawthorn, Swinburne, VIC 3122, Australia

E. Pelinovsky and C. Kharif (eds.), *Extreme Ocean Waves*,
DOI 10.1007/978-3-319-21575-4_3

and hydrodynamics (Chabchoub et al. 2011, 2012; Onorato et al. 2013a; Chalikov 2009) (see a general overview in Onorato et al. 2013b). At first order, an idealized wave field can be approximated by a linear superposition of waves with randomly distributed amplitudes and phases. Under these circumstances, statistical properties of water surface elevations can be conveniently approximated by the Normal or Gaussian distribution (Ochi 1998).

Nonlinear contributions, however, cannot be totally disregarded if the steepness is sufficiently high. This routinely happens in the real ocean, where the most obvious manifestation of nonlinearity is a sharpening of wave crests and flattening of troughs, which results in a weak departure from Normality as crests becomes higher and trough shallower than linear theory would predict. In order to capture such an effect, second-order corrections are normally applied to the linear solution (Longuet-Higgins 1963) and theoretical (second-order based) distributions extracted (e.g. Tayfun 1980; Forristall 2000).

Interestingly enough, if wave energy is concentrated around a dominant component (narrow banded assumption) and confined within a rather narrow range of directions (the wave field is approximately unidirectional), higher-order nonlinear effects, such as modulational wave instability, take place. The instability of freely propagating deep water wave trains to side band perturbations is a well-known phenomenon (e.g., Janssen 2003) and causes particularly strong focusing of wave energy when the Benjamin-Feir Index (*BFI*), a ratio of steepness to dispersion (Onorato et al. 2001; Janssen 2003) is $O(1)$. As a consequence, a small part of a wave train grows at the expense of the surrounding waves, which eventually turns into an extreme wave. In a random sea state, this results in strong deviations from Gaussian-and second-order-based statistics (Janssen 2003; Onorato et al. 2001, 2006b; Mori and Yasuda 2002; Mori et al. 2007; Socquet-Juglard et al. 2005; Trulsen and Dysthe 1997; Mori and Janssen 2006; Toffoli et al. 2009). It should be mentioned, however, that the strength of higher-order effects (and modulational instability in particular) reduces with the increase of the directional spreading, attenuating deviations from Normality (Onorato et al. 2009a, b; Toffoli et al. 2010; Waseda et al. 2009).

These results apply for a condition of deep water, where the relative depth $k_0 h \rightarrow \infty$ with k_0 the wavenumber associated to the dominant wave component and h the water depth. Many observations of extreme waves in the ocean, nonetheless, have been reported in conditions of arbitrary water depth with $k_0 h < 2$ (Haver and Andersen 2000; Trulsen 2007; Babanin et al. 2011; Chien et al. 2002). Under these circumstances, waves interact with the nearby sea floor, generating a current that subtracts energy from nonlinear focussing. Correspondingly, modulation of wave trains to side band perturbations attenuates and eventually vanishes for $k_0 h = 1.36$. Whereas this is valid for collinear perturbations, however, a plane wave can still destabilize if subjected to oblique (directional) disturbances (Slunyaev et al. 2002; Benjamin 1967; Whitham 1974; Janssen and Onorato 2007; Mori and Yasuda 2002; Kristiansen et al. 2005; Francius and Kharif 2006; McLean 1982).

An experimental verification of the destabilization of wave trains due to oblique perturbations in water of finite depth was recently discussed in Toffoli et al. (2013). Experiments also corroborated that oblique perturbations lead to a notable amplitude growth. Therefore, it would be feasible to assume that statistical properties of surface gravity waves and specifically the occurrence of extremes could be affected by modulational instability processes in finite water depth for directional wave fields. In this regard, some attempts to assess the effect of wave instability, directionality and water depth on wave statistics was presented in Toffoli et al. (2007, 2009), among others. Results, however, were not conclusive and hence it remains unclear, whether the observed amplitude growth by oblique perturbations is sufficient to induce substantial deviations from Gaussian-and second-order-based statistical distributions in finite water depth.

Laboratory experiments in a large wave basin are here used to discusses the effect of high-order nonlinearity and particularly modulational instability on wave statistics and probability of extremes. Direct numerical simulations of the potential Euler equations and field measurements are used to support data interpretation. In the following section, the data sets and the initial conditions for simulations and experiments are discussed. In Sect. 3, spectral evolution in the wave basin is discussed to asses the extent of wave dissipation (by bottom friction and/or wave breaking). In Sect. 4, statistical properties of the water surface are presented. Concluding remarks are reported in the last Section.

2 Data Sets

2.1 Laboratory Experiments

A set of experiments have been carried out in the ocean wave basin at MARINTEK (Norway). The basin is 70 m wide and 50 m long (see Fig. 1) and is equipped with an adjustable floor. For the present study, water depth was set equal to 0.78 m. The basin is also equipped with a directional wavemaker along the 70 m side, which consists altogether of 144 individually controlled flaps. Reflection is controlled by two absorbing beaches at the opposite side of the wavemaker and on its right-hand side. On the left-hand side, a second wave maker is present, which acts as a vertical wall. Overall, reflection in the middle of the basin is negligible (cf. Onorato et al. 2009a).

Wave measurements were concentrated along the central axis of the basin (see Fig. 1) where a total of 27 resistance wave gauges operating at a sampling frequency of 200 Hz were deployed to trace the spatial evolution of the wave field. Three 3-probe arrays and one 8-probe array were installed to monitor the evolution of the directional wave spectrum too.

Initial conditions at each paddle of the wavemaker were imposed in the form time series of irregular waves. These were generated as an inverse Fast Fourier Transform of an input directional wave spectrum $E(\omega, \vartheta) = S(\omega)D(\vartheta)$, where $S(\omega)$ is the frequency spectrum, $D(\vartheta)$ is the directional function, ω is the angular frequency and ϑ is the direction. Amplitudes were randomly chosen from the Rayleigh distribution,

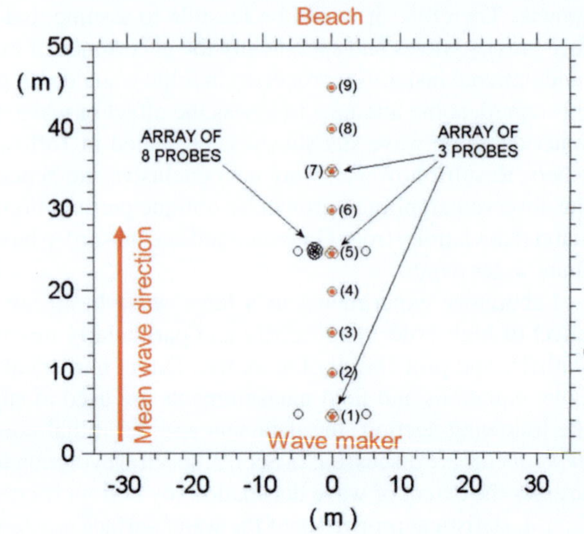

Fig. 1 Experimental set up in the ocean wave basin at MARINTEK

while the random phases were assumed to be uniformly distributed in the interval [0, 2π). Hence, an initial Gaussian wave field is here used as input. Note that second- and higher-order nonlinearity (if any) develops naturally as waves start propagating along the basin.

For convenience, $S(\omega)$ is expressed in the form of a JONSWAP spectrum (Komen et al. 1994):

$$S(\omega) = \frac{2\pi\alpha g^2}{\omega^5} \exp\left[-\frac{5}{4}\left(\frac{\omega}{\omega_p}\right)^{-4}\right] \gamma^{\exp[-(\omega-\omega_p)^2/(2\sigma^2\omega_p^2)]}, \tag{1}$$

where α is the Phillips parameter, γ is the peak enhancement factor and ω_p is the peak frequency; the parameter σ is equal to 0.07 if $\omega \leq \omega_p$ and 0.09 if $\omega > \omega_p$. For the present study, three spectral configurations were selected; spectral parameters are detailed in Table 1. For all cases, the peak enhancement factor γ is kept constant and equal to 6. Several peak periods T_p are used. As the water depth is uniform, this ensures that different spectral configurations operate in a different regimes of relative water depth (k_0h). Significant wave height H_s was chosen in such a way that the wave steepness k_0a, where $a = H_s/2$, is constantly equal to 0.13. This is a fairly high value of wave steepness, which was often recorded during ship accidents reported as being due to bad weather conditions (Toffoli et al. 2005). As both k_0h and k_0a affect the degree of nonlinearity in the system, a general steepness parameter μ was applied (see Table 1). It can be written as follows (see e.g. Toffoli et al. 2007 and references therein):

Table 1 Input spectral parameters

Exp. ID	T_p [s]	L_p [m]	H_s [m]	kh	α	γ	ka	μ
A	1.35	2.76	0.12	1.78	0.013	6	0.13	0.19
B	1.64	3.81	0.16	1.30	0.018	6	0.13	0.25
C	1.77	4.27	0.18	1.15	0.011	6	0.13	0.28

$$\mu = \frac{1}{2} H_s\, k_0\, coth(k_0 h) \left[1 + \frac{3}{a\ sinh^2\,(k_0 h)} \right]. \tag{2}$$

Note that μ tends to the wave steepness $k_0 a$ for deep water conditions, while it tends to the Ursell number for the shallow-water limit (i.e., $k_0 h \rightarrow 0$).

In order to model the directional spreading, a frequency-independent $D(\vartheta) = cos^N(\vartheta)$ function was applied. Two different values of the directional spreading coefficient N were used for every JONSWAP configurations: $N = 840$ and $N = 2$, representing unidirectional and directional sea states, respectively.

It is worth mentioning that the Benjamin-Feir Index is attenuated by both wave directionality and water depth. For unidirectional wave fields in deep water, the BFI associated to the selected spectra is $O(1)$, while it drops to about 10^{-1} for the shallowest and most directional condition that is considered herein (see Mori et al. 2011; Janssen and Onorato 2007 for details on BFI calculation). Although higher value of *BFI* are possible in principle, the necessary steepness and dispersion (spectral bandwidth) would not be reasonable for ocean and coastal waves.

In order to ensure enough data to perform statistically significant analysis, four time series of 20 min were recorded during the experiment. For each set of measurement, different random amplitudes and phases were applied.

2.2 Numerical Simulations

Direct numerical simulations were carried out in an attempt to replicate experimental results. Computations consisted in solving the potential Euler equations of motion. Assuming an irrotational, inviscid and incompressible fluid flow, a velocity potential $\phi(x, y, z, t)$ that satisfies the Laplace's equation everywhere in the fluid can be defined. For the present study, a constant water depth is also assumed. At the bottom ($z = -h$) the boundary condition is such that the vertical velocity is zero (i.e., $\phi_z|_h = 0$). At the free surface $z = \eta(x, y, t)$, the kinematic and dynamic boundary conditions are satisfied for the free surface elevation and the velocity potential at the free surface $\psi(x, y, t) = \phi(x, y, \eta(x, y, t), t)$. Using the free surface variables these conditions can be written as follows (see, e.g., Zakharov 1968):

$$\psi_t + g\eta + \frac{1}{2}\left(\psi_x^2 + \psi_y^2\right) - \frac{1}{2}W^2\left(1 + \eta_x^2 + \eta_y^2\right) = 0, \tag{3}$$

$$\eta_t + \psi_x \eta_x + \psi_y \eta_y - W\left(1 + \eta_x^2 + \eta_y^2\right) = 0, \quad (4)$$

where the subscripts indicate the partial derivatives, and $W(x, y, t) = \phi_z|_\eta$ represents the vertical velocity at the free surface.

The temporal evolution of the surface elevation of an initial surface $\eta(x, y, t = 0)$ can be estimated directly from the system of equations (3) and (4) by solving the system of equations with a Higher-Order Spectral Method (HOSM) as derived in West et al. (1987). A substantial advantage of HOSM, compared with other methods, is that it is able to simulate the evolution of a wave field in a large number of random realizations of the surface elevation in a reasonable computational time, without limitations in terms of the spectral bandwidth.

This approach is based on a pseudo-spectral method, which uses a series expansion in the wave steepness (ε) of the vertical velocity $W(x, y, t)$ about the free surface. Here, we considered a third-order expansion. The former allows the inclusion of four waves interactions (Tanaka 2001b), which is directly responsible for modulational instability. It is important to mention, however, that the model is not fully nonlinear. Therefore, effect of order higher than modulational instability are not accounted for. After evaluating the vertical velocity at the free surface at third order, the free surface velocity potential $\psi(x, y, t)$ and the surface elevation $\eta(x, y, t)$ can be integrated in time from Eqs. (3) and (4). The time integration is performed by means of a fourth-order Runge–Kutta method with a constant time step equivalent to $T_p/200$. All aliasing errors generated in the nonlinear terms are removed (West et al. 1987; Tanaka 2001b). Note, however, that no additional terms were included to take into account wave dissipation. A concise review of HOSM can be found in Tanaka (2001a).

The initial conditions for the model is a linear surface derived with a inverse Fast Fourier Transform from an input spectrum. In this regard, spectral conditions are identical to the ones used in the laboratory experiments (see Table 1). The dimension of the physical domain was defined by a mesh of 256 $\times$ 256 points with a resolution in both dimensions equivalent to $\Delta x = \Delta y = L_p/4$ so that a dominant wave is discredited by approximately 25 grid points. The model was run to ensure a temporal evolution of the initial surface equivalent to 60 peak periods. To ensure enough data for statistical analysis, furthermore, 100 random realizations were carried out for each spectral configuration.

It is important to remark that numerical simulations provide a temporal evolution of the surface elevation, while a spatial evolution is observed in the ocean basin. To compare the results, we assume that space and time can be converted according to the group velocity. This approach has been used to compare HOSM simulations with laboratory experiments in Toffoli et al. (2010) with satisfactory results.

For benchmarking second-order effects, numerical simulations from a second-order wave model were also used to derive statistical information. Second-order simulations are identical to those used in Toffoli et al. (2007).

2.3 Field Measurements: The Lake George Data Set

Field measurements of surface elevations gathered at the Lake George field experimental site (Australia) between September 1997 and August 2000 were used to explore statistical properties of finite water depth waves (see Young et al. 2005; Toffoli et al. 2007 for details). Records were obtained by using an array of eight capacitance gauges, acquiring at a sampling frequency of 25 Hz. This set of probes was located under a measurement bridge in the form of a centered pentagon with a diameter of 15 cm sufficiently far from the structure in order to avoid any disturbances.

The lake bed in the region of the observation site was flat and the water depth gradually changed from 1.1 to 0.4 m due to a natural process of drying out. This ensured a fairly large range of relative depths kh. Records were sorted out to ensure that data with general steepness μ as used in the experiment were available for analysis (see Toffoli et al. 2007).

As field measurements consist in wind generated waves, the directional spreading is normally broad (Young et al. 2005; Toffoli et al. 2007). To some extent, the wave directional spreading is consistent with experiments and simulations that were obtained with $N = 2$.

3 Evolution of Significant Wave Height and Wave Energy Spectrum

It is instructive to observe how the wave energy spectrum and the concurrent significant wave height evolve in space within the wave basin. In Figs. 2 and 3, the spatial evolution of H_s and frequency spectrum is presented.

Initial conditions were designed to minimize breaking probability, while maintaining a sufficiently large value of steepness (cf. Onorato et al. 2009a; Toffoli et al. 2010). For unidirectional sea states (upper panel in Fig. 2), in fact, the significant wave height does not show any notable attenuation along the basin for any k_0h. This, apparently, excludes any substantial effects due to bottom friction. Directional sea states, however, appear to be more prone to dissipation. Although energy loss is limited for the deepest water depth condition (see case $\mu = 0.19$ in the lower panel of Fig. 2), it becomes substantial for shallower depths, where approximately 25 % of wave energy is lost along the basin. Based on visual observation, this dissipation is primarily attributed to a more frequent occurrence of wave breaking.

As dissipation is limited for unidirectional waves, the wave spectrum does not show any substantial changes. For the conditions of deepest water ($kh = 1.7$ and $\mu = 0.19$), nonetheless, the wave spectrum undergoes an expected spectral downshifting due to an energy redistribution from higher to lower frequencies, as a result of modulational instability (Onorato et al. 2009a). For lower relative water depths, modulational instability attenuates (and eventually vanishes for $k_0h = 1.36$) with a consequent suppression of peak downshifting (see upper panel in Fig. 3). This is also consistent with the evolution of numerical wave fields for $k_0h \approx 1.36$ (Janssen

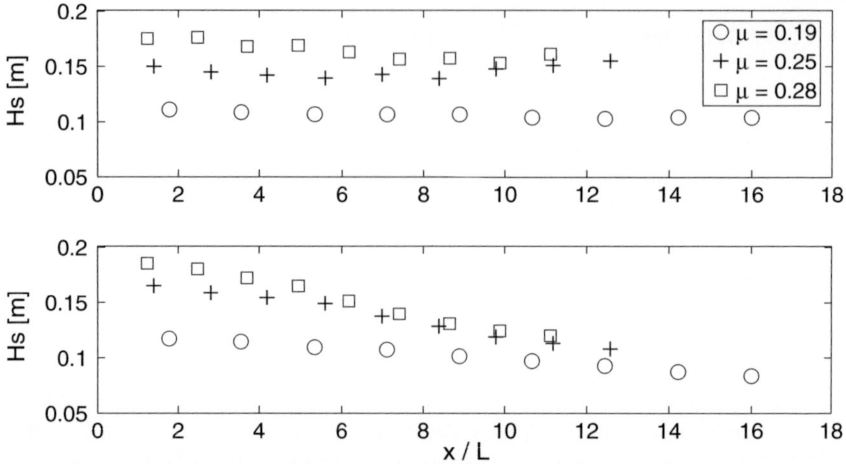

Fig. 2 Spatial evolution of significant wave height H_s in the ocean wave basin: unidirectional wave fields ($N = 840$) in the *upper panel* and directional wave fields ($N = 2$) in the *lower panel*

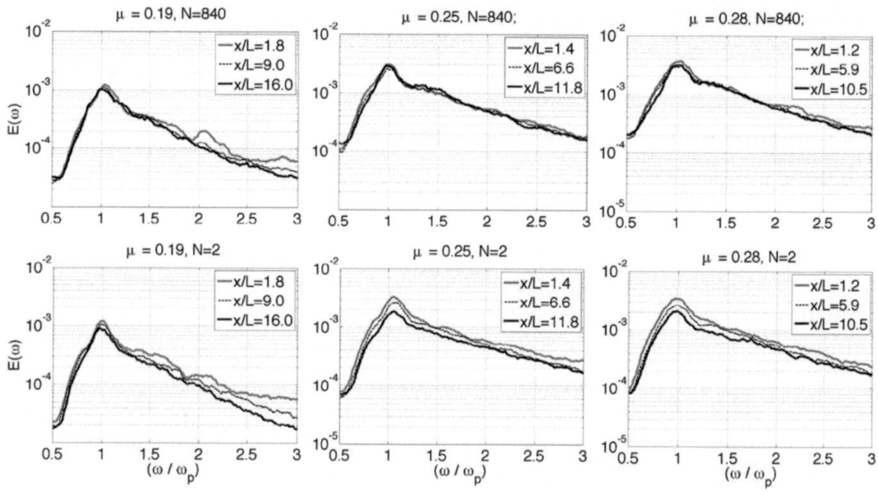

Fig. 3 Spatial evolution of significant wave height H_s in the ocean wave basin

and Onorato 2007; Toffoli et al. 2009). Similarly, the same behavior of the spectral peak is observed for directional wave fields (see lower panels in Fig. 3). For the latter, however, a significant energy loss around the spectral peak is detectable as a result of wave breaking.

4 Wave Statistics

4.1 Skewness and Kurtosis

Statistical properties of the surface elevation can be conveniently summaries by the third-and fourth-order moments of the probability density function, namely the skewness and the kurtosis, respectively. Whereas the former is an indicator of vertical asymmetry in the signal and hence of the strength of second-order effects (e.g., Ochi 1998), the latter provides information on the percentage of extremes in the record and it is conveniently used to estimate the effect of modulational instability on the occurrence of extremes (Onorato et al. 2001). The initial linear surfaces for the experimental and numerical tests are Normally distributed and hence, the skewness is zero (i.e., the wave profile is vertically symmetric) and the kurtosis is equal to 3. As waves propagate along the tank, nonlinear interactions take places, modifying the statistical properties of the wave field. The evolution of skewness and kurtosis, in this regard, is presented in Figs. 4, 5 and 6 for different values of the general nonlinear parameter μ.

In a condition of fairly deep water ($k_0h = 1.78$), development of modulational instability and concurrent formation of very large waves is observed for long crested wave fields ($N = 840$). Numerical simulations provides a good approximation of the experimental results.

As expected, the broadening of the wave spectrum suppresses effect related to modulational instability. While wave instability can still be active, amplitude growth is no longer notable and the kurtosis does not diverge from expected linear (kurtosis $= 3$) and second-order-based (kurtosis ≈ 3.1) statistics (cf. Onorato et al.

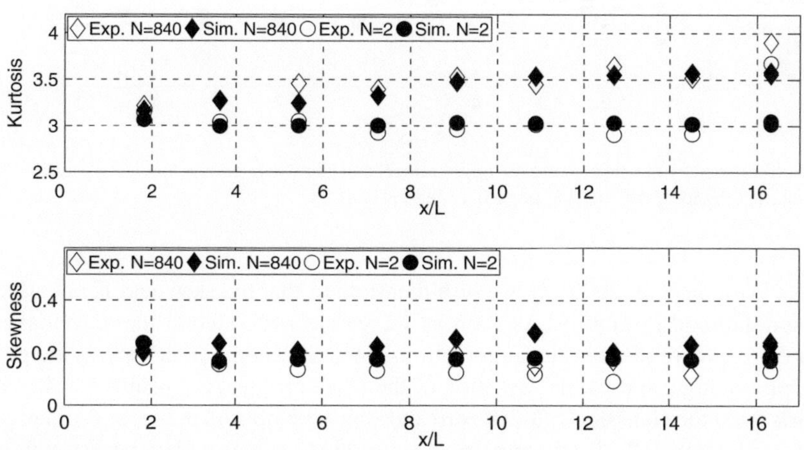

Fig. 4 Temporal evolution of kurtosis (*upper panel*) and skewness (*lower panel*) for an initial wave field characterized by $kh = 1.78$, $ak = 0.13$, and $\mu = 0.19$

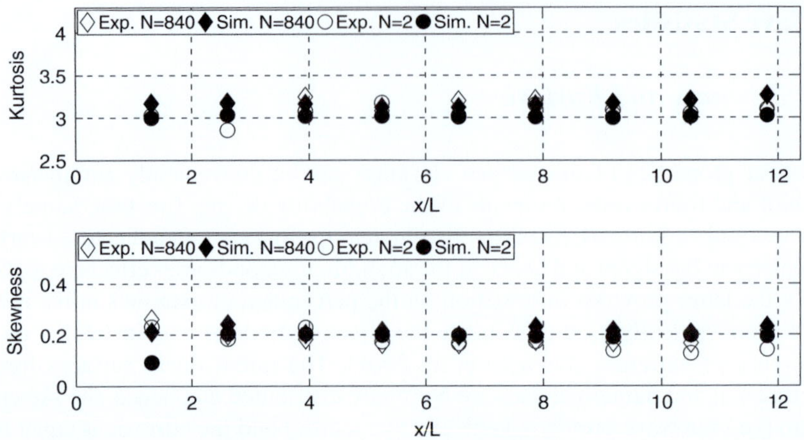

Fig. 5 Temporal evolution of kurtosis (*upper panel*) and skewness (*lower panel*) for an initial wave field characterized by $kh = 1.30$, $ak = 0.13$, and $\mu = 0.25$

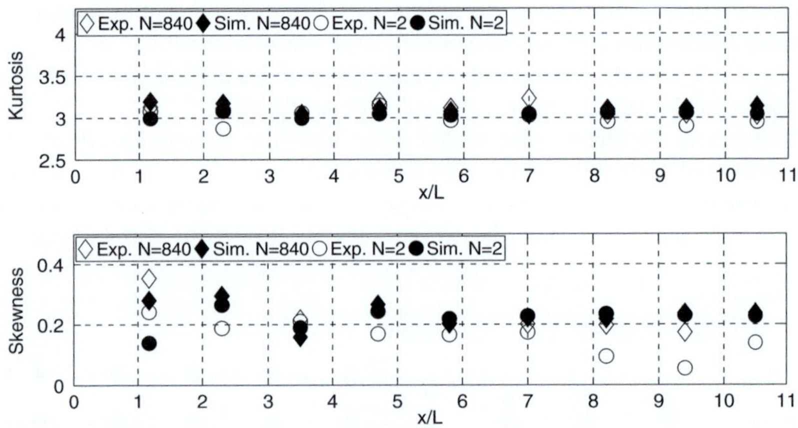

Fig. 6 Temporal evolution of kurtosis (*upper panel*) and skewness (*lower panel*) for an initial wave field characterized by $kh = 1.15$, $ak = 0.13$, and $\mu = 0.28$

2009a; Waseda et al. 2009). It is worth mentioning that the skewness is not affected by modulational instability. As soon as waves are generated at the wavemaker or allowed to evolve in numerical simulations, bound (second order) waves are generated, producing a vertical deformation of the wave profile. As a result, the skewness departs from the Gaussian value of zero and stabilizes around the second-order value of approximately 0.2. Again, numerical simulations agrees with experimental data.

As water depth decreases, modulational instability gradually weakens and eventually vanishes for $kh = 1.36$. Experimental and numerical unidirectional wave fields, in this respect, no longer show a clear growing pattern for the kurtosis, which remains only slightly above the value of 3. Despite verified effects due to oblique perturbations, no substantial changes in the kurtosis are observed for directional wave fields too. In terms of skewness no significant differences are observed with respect to deeper water conditions. As the water depth is not too shallow, the strength of second-order contributions does not intensify notably. It is important to mention, nonetheless, that the finite extension of the basin may not allow instability to fully develop, especially due to the very low values of *BFI*. Simulations of longer evolutions (equivalent to about 500 wavelengths), though, does not show any significance deviation of wave statistics from second-order-based values.

To summarize the behavior of kurtosis, the latter is presented as a function of the relative water depth k_0h in Fig. 7 (values in the proximity of the center of the basin are considered). Numerical and field data are shown for comparison too. Whereas kurtosis attenuates with reduction of relative water depth in unidirectional wave fields ($N = 840$), a growing trend (i.e., kurtosis increases with reducing k_0h) is observed for directional sea states ($N = 2$). Overall, experiments agree with numerical simulations qualitatively. Nevertheless, simulations underestimate the kurtosis in directional wave fields with $k_0h < 1.4$. In this regard, we remark that numerical simulations account for a third-order expansion ($M = 3$) and hence, only modulational instability is considered. Neglected higher-order effects appears to play a significant role.

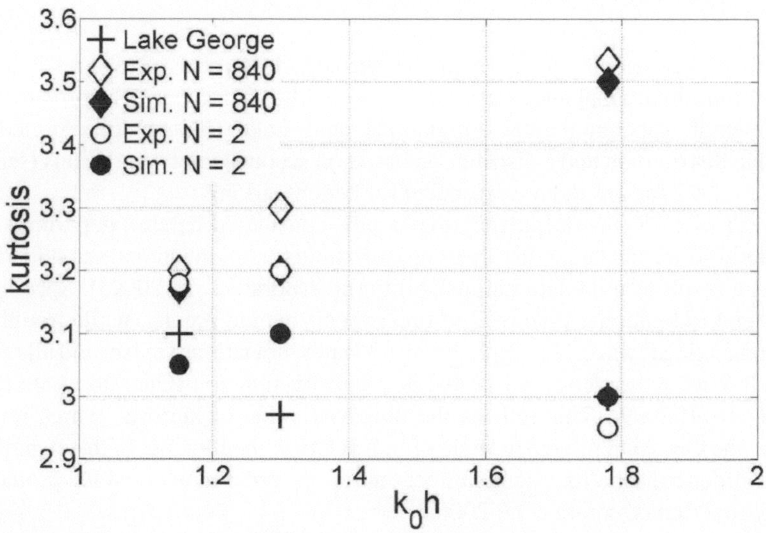

Fig. 7 Kurtosis as a function relative water depth k_0h. Note the coincidence of the Lake George experimental data with the simulation ($N = 2$) at $k_0h = 1.78$

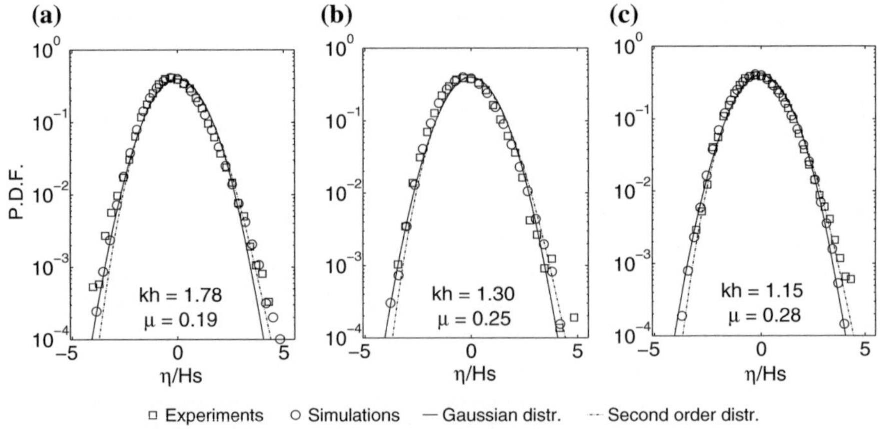

Fig. 8 Probability density function for unidirectional sea states

Field measurements of wind generated wave fields at Lake George also indicates a increase of kurtosis for small relative depth. In contrast with experiments and numerical simulations, however, a deviation from Gaussianity becomes only evident for $k_0 h < 1.2$, while random waves remains Gaussian for deeper water depth.

4.2 Probability Density Function

The probability density function of the surface elevation is presented in Figs. 8 and 9 for unidirectional and directional wave fields, respectively. Records from the experimental tests, numerical simulations, and Lake George are reported; the Gaussian distribution and a distribution based on second-order simulations (see Toffoli et al. 2007 for details) are presented for benchmarking.

For $kh = 1.78$ (the deepest relative depths considered herein), departures from Gaussian and second-order distributions are visible for unidirectional sea states ($N = 840$) as a result of modulational instability (cf. Onorato et al. 2009a). Specifically, crests tend to be higher than second-order theory would predict, while troughs are deeper than linear waves. For directional sea states, deviations vanish, and all records fits well a second-order-based probability density function (Onorato et al. 2009a; Waseda et al. 2009). This reflects the observed value of kurtosis, which remains close to the Gaussian reference value of 3. It is worth mentioning, in this regard, that second-order contributions do not affect the fourth- order moment of the probability density function (Onorato et al. 2006a).

With the decrease of relative water depth, deviations at the upper and lower tails of the probability density function attenuates for unidirectional sea states. For $kh = 1.30$ and 1.15, positive elevations fits the second-order-based distribution, while negative

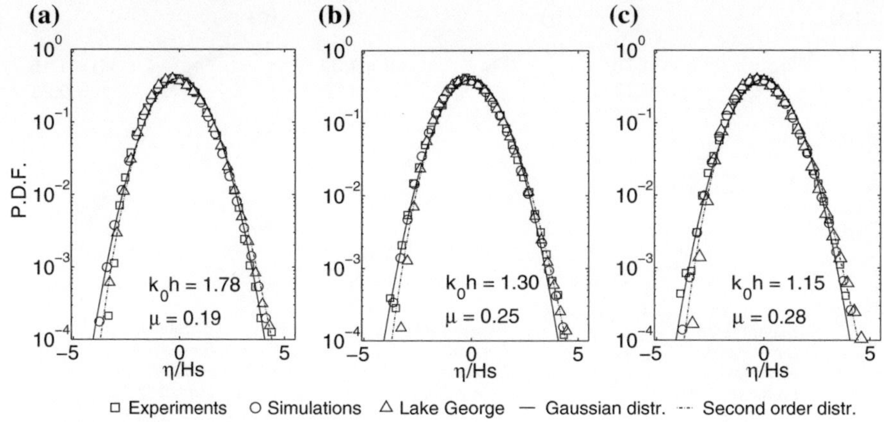

Fig. 9 Probability density function for directional sea states

elevations remains slightly deeper than second-order predictions and better fit the lower tails of the Gaussian distribution. This results in slightly larger wave heights when compared with second-order waves, justifying values of kurtosis higher than three (normally, kurtosis ≈ 3.2 for unidirectional random wave fields in finite water depth). For directional wave fields, a good fits of all records with second-order-based distributions is again observed for the upper tail of the distribution. For the Lake George data set, also the lower tail of the distribution seems to fit relatively well the second-order distribution. Experimental and numerical data, on the contrary, shows a weak deviation of the lower tail from second-order distribution, indicating that troughs become deeper. The lower tail of the Gaussian distribution offers a better fit of experimental and numerical directional wave fields. As aforementioned, the deepening of trough support the observed increase of kurtosis with decreasing relative depth (from a kurtosis of about 3 for $k_0h = 1.78$ to approximately 3.2 for $k_0h < 1.78$, see Fig. 7).

4.3 Wave Crest Distribution

In present section, a comparison of wave crest distributions as obtained from experimental records, numerical simulations, and field measurements is discussed. Wave crest distributions in the form of exceedance probability are shown in Figs. 10 and 11 for unidirectional and directional sea states, respectively. The Rayleigh distribution and the second-order Forristall 2D distribution (Forristall 2000) are included for benchmarking linear and second-order statistics.

It is important to remark that numerical simulations provide a surface and not a time series. Extraction of wave crests from a surface $\eta(x, y, t)$ is not trivial. The problem is here bypassed by recording the temporal evolution at one numerical

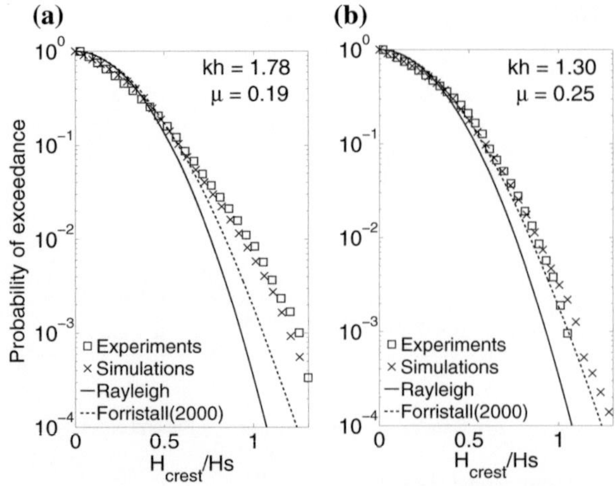

Fig. 10 Wave crest distribution for unidirectional sea states

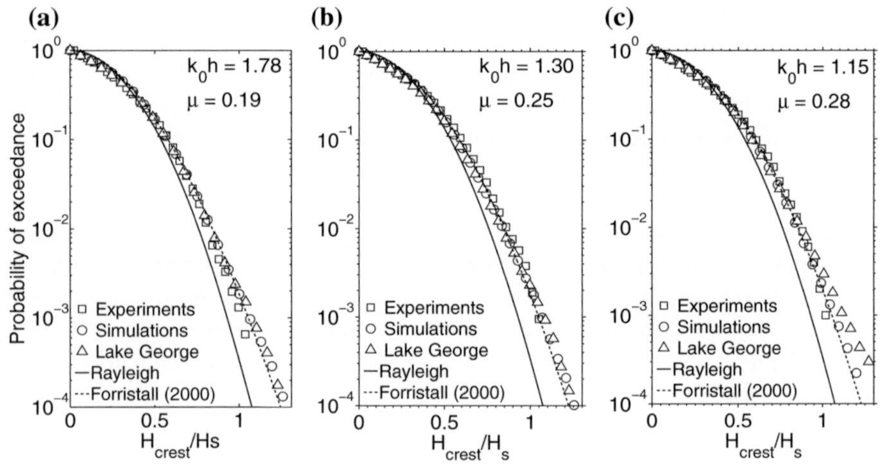

Fig. 11 Wave crest distribution for directional sea states

probes, which is located in the middle of the physical domain. Calculation of wave crests can then performed using a standard zero-downcrossing method.

For the deepest water depth and narrowest directional spreading, deviations from second-order statistics are substantial, despite the fact that waves already start feeling the sea floor for $k_0h = 1.78$. Extreme waves with crests greater than 1.3 times H_s have a probability of occurrence one order of magnitude higher than second-order theory would predict. All records in directional sea states, on the contrary, show a good agreement with second-order statistics as modulational instability does not have any notable effect on wave amplitude growth. Nonetheless, crests in the tail of the exceedance probability ($H_{crest}/H_s > 1$ and probability levels lower than 0.001) show a weak deviation from second-order predictions.

For shallower relative water depth ($k_0h = 1.30$ and 1.15), departure from second-order distribution in unidirectional sea states vanishes and crests fits the Forristall (2000) distribution. Despite the survivability of modulational instability and possible wave amplitude growth due to oblique perturbations (Toffoli et al. 2013), wave crest distributions for directional sea states do not deviates notably from second-order distributions, apart from a weak departure for probability lower than 0.001. Interestingly enough, field measurements at Lake George show a more substantial departure from second-order statistics in the tail of the distribution: this deviation was already discussed in Toffoli et al. (2007) and found to be statistically significant (at least within a 95 % confidence interval). In this regard, however, it is worth mentioning that experimental data are not sufficient to assess probability levels as low as 0.0001, where this deviation is observed. The fact that numerical simulations are not capable of capturing such a deviation, it seems to corroborate that nonlinear effects higher than third-order (modulational instability) are not negligible under the present circumstances.

5 Conclusion

Laboratory experiments in a large wave basin, numerical simulations with a truncated form of the potential Euler equations and field experiments at the Lake George experimental site (Australia) have been discussed to assess the role of third-order nonlinearity, and particularly modulational instability, on wave statistics. Specifically, the role of oblique (directional) perturbations on the instability of wave packets and their ability to trigger wave amplitude growth in water of finite depth was investigated.

Initial conditions for experimental tests and numerical simulations were configured to replicate wave conditions at Lake George. To this end, spectral conditions with a constant wave steepness $k_0a = 0.13$ and different relative water depths ($k_oh = 1.78$, 1.30, and 1.15) were selected. Directional spreading was modeled with a $cos^N(\vartheta)$ function. An almost unidirectional distribution ($N = 840$) and a broad directional distribution typical for wind generated waves ($N = 2$) were applied.

Experimental and numerical records are in good agreement and confirm substantial departures from linear and second-order-based statistics, provides the wave field is unidirectional and relative water depth is sufficiently deep ($k_0h = 1.78$ in the case of the present exercise). The observed deviation, however, are suppressed by the directional spreading of the wave energy, which force wave statistics to fit second-order-based distributions. This is consistent with previous numerical and experimental studies (see Socquet-Juglard et al. 2005; Onorato et al. 2009a; Waseda et al. 2009 among others).

For shallower relative depths ($k_0h = 1.30$ and 1.15), the interaction with the sea floor subtracts energy to modulational instability processes, inhibiting any deviations from second-order statistics. Despite the observed amplitude growth in regular wave packets due to directional perturbations Toffoli et al. (2013), wave statistics of directional wave fields does not deviates notably from standard second-order distributions. It should be mentioned, nonetheless, that field measurements at Lake George indicates that the most extreme wave crests (probability levels lower than 0.001) are slightly under estimated by numerical simulations. This is not captured by numerical simulations, which only includes third-order effects. Further investigation on the effects of higher-order nonlinear contributions is called for to better understand the statistical properties of waves in finite water depth.

Acknowledgments This work has been supported by the European Communitys Sixth Framework Programme through the grant to the budget of the Integrated Infrastructure Initiative HYDRALAB IV within the Transnational Access Activities, Contract No. 022441. F.W.O. project G.0333.09 and E.U. project EXTREME SEAS (contract SCP8-GA-2009-234175) are acknowledged. The authors also acknowledge the Hercules Foundation and the Flemish Government department EWI for providing access to the Flemish Supercomputer Center.

References

Babanin AV, Hsu T-W, Roland A, Ou S-H, Doong D-J, Kao CC (2011) Spectral wave modelling of Typhoon Krosa. Nat Hazards Earth Syst Sci 11(2):501–511

Bailung H, Sharma SK, Nakamura Y (2011) Observation of peregrine solitons in a multicomponent plasma with negative ions. Phys Rev Lett 107:255005

Benjamin TB (1967) Instability of periodic wave trains in nonlinear dispersive systems. Proc R Soc Lond A299:59–75

Chabchoub A, Hoffmann NP, Akhmediev N (2011) Rogue wave observation in a water wave tank. Phys Rev Lett 106(20):204502

Chabchoub A, Hoffmann N, Onorato M, Akhmediev N (2012) Super rogue waves: observation of a higher-order breather in water waves. Phys Rev X 2(1):011015

Chalikov D (2009) Freak waves: their occurrence and probability. Phys Fluids 21(076602):1–4

Chien H, Kao C-C, Chuang LZH (2002) On the characteristics of observed coastal freak waves. Coast Eng J 44(04):301–319

Forristall GZ (2000) Wave crests distributions: observations and second-order theory. J Phys Ocean 30:1931–1943

Francius M, Kharif C (2006) Three dimensional instabilities of periodic gravity waves in shallow water. J Fluid Mech 561:417–437

Haver S, Andersen J (2000) Freak waves: rare realizations of a typical population or typical realizations of a rare population? In: Proceedings of the 10th international offshore and polar engineering (ISOPE) conference, Seattle, USA, May 2000

Janssen PAEM (2003) Nonlinear four-wave interaction and freak waves. J Phys Ocean 33(4):863–884

Janssen PAEM, Onorato M (2007) The intermediate water depth limit of the Zakharov equation and consequences for wave prediction. J Phys Oceanogr 37:2389–2400

Kibler B, Fatome J, Finot C, Millot G, Dias F, Genty G, Akhmediev N, Dudley JM (2010) The peregrine soliton in nonlinear fibre optics. Nat Phys 6(10):790–795

Komen GJ, Cavaleri L, Donelan M, Hasselmann K, Hasselmann H, Janssen PAEM (1994) Dynamics and modeling of ocean waves. Cambridge University Press, Cambridge

Kristiansen Ø, Fructus D, Clamond D, Grue J (2005) Simulations of crescent water wave patterns on finite depth. Phys Fluids 17:064101

Longuet-Higgins MS (1963) The effect of non-linearities on statistical distribution in the theory of sea waves. J Fluid Mech 17:459–480
McLean JW (1982) Instabilities of finite-amplitude gravity waves on water of finite depth. J Fluid Mech 114(1):331–341
Mori N, Onorato M, Janssen PAEM (2011) On the estimation of the kurtosis in directional sea states for freak wave forecasting. J Phys Ocean 41:1484–1497
Mori N, Janssen PAEM (2006) On kurtosis and occurrence probability of freak waves. J Phys Ocean 36:1471–1483
Mori N, Onorato M, Janssen PAEM, Osborne AR, Serio M (2007) On the extreme statistics of long–crested deep water waves: theory and experiments. J Geophys Res 112(C09011). doi:10.1029/2006JC004024
Mori N, Yasuda T (2002) Effects of high-order nonlinear interactions on unidirectional wave trains. Ocean Eng 29:1233–1245
Ochi MK (1998) Ocean waves: the stochastic approach. Cambridge University Press, Cambridge
Onorato M, Osborne AR, Serio M, Bertone S (2001) Freak wave in random oceanic sea states. Phys Rev Lett 86(25):5831–5834
Onorato M, Osborne A, Serio M, Cavaleri L, Brandini C, Stansberg CT (2006a) Extreme waves, modulational instability and second order theory: wave flume experiments on irregular waves. Eur J Mech B/Fluids 25:586–601
Onorato M, Osborne AR, Serio M (2006b) Modulation instability in crossing sea states: a possible mechanism for the formation of freak waves. Phys Rev Lett 96(014503)
Onorato M, Cavaleri L, Fouques S, Gramstad O, Janssen PAEM, Monbaliu J, Osborne AR, Pakozdi C, Serio M, Stansberg CT, Toffoli A, Trulsen K (2009a) Statistical properties of mechanically generated surface gravity waves: a laboratory experiment in a 3D wave basin. J Fluid Mech 627:235–257
Onorato M, Proment D, Clauss G, Klein M (2013a) Rogue waves: from nonlinear Schrödinger breather solutions to sea-keeping test. PLoS ONE 8(2):e54629
Onorato M, Residori S, Bortolozzo U, Montina A, Arecchi FT (2013b) Rogue waves and their generating mechanisms in different physical contexts. Phys Rep 528(2):47–89
Onorato M, Waseda T, Toffoli A, Cavaleri L, Gramstad O, Janssen PAEM, Kinoshita T, Monbaliu J, Mori N, Osborne AR, Serio M, Stansberg CT, Tamura H, Trulsen K (2009b) Statistical properties of directional ocean waves: the role of the modulational instability in the formation of extreme events. Phys Rev Lett **102**(114502)
Slunyaev A, Kharif C, Pelinovsky E, Talipova T (2002) Nonlinear wave focusing on water of finite depth. Phys D 173(1–2):77–96
Socquet-Juglard H, Dysthe K, Trulsen K, Krogstad HE, Liu J (2005) Distribution of surface gravity waves during spectral changes. J Fluid Mech 542:195–216
Solli DR, Ropers C, Koonath P, Jalali B (2007) Optical rogue waves. Nature 450:1054–1057
Tanaka M (2001a) A method of studying nonlinear random field of surface gravity waves by direct numerical simulations. Fluid Dyn Res 28:41–60
Tanaka M (2001b) Verification of Hasselmann's energy transfer among surface gravity waves by direct numerical simulations of primitive equations. J Fluid Mech 444:199–221
Tayfun MA (1980) Narrow-band nonlinear sea waves. J Geophys Res 85(C3):1548–1552
Toffoli A, Lefèvre JM, Bitner-Gregersen E, Monbaliu J (2005) Towards the identification of warning criteria: analysis of a ship accident database. Appl Ocean Res 27:281–291
Toffoli A, Onorato M, Babanin AV, Bitner-Gregersen E, Osborne AR, Monbaliu J (2007) Second-order theory and set-up in surface gravity waves: a comparison with experimental data. J Phys Ocean 37:2726–2739
Toffoli A, Benoit M, Onorato M, Bitner-Gregersen EM (2009) The effect of third-order nonlinearity on statistical properties of random directional waves in finite depth. Nonlinear Process Geophys 16:131–139
Toffoli A, Babanin AV, Onorato M, Waseda T (2010) Maximum steepness of oceanic waves: field and laboratory experiments. Geophys Res Lett 37:L05603

Toffoli A, Gramstad O, Trulsen K, Monbaliu J, Bitner-Gregersen EM, Onorato M (2010) Evolution of weakly nonlinear random directional waves: laboratory experiments and numerical simulations. J Fluid Mech 664:313–336

Toffoli A, Fernandez L, Monbaliu J, Benoit M, Gagnaire-Renou E, Lefevre JM, Cavaleri L, Proment D, Pakozdi C, Stansberg CT, Waseda T, Onorato M (2013) Experimental evidence of the modulation of a plane wave to oblique perturbations and generation of rogue waves in finite water depth. Phys Fluids 25:091701

Trulsen K (2007) Weakly nonlinear sea surface waves—freak waves and deterministic forecasting. In: Geometric modelling, numerical simulation, and optimization. Springer, Berlin, pp 191–209

Trulsen K, Dysthe KB (1997) Freak waves—a three-dimensional wave simulation. In: Proceedings of the 21st symposium on naval hydrodynamicsWashington, DC, 24-28 January 1997. National Academy Press, pp 550-560

Waseda T, Kinoshita T, Tamura H (2009) Evolution of a random directional wave and freak wave occurrence. J Phys Oceanogr 39:621–639

West BJ, Brueckner KA, Jand RS, Milder DM, Milton RL (1987) A new method for surface hydrodynamics. J Geophys Res 92(C11):11803–11824

Whitham GB (1974) Linear and nonlinear waves. Wiley Interscience, New York

Young IR, Banner ML, Donelan MA, Babanin AV, Melville WK, Veron F, McCormick C (2005) An integrated system for the study of wind waves source terms in finite depth water. J Atmos Ocean Technol 22:814–828

Zakharov V (1968) Stability of period waves of finite amplitude on surface of a deep fluid. J Appl Mech Tech Phys 9:190–194

Modeling of Rogue Wave Shapes in Shallow Water

T. Talipova, C. Kharif and J.P. Giovanangeli

Abstract Various shapes of rogue waves are discussed within the framework of the mechanism of nonlinear focusing of transient frequency modulated wave groups. A particular attention is paid to the formation of troughs in front of high crests. The conditions of appearance of the "three sisters" are discussed too. It is important to emphasize that this mechanism is not too sensitive to the variation of the shape of transient frequency modulated wave groups. The variable-polarity shape of a rogue wave is more probable than only one crest or one trough, because the generation of the latter ones needs a specific phase relation between individual waves in the group.

1 Introduction

The interest in occurrence of abnormal huge waves on the sea surface has arisen a long time ago and the physical mechanisms generating these giant water waves are now well understood and documented (Kharif and Pelinovsky 2003; Slunyaev et al. 2013). Rogue waves are observed everywhere, both in deep and shallow waters and sometimes even on beaches. The theoretical background for internal rogue waves had been done in (Grimshaw et al. 2010a, b; Talipova et al. 2011). The shapes of rogue waves are various. Sometimes they look like solitary waves, sometimes they appear as a group of waves (the "three sisters") or as a wall of water (Mallory 1974;

T. Talipova (✉)
Institute of Applied Physics RAS, 46 Uljanov Str., Nizhny Novgorod
603950, Russia
e-mail: tata@hydro.appl.sci-nnov.ru

T. Talipova
Nizhny Novgorod Technical State University, 24 Minina str., Nizhny Novgorod
603950, Russia

C. Kharif
IRPHE, Ecole Centrale Marseille, 49 rue F. Joliot-Curie, 13384 Marseille Cedex 13, France
e-mail: ckharif@centrale-marseille.fr

J.P. Giovanangeli
IRPHE, CNRS, 49 rue F. Joliot-Curie, 13384 Marseille Cedex 13, France

E. Pelinovsky and C. Kharif (eds.), *Extreme Ocean Waves*,
DOI 10.1007/978-3-319-21575-4_4

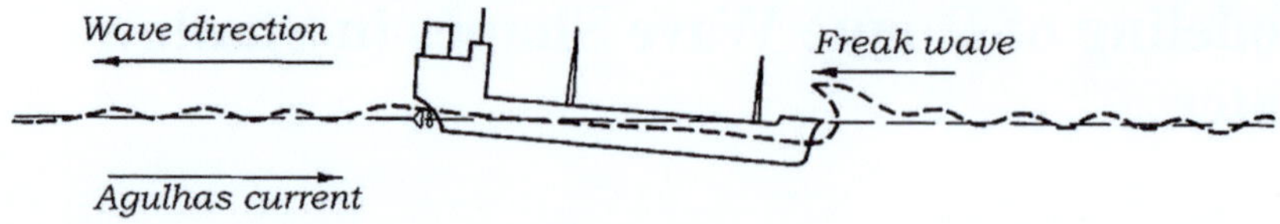

Fig. 1 Rogue wave collision with the "Taganrorsky Zaliv" (from the book by Lavrenov 2003)

Fig. 2 The Draupner New Year wave (from the paper by Haver and Jan Andersen 2000)

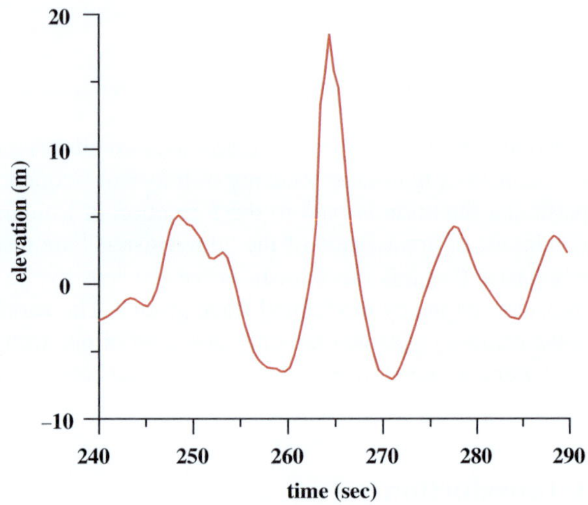

Torum and Gudmestad 1990; Olagnon and Athanassoulis 2001; Chien et al. 2002; Rosenthal 2003). In some descriptions (see, Lavrenov 2003), a long shallow trough occurs in front of a very high crest (Fig. 1), and such a wave can be very dangerous for shipping.

Indeed, there is no unique representation of rogue wave shapes. In theory, until now main attention has been paid to the possible values reached by the amplitude or height of freak waves, but not to their shapes. One attempt to explain the shape of the Draupner New Year wave (Fig. 2) from various nonlinear water wave theories has been made in the paper by Walker et al. (2004).

Here, we discuss theoretical shapes of rogue waves in a basin of moderate depth due to the focusing of transient wave groups. As it is discussed in a review paper (Kharif and Pelinovsky 2003), various mechanisms of wave group focusing may be suggested by using (i) water wave amplitude and frequency variations in space due to wind action, (ii) nonlinear modulational instability, and (iii) sea current or sea bottom inhomogeneity. The simplest explanation of rogue wave occurrence due to transient group focusing may be described as follows (Kharif et al. 2001; Slunyaev et al. 2002). If initially short wave groups are located in front of longer wave groups having larger group velocities, then during the stage of evolution, longer waves will overtake shorter waves. A huge wave can occur at some fixed time because of the superposition of waves merging at a given location. Afterwards, the longer waves will be in front of the shorter waves and the amplitude of the highest wave will decrease.

Such a mechanism has been reproduced in various laboratory tanks (Baldock and Swan 1996; Johannessen and Swan 2001; Brown and Jensen 2001; Clauss 2002; Shemer et al. 2006; Giovanangeli et al. 2005; Touboul et al. 2006; Kharif et al. 2008).

The elements of the nonlinear dispersive theory of wave focusing are given in Sect. 2 and the results of the numerical model are described in Sect. 3.

2 Theoretical Model

The dynamics of nonlinear long surface water waves on constant depth may be described by the Korteweg-de Vries equation (Dingemans 1996)

$$\frac{\partial \eta}{\partial t} + c\left(1 + \frac{3\eta}{2h}\right)\frac{\partial \eta}{\partial x} + \frac{ch^2}{6}\frac{\partial^3 \eta}{\partial x^3} = 0, \tag{1}$$

where η is the water surface elevation, h is the undisturbed water depth, $c = \sqrt{gh}$ is the linear speed of long surface wave and g is the gravity acceleration. Equation (1) may be reduced to dimensionless form (3) by the following transformations (2)

$$\zeta = \frac{\eta}{h}, \quad \tau = \frac{c}{h}t, \quad y = \frac{x - ct}{h} \tag{2}$$

$$\frac{\partial \zeta}{\partial \tau} + \frac{3}{2}\zeta\frac{\partial \zeta}{\partial y} + \frac{1}{6}\frac{\partial^3 \zeta}{\partial y^3} = 0. \tag{3}$$

The effective process to generate transient wave group focusing into a rogue wave was suggested in a recent paper (Pelinovsky et al. 2000). It is based on the invariance of the Korteweg-de Vries equation (3) with respect to reversal of time and abscissa. It means that we may choose the expected form of freak wave $\zeta_{\mathrm{fr}}(x)$ as the initial condition for Eq. (3) and solve it for any time $t = T$. Solutions found analytically or numerically after reversal of abscissa $\zeta(-x)$ describes the wave train which evolution may lead to the occurrence of waves of abnormal amplitude with the chosen shape $\zeta_{\mathrm{fr}}(x)$ and at time $t = T$. From Eq. (3) solved within the framework of a deterministic approach, with zero boundary conditions when $|x|$ goes to ∞ and the shape of the abnormal wave described by positive pulse with amplitude A_0 and length L, we show that the process is controlled by the Ursell parameter (Kharif et al. 2000). Furthermore, it is shown in the paper by Pelinovsky et al. (2000) that for a single rogue wave the Ursell parameter satisfies the following condition

$$\mathrm{Ur} = A_0 L^2 \ll 1 \tag{4}$$

The very steep wave appears due to the focusing of a group of waves of moderate amplitude. For the sake of simplicity, this wave may be approximated by the δ-function

$$\varsigma_{\mathrm{f}}(y) = Q\delta(y). \tag{5}$$

The coefficient Q in (5) is equal to wave "mass"

$$M_{\mathrm{f}} = \int_{-\infty}^{\infty} \varsigma_{\mathrm{f}}(y)\mathrm{d}y = Q. \tag{6}$$

The potential energy of this wave is infinite formally. Within the framework of equation (3) which may be solved by using the method of inverse scattering transform (Drazin and Johnson 1993), the delta pulse (5) evolves into a solitary wave

$$\varsigma_{\mathrm{s}} = A_{\mathrm{s}}\mathrm{sech}^2\left[\gamma\left(y - (1 + A_{\mathrm{s}}/2)\tau\right)\right], \tag{7}$$

with dimensionless amplitude A_{s} and inverse width γ

$$A_{\mathrm{s}} = \frac{3}{4}Q^2 \quad \gamma = \sqrt{\frac{3}{4}A_{\mathrm{s}}} = \frac{3}{4}Q \tag{8}$$

There is a dispersive tail spreading in space and damping in time. The solitary wave mass M_{s} and its energy E_{s} are conserved in time and equal accordingly to

$$M_{\mathrm{s}} = \frac{A_{\mathrm{s}}}{\gamma}\int_{-\infty}^{\infty} \mathrm{sech}^2 z\mathrm{d}z = \frac{2A_{\mathrm{s}}}{\gamma} = 2Q \tag{9}$$

$$E_{\mathrm{s}} = \frac{A_{\mathrm{s}}^2}{\gamma}\int_{-\infty}^{\infty} \mathrm{sech}^4 z\mathrm{d}z = \frac{4A_{\mathrm{s}}^2}{3\gamma} = Q^3. \tag{10}$$

We emphasize that the solitary wave mass is larger than twice the rogue wave mass, therefore incipient dispersive tail has negative mass

$$M_{\mathrm{t}} = -Q. \tag{11}$$

The energy of dispersive tail goes to infinity also as the energy of the initial delta pulse. Hence, if the solitary wave is deleted from the wavefield, the energy of dispersive tail is large enough to produce a wave of abnormal amplitude. Since dispersive tail mass is negative, it is reasonable to assume that the deep negative trough prevails in the rogue wave generation. Dispersive wave tail, especially with small amplitude, within the framework of the Korteweg-de Vries equation, evolves like the Airy function, and because its mass M_{t}, accordingly to (11), is proportional to the mass of expected rogue wave Q, the waves in the dispersive tail contain the information about both time (or position) of rogue wave occurrence and rogue amplitude due to self-similarity of the Airy function.

When the initial rogue wave disturbance has negative polarity, solitary waves are not generated irrespective of the Ursell parameter value, and the whole energy goes into damping dispersive wave train. Let us mention that within the framework of an idealized problem, solitary waves prevent the formation of rogue waves whose amplitude has to be higher not only than the amplitude of the dispersive tail but also higher than solitary wave amplitudes, constant in time. Hence, it is reasonable to suggest that without solitary waves into dispersive tail the formation of rogue wave of variable-polarity is more probable. In this case the condition about the Ursell parameter is satisfied.

3 Numerical Model

The numerical integration of the Korteweg-de Vries equation (3) is based on a finite difference scheme which satisfies the Courant criterion. The main goal of the numerical simulations is to analyze the conditions of variable-polarity rogue wave generation from transient wave groups without solitary waves.

Following Pelinovsky et al. (2000), we generate numerically transient wave groups from a short Gaussian pulse given by $A_f \exp(-y^2/L^2)$. The corresponding Ursell parameter is sufficiently small. The transient group corresponds to a solitary wave plus a damping dispersive wave train. After reverse of abscissa this transient wave group focuses again into the rogue wave with the Gaussian pulse shape. This process is shown in Fig. 3 for two values of the Gaussian pulse amplitude 0.2 and 0.4 and the same width $L = 0.55$. In this case the rogue wave occurs at $\tau = 2000$. Amplitudes

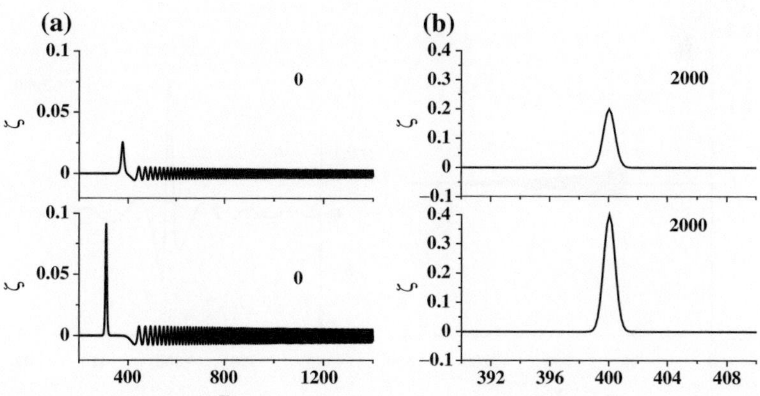

Fig. 3 Transient wave groups (**a**), leading to the formation of a Gaussian pulse of positive polarity (**b**) with amplitude values 0.2 (*black*) and 0.4 (*red*). The width is 0.55

of generated rogue waves in both cases (Fig. 3b) are more than 4 times larger than the amplitude of the corresponding solitary waves in initial wave transient groups (Fig. 3a), and it is more than the criterion needed for the freak wave occurrence: the amplitude of the freak wave has to be more than twice the amplitude of background waves.

Note that amplitudes of generated solitary waves in both runs differ from one to another by a factor 4 (0.025–0.092), whereas the amplitudes of dispersive tails differ by a factor 2. So, this simple numerical experiment confirms our theoretical conclusions that influence of amplitude of the dispersive tail on the amplitude of the rogue wave is strong (practically linear when the Ursell parameter is very weak).

Additional numerical simulations were run, corresponding to truncated transient wave groups: the solitary wave has been ignored. Hence, we consider the mechanism of rogue wave formation directly from the dispersive tail only. Results of these runs

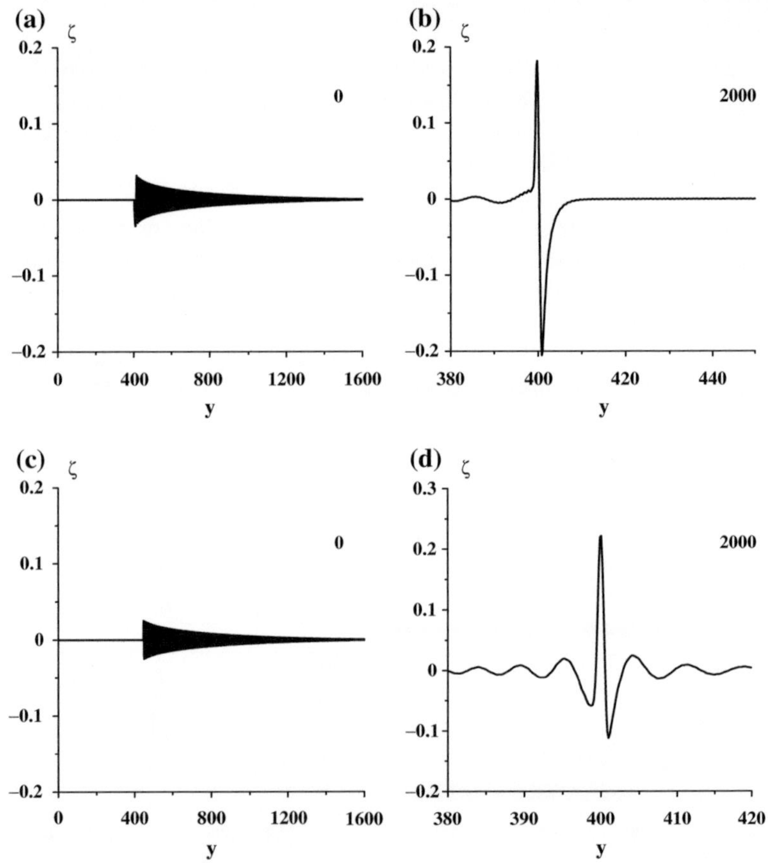

Fig. 4 **a** Initial dispersive wave train; **b** rogue wave generated by dispersive focusing; **c** initial wave train where one negative half-wave is deleted; **d** rogue wave generated by focusing of this wave train

are shown in Fig. 4. Due to dispersive focusing of the tail (Fig. 4a), the variable-polarity high-amplitude wave is generated (Fig. 4b) and its height (from trough to crest) is equal to 0.4 that is the same height than that of the rogue wave generated from the full wave group including the solitary wave. The excess of wave height above the initial height of dispersive tail is about 6.7, so such a wave satisfies the amplitude criterion of rogue wave occurrence. It is evident that the negative trough of the rogue wave is longer than the positive crest, and the negative total mass described into Sect. 2 is conserved.

So the evolution of the dispersive tail allows us to explain the appearance of the long trough (which has a specific shape within the framework of an idealized model) ahead of the positive pulse as it is described in the book by Lavrenov (2003).

The amplitude of oscillations in the dispersive tail varies significantly with wave position, and the mass distribution here is very nonuniform. So, if we delete the last high-energy negative half-wave (Fig. 4c), the mass of the tail is modified significantly, and the rogue wave which focuses from such a tail after reversing of abscissa consists of one high peak and moderately deep troughs (Fig. 4d). It is interesting to note that the wave shape in Fig. 4d is similar to the New Year wave (Fig. 2). Despite the fact that the rogue wave height becomes smaller (0.33 against 0.4 in the previous case), the excess of wave height above the initial height of the dispersive tail is about 6.7 as in the previous case. Thus, the mass of dispersive wave train influences significantly the shape of the rogue wave but in any case we obtain the variable-polarity rogue wave. A second series of numerical simulations has been performed corresponding to a Gaussian pulse of negative polarity. Its focusing leads to occurrence of abnormal deep trough on the sea surface (Fig. 5). It is well known that during the evolution of such a pulse, solitary waves do not occur and the shape of the transient wave group is close to the Airy function profile, especially for small values of the amplitude. The maximal wave height (from trough to crest) in the tail in Fig. 4a is 0.1, while the pulse amplitude is 0.2. So, the amplitude criterion of rogue wave is satisfied in this case too.

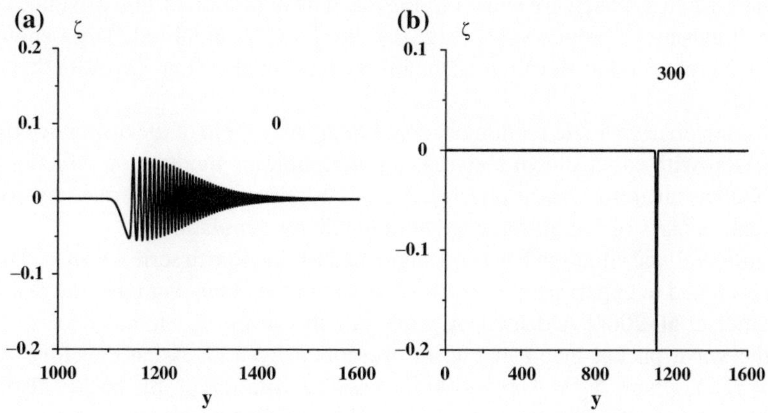

Fig. 5 **a** Initial wave train, **b** its transformation into a Gaussian pulse of negative polarity

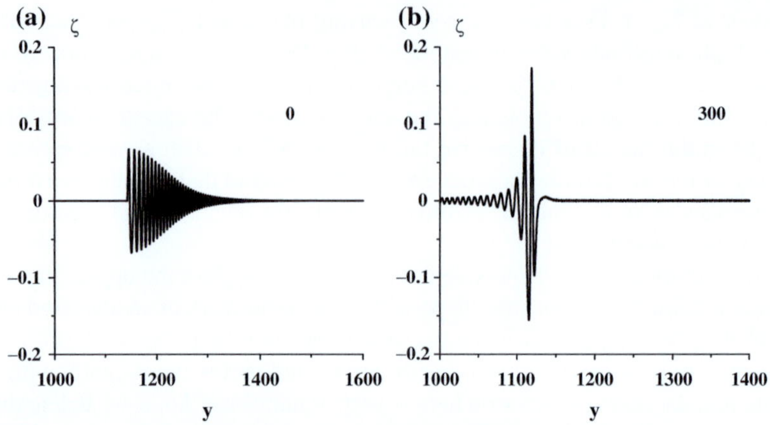

Fig. 6 **a** Initial wave train, **b** transformation into "three sisters"

The removal of back long negative half-wave from the wave train is shown in Fig. 5a (see Fig. 6a). The generation of a trough in front of the high positive pulse is observed (Fig. 6b). However, in this situation the trough with larger amplitude is behind the crest and following crests also. This is close to the wave packet often called in the literature as "three sisters." The decrease in elevation between the first crest and following trough is equal to 0.33 that is more than three times the height of the initial dispersive wave train, and the amplitude criterion is satisfied. Thus, our assumption that any dispersive wave train without solitons may generate the variable-polarity rogue wave is confirmed by the evolution of this class of transient wave group also.

For the third series of numerical experiments, the rogue wave generation from a transient wave group has been chosen as a wave with a shape close to that shown in the book by Lavrenov (2003) (see Fig. 1). For this case, the solitary wave and dispersive wave tail used are shown in Fig. 6a. The evolution of this wavefield is the chosen "Lavrenov's" rogue wave (Fig. 7b). We fixed its amplitude large enough to obtain a decrease in the elevation of the initial wave packet (Fig. 7a) more than three times.

The main characteristic feature observed in the wavefield of the dispersive train in comparison with cases shown above, is a nonmonotonic modulation which may be interpreted as an almost linear interference of both wave trains generated by positive and negative parts of initial rogue wave in the direct simulation.

A more realistic situation has been suggested for the fourth series of runs. This situation is closed to experimental results obtained in the Hannover tank and described by Shemer et al. 2006. A dispersive wave tail shown in Fig. 8a, has been obtained from the wave packet given in Fig. 4a multiplied by a Gaussian envelope $A_g \exp(-(y-b)^2/L^2)$, where $A_g = 1$, $b = 800$, $L = 200$. Evolution of this packet also leads to the generation of the "three sisters" (Fig. 8b) and the maximal wave height of this group is ten times larger than the maximal height of the initial wave packet.

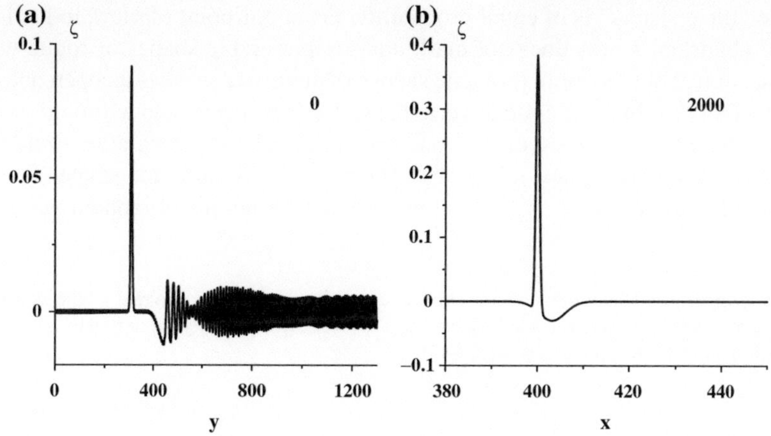

Fig. 7 **a** Initial wave packet, **b** transformation into “Lavrenov’s” wave

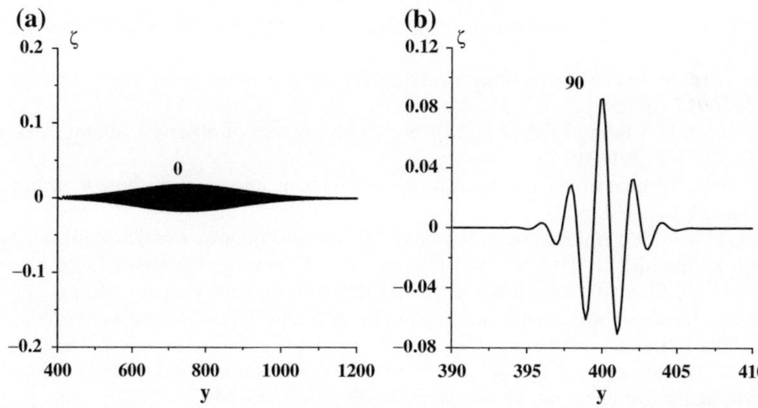

Fig. 8 **a** Initial wavefield, **b** transformation into “three sisters”

4 Conclusions

Within the framework of nonlinear-dispersive mechanism, relevant variety of shapes of rogue waves may be obtained, including the “Lavrenov’s” wave which consists of a huge crest and a long trough in front of it. It is important to emphasize that this mechanism is not too sensitive to the variation of the shape of transient wave groups. The optimal focusing of transient wave groups which requires a special phase relation gives the best conditions for rogue wave occurrence with huge amplitude. Nevertheless, the amplitude criterion is satisfied for conditions of strong deformations of the wave group, initially leading to optimal focusing, as it is shown in this work. It is clear from this simple theory that we can always get any natural form of abnormal wave. Within the framework of this model, the generation of the “Lavrenov’s” wave and the “three sisters” is of equal probability. From our point of view, today in situ data of abnormal waves does not mark out any preferable shapes of rogue waves. The question about the more probable shapes of abnormal wave is an open question. It seems that the shape of a rogue wave in the form of a crest and a through is more probable than only one crest or only one trough, because the generation of the latter ones needs a specific phase relation. In future, we will study the shapes of rogue waves within the framework of direct numerical simulations of random wind wave fields.

Acknowledgments The authors devote this paper in memory of Igor Lavrenov who was one of the pioneers to investigate rogue wave dynamics. This work was partially supported by Volkswagen Foundation and RFBR grant 14-05-91370-CT_a.

References

Baldock TE, Swan C (1996) Extreme waves in shallow and intermediate water depths. Coast Eng 27:21–46

Brown MG, Jensen A (2001) Experiments on focusing unidirectional water waves. J Geophys Res 106(C8):16917–16928

Chien H, Kao C-C, Chuang LZH (2002) On the characteristics of observed coastal freak waves. Coast Eng J 44(4):301–319

Clauss G (2002) Dramas of the sea: episodic waves and their impact on offshore structures. Appl Ocean Res 24:147–161

Dingemans MW (1996) Water wave propagation over uneven bottoms. World Scientific, Singapore

Drazin PG, Johnson RS (1993) Solitons: an introduction. Cambridge University Press, Cambridge

Giovanangeli JP, Kharif C, Pelinovsky E (2005) Experimental study of the wind effect on the focusing of transient wave groups. In: Olagnon M, Prevosto M (eds) Rogue waves 2004, (Brest, France, 2000). Ifremer, France

Grimshaw R, Pelinovsky E, Talipova T, Kurkina O (2010a) Internal solitary waves: propagation, deformation and disintegration. Nonlinear Proc Geoph 17:633–649

Grimshaw R, Pelinovsky E, Talipova T, Sergeeva A (2010b): Rogue internal waves in the ocean: long wave model. Eur Phys J special topics 185:195–208

Haver S, Jan Andersen O (2000) Freak waves: rare realizations of a typical population or typical realizations of a rare population? In: Proceedings of the tenth international offshore and polar engineering conference, Seattle, USA

Johannessen TB, Swan C (2001) A laboratory study of the focusing of transient and directionally spread surface water waves. Proc R Soc Lond Ser A 457:971–1006

Mallory JK (1974) Abnormal waves on the south-east of South Africa. Inst Hydrod Rev 51:89–129

Kharif C, Pelinovsky E (2003) Physical mechanisms of the rogue wave phenomenon. Eur J Mech B Fluids 22:603–634

Kharif C, Pelinovsky E, Talipova T (2000) Formation de vagues géantes en eau peu profonde. Comptes Rendus de l’Academie des Sciences 328(serie IIb, 11):801–807

Kharif C, Pelinovsky E, Talipova T, Slunyaev A (2001) Focusing of nonlinear wave groups in deep water. JETP Lett 73(4):190–195

Kharif C, Giovanangeli JP, Touboul J, Grare L, Pelinovsky EN (2008) Influence of wind on extreme wave events: experimental and numerical approaches. J Fluid Mech 594:209–247

Lavrenov IV (2003) Wind waves in ocean. Springer, New York

Olagnon M, Athanassoulis GA (eds) (2001) In: Rogue waves 2000, Ifremer, France

Pelinovsky E, Talipova T, Kharif C (2000) Nonlinear dispersive mechanism of the freak wave formation in shallow water. Phys D 147(1–2):83–94

Rosenthal W (2003) MaxWave. forecast and impact on marine structures. GKSS Research Center, Geesthacht, Germany

Shemer L, Goulitski K, Kit E (2006) Steep transient waves in tank—experiments and simulations. In: Proceedings of OMAE, 4–9 June 2006, Hamburg, Germany

Slunyaev A, Pelinovsky E, Sergeeva A, Chabchoub A, Hoffmann N, Onorato M, Akhmediev N, (2013): Super-rogue waves in simulations based on weakly nonlinear and fully nonlinear hydrodynamic equations. Phys. Rev. E 88:012 909

Slunyaev A, Kharif C, Pelinovsky E, Talipova T (2002) Nonlinear wave focusing on water of finite depth. Phys D 173(1–2):77–96

Talipova TG, Pelinovsky EN, Kharif CH (2011): Modulational instability of long internal waves of moderate amplitudes in stratified and horizontally inhomogeneous ocean. JETP Letters 94(3):199–203

Torum A, Gudmestad OT (eds) (1990) In: Water wave kinematics. Kluwer, Dordrecht

Touboul J, Giovanangeli JP, Kharif Ch, Pelinovsky E (2006) Experiments and simulations of freak waves under the action of wind. Eur J Mech B:Fluids 25(5):662–676

Walker DAG, Taylor PH, Eatock Taylor R (2004) The shape of large surface waves on the open sea and the Draupner new year wave. Appl Ocean Res 26:73–83

Non-Gaussian Properties of Shallow Water Waves in Crossing Seas

Alessandro Toffoli, Miguel Onorato, A.R. Osborne and Jaak Monbaliu

Abstract The Kadomtsev–Petviashvili equation, an extension of the Korteweg–de Vries equation in two horizontal dimensions, is here used to study the statistical properties of random shallow water waves in constant depth for crossing sea states. Numerical simulations indicate that the interaction of two crossing wave trains generates steep and high-amplitude peaks, thus enhancing the deviation of the surface elevation from the Gaussian statistics. The analysis of the skewness and the kurtosis shows that the statistical properties depend on the angle between the two wave trains.

1 Introduction

Sea states characterized by two spectral peaks with different mean directions, also known as mixed or crossing seas, are quite common in nature. Such conditions, for example, can be easily observed when the wind direction suddenly changes or a swell system coming from a distant source interferes with the local sea. A recent study by Toffoli et al. (2005), based on data collected from January 1995 to April 1999 by the Lloyd's Marine Information Service (Bitner and Eknes 2001), has also revealed that a large percentage of ship accidents, reported as being due to bad weather conditions, has occurred in crossing sea states (see also Donelan and Magnusson 2005; Greenslade 2001).

In that respect, mixed sea conditions are suspected to enhance the probability of occurrence for extreme wave events (see, e.g., Lehner et al. 2005). For the case of

A. Toffoli (✉) · J. Monbaliu
Swinburne University of Technology, Melbourne, Australia
e-mail: toffoli.alessandro@gmail.com

J. Monbaliu
e-mail: jaak.monbaliu@bwk.kuleuven.be

M. Onorato · A.R. Osborne
Universitá di Torino, Torino, Italy
e-mail: onorato@ph.unito.it

A.R. Osborne
e-mail: al.osborne@gmail.com

E. Pelinovsky and C. Kharif (eds.), *Extreme Ocean Waves*,
DOI 10.1007/978-3-319-21575-4_5

deep water, theoretical studies on the influence of a second spectral peak (Onorato et al. 2006; Shukla et al. 2006), which propagates with a mean direction that differs from the principal wave system, showed that waves can be unstable in mixed seas. Moreover, using second-order wave theory to simulate random time series of the sea surface, Toffoli et al. (2006) have shown that crossing sea state conditions may modify the form of the wave crest distribution as an excess of large waves may occur when two wave systems coexist.

In the case of shallow water, Peterson et al. (2003) and Soomere and Engelbrecht (2005, 2006) have studied the interaction of multi-directional wave trains by considering soliton-like solutions of the Kadomtsev–Petviashvili equation (Kadomtsev and Petviashvili 1970), which can be thought as an extension of the Korteweg–de Vries equation in the case of propagation in the (x, y) plane. The most interesting feature that arises from the interaction of two non-collinear solitons results in the formation of a steep and high peak (Miles 1977a, b). In particular, not only does its amplitude exceed the value predicted by linear theory, but also assumes values up to four times the amplitude of the incoming solitons.

Although directional, shallow water wave fields have recently been investigated (see, e.g., Herbers et al. 2007; Janssen et al. 2006), the evolution of shallow water waves in crossing sea state conditions has not yet been addressed properly. The aim of the present study is to discuss the case of possible nonlinear interactions that may arise in shallow water when two sea states, with a certain frequency distribution and directional spreading, coexist. We mainly concentrate on the occurrence of extreme wave events and hence on the form of the probability density function of the surface elevation and its third- and fourth-order moments.

In order to accomplish this task, the Kadomtsev–Petviashvili equation has been simulated numerically. Our choice on such equation is not due to the fact that we believe that the Kadomtsev–Petviashvili equation is the most accurate theoretical approach to describe shallow water waves (higher order models based on the Boussinesq equations are surely more appropriate to describe different effects that take place in shallow water, Herbers et al. 2007). We are motivated by the fact that the Kadomtsev–Petviashvili equation is the leading order equation in the shallow water expansion of the Euler equations where directionality is included. Our aim is to highlight a nonlinear mechanism described by the Kadomtsev–Petviashvili equation, which brings the statistical properties of the surface elevations far from the gaussian behavior. The numerical solution of the Kadomtsev–Petviashvili equation, moreover, is fast. Therefore, this facilitates the performance of many numerical experiments considering different random phases, different degrees of nonlinearity, and different angles between the incoming wave systems. It should be mentioned, furthermore, that the bottom topography has an important role in the evolution of shallow water waves (Janssen et al. 2006). For convenience, however, we only discuss the case of flat bottom.

The paper is organized as follows: we first begin with a general description of the Kadomtsev–Petviashvili equation. In the section following that we describe the numerical experiment. In Sect. 4, the skewness and the kurtosis of the simulated surface elevation are investigated as a function of the angle between the two wave systems. A direct analysis of the sea surface and the concurrent probability density function is presented in Sect. 5. In Sect. 6, we discuss the statistical distribution of the wave height. Some concluding remarks are presented in Sect. 7.

2 The Kadomtsev–Petviashvili Equation

The Korteweg–de Vries (KdV) equation describes nonlinear waves that propagate in the x-direction. A periodic solution of such an equation is expressed by means of a Jacobian elliptic function, which is known as cnoidal wave (see, e.g., Johnson 1997). To express the degree of nonlinearity of KdV-like wave trains, the Ursell number (Ursell 1953) is often used:

$$Ur = \frac{3}{16}\frac{a\lambda^2}{\pi^2 h^3} = \frac{3}{4}\frac{ka}{k^3h^3}. \tag{1}$$

Small Ursell numbers indicate linear wave motion ($Ur \ll 1$); large values address to increasing nonlinear effects (Osborne and Petti 1994).

One of the major shortcomings of cnoidal wave theory is its one-dimensional nature, i.e., it describes long-crested waves. A real sea surface, however, has two horizontal dimensions, and waves can be both long and short crested. The y-dependence, for example, would not be trivial in the case of two crossing waves. In order to describe the evolution of weakly directional waves in shallow water, a two-dimensional generalization of the KdV equation was derived by Kadomtsev and Petviashvili (1970); hereafter, we shall refer to this equation as the KP equation. For an irrotational, inviscid fluid in constant depth, the KP equation can be obtained from the Euler equations, if the following assumptions are satisfied (see, e.g., Johnson 1997; Segur and Finkel 1985 for a review): (I) waves are of small amplitude, i.e., $\delta = a/h \ll 1$; (II) the water is shallow, i.e., $\beta = (kh)^2 \ll 1$; (III) the waves are weakly directional, i.e., $\zeta = (k_y/k_x)^2 \ll 1$, where k_x and k_y are the components of the wavenumber vector $\mathbf{k}$; (IV) these three small effects are comparable, i.e., $\zeta \simeq O(\delta) \simeq O(\beta)$; and (V) waves propagate only toward positive values of x. In dimensional form, the KP equation can be written as follows:

$$\frac{\partial}{\partial x}\left(\frac{\partial \eta}{\partial t} + \frac{3}{2}\frac{C_0}{h}\eta\frac{\partial \eta}{\partial x} + \frac{1}{6}h^2 C_0 \frac{\partial^3 x}{\partial x^3}\right) + \frac{1}{2}C_0\frac{\partial^2 \eta}{\partial y^2} = 0, \tag{2}$$

where $C_0 = \sqrt{gh}$ is the phase velocity and h is the water depth.

Fornberg and Whitham (1978) developed a method for solving KdV-type wave equations. The method uses a pseudo-spectral treatment of the space dependence together with a leap-frog scheme in time. Here, we have extended this method to the KP equation in order to perform numerical simulations. Boundary conditions are assumed to be periodic, and derivatives are performed using the Fast Fourier

Transform (FFT) algorithm. The number of grid points in physical space was set to 128 × 64. The numerical method has been successfully tested against analytical soliton solutions of the KP equation.

3 Numerical Experiment

A two-dimensional wave field is commonly described by a directional wave spectrum: $E(\omega, \vartheta) = S(\omega)D(\omega, \vartheta)$, where ω is the angular frequency and ϑ is the direction. Here, we use a JONSWAP spectral formulation (Komen et al. 1994) with dominant wavelength $\lambda_p = 76.5\ m$, peak enhancement factor $\gamma = 3.3$, and Phillips parameter $\alpha = 0.004$ to model the wave energy in the frequency domain ($S(\omega)$); this corresponds to a significant wave height H_s of $1.7\ m$ (calculated as four times the standard deviation). A $cos^{2s}(\vartheta/2)$ spreading function (see, e.g., Hauser et al. 2005) is then applied to model the energy in the directional domain ($D(\omega, \vartheta)$). The spreading coefficient s is expressed as a function of the angular frequency (Goda 2000):

$$s(\omega) = \left(\frac{\omega}{\omega_p}\right)^5 s_{max} \qquad if \quad \omega \leq \omega_p \tag{3}$$

$$s(\omega) = \left(\frac{\omega}{\omega_p}\right)^{-2.5} s_{max} \qquad if \quad \omega > \omega_p, \tag{4}$$

where ω_p is the peak frequency. For the present study, we have chosen the spreading coefficient such that at the peak frequency $s(\omega_p) = 25$, which corresponds to the directional spreading of a short-decay swell (Goda 2000). Such a spectrum, however, is unimodal, i.e., single peaked. In order to describe crossing sea conditions, which represent bimodal sea states, the wave field is defined using two identical, aforementioned, single-peaked spectra with different mean directions, i.e., $\vartheta_1 \neq \vartheta_2$. The resulting sea state, $\tilde{E}(\omega, \vartheta) = E(\omega, \vartheta - \vartheta_1) + E(\omega, \vartheta - \vartheta_2)$, is characterized by $\lambda_p = 76.5\ m$ and $H_s = 2.4\ m$. Note that in case $\vartheta_1 = \vartheta_2$, such a spectrum reduces to an unimodal energy distribution.

By means of the linear dispersion relation, the directional spectrum $\tilde{E}(\omega, \vartheta)$ is expressed as a function of the wavenumber vector $\mathbf{k} \equiv (k_x, k_y)$ (Tanaka 2001). For a typical, bimodal, wavenumber spectrum $\tilde{E}(\mathbf{k})$ used in this study, the two spectral peaks are assumed to be symmetric with respect to the x-direction, such that $\vartheta_1 = -\vartheta_2 = \vartheta^*$.

The spectrum $\tilde{E}(\mathbf{k})$ is then used to generate a random sea surface $\eta(x, y)$ at an initial time $t = 0\ s$ as a linear superposition of Fourier modes. It is straightforward to show that a linear surface, which will be our initial condition for the simulations, can be expressed as follows:

$$\eta(\mathbf{x}) = \sum_{i=1}^{N}\sum_{j=1}^{M} a_{ij}\ cos[(k_x)_i\ x + (k_y)_j\ y + \epsilon_{ij}], \tag{5}$$

where $\mathbf{x} \equiv (x, y)$ is the position vector; ϵ_{ij} is the random phase, which is selected from a uniform distribution in the interval $[0, 2\pi]$; N and M represent the number of wavenumbers in the x- and y-direction, respectively; and a_{ij} is the spectral amplitude, which is calculated as follows:

$$a_{ij} = \sqrt{2\ \tilde{E}[(k_x)_i, (k_y)_j]\ \Delta k_x\ \Delta k_y}. \tag{6}$$

Note that the use of a determinist amplitude may not include all natural variability of waves. However, if a directional wave field is simulated, the addition of different directional components automatically restores this variability (Forristall 2000).

The summations in Eq. (5) are performed by means of a two-dimensional inverse Fast Fourier Transform. The input spectrum $\tilde{E}(\mathbf{k})$ is defined using equally spaced wavenumbers in both x- and y-direction, i.e., Δk_x and Δk_y are constants; this automatically generates periodic boundaries, which are required for numerical simulations of the KP equation. The spatial domain is chosen such that the wave field measures 400 m in the x-direction and 200 m in the y-direction.

We now consider this linear surface in a shallow water environment at three different water depths. This defines three different degrees of nonlinearity as measured by the Ursell number, Eq. (1), where $a = H_s/2$ and $k = 2\pi/\lambda_p$. The different water depths and the concurrent Ursell numbers are presented in Table 1. Under these conditions, the linear surface is then used as input in the KP model; the surface is let evolve in time according to Eq. (2) until it reaches a statistically stationary condition. The experiment is repeated for different values of the angles ϑ^* between a minimum of 0° (unimodal spectrum) up to a maximum of 45° with a step size of 2.5°. The spectral energy is kept constant, i.e., $H_s = const$ and $\lambda_p = const$, for all repetitions of the numerical experiment.

For each time step, $\Delta t = 0.001\ s$, the model produces a two-dimensional surface as output; the skewness and the kurtosis have been computed from the space series associated with each instant in time. For all spectral densities, the numerical experiment is repeated 500 times with different random phases; this should in principle stabilize the statistical information from the simulated samples. Furthermore, time series have been recorded at four different grid points of the physical domain, starting after the surface has reached a statistically stable condition.

Table 1 Water depths and degrees of nonlinearity: water depths h [m]; relative water depth kh; and Ursell number Ur

h [m]	kh	Ur
7.00	0.57	0.39
6.00	0.49	0.62
5.00	0.41	1.10

4 Skewness and Kurtosis of the Simulated Surface Elevation

When the water depth decreases, the nonlinear interaction becomes more and more relevant. The most obvious manifestation of this nonlinearity is the sharpening of the wave crests and the flattening of the wave troughs, which results in a departure from the Gaussian statistics. The deviation is usually expressed in terms of the skewness, λ_3, and kurtosis, λ_4, which represent the third- and fourth-order moments of the probability density function, respectively. Whereas the first expresses the vertical asymmetry of the wave profile, the second is an indication of extreme events. For waves in shallow water, $kh \ll 0$, it is therefore expected that $\lambda_3 > 0$ and $\lambda_4 > 3$, where 0 and 3 are the values of skewness and kurtosis for a Gaussian random process.

In Fig. 1, we show the skewness and kurtosis of the simulated surface elevation $\eta(x, y)$ as a function of time for a single-peaked sea state, i.e., $\vartheta^* = 0°$ and a degree of nonlinearity $Ur = 0.62$. Note that, since the physical phenomenon is considered a random process, the statistical properties of the wave field are retrieved by calculating average values at each time step over the collection of simulated samples (ensemble average).

The input, linear, surface evolves at each instant of time according to the KP equation; the statistical properties of the wave surface show a rapid and significant deviation from the Gaussian statistics. After about 600 s, the values of the skewness and kurtosis reach a stationary condition (cf. Pelinovsky and Sergeeva 2006). For the considered degree of nonlinearity ($Ur = 0.62$), for example, the moments tend to stabilize around the following values: $\lambda_3 \simeq 0.60$ and $\lambda_4 \simeq 3.39$. As one would

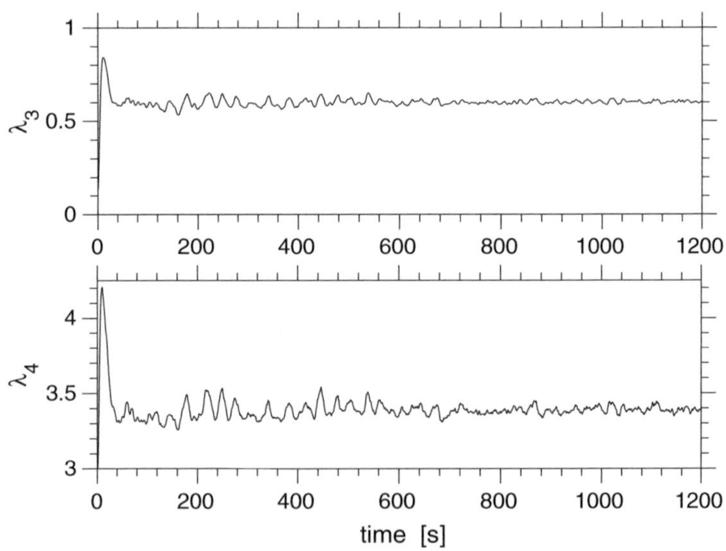

Fig. 1 Skewness (*upper panel*) and kurtosis (*lower panel*) as a function of time from numerical simulations with $\vartheta_1 = -\vartheta_2 = \vartheta^* = 0°$ and $Ur = 0.62$

expect, the skewness and kurtosis vary in magnitude as the nonlinearity of the system changes: for $Ur = 0.39$, $\lambda_3 \simeq 0.44$ and $\lambda_4 \simeq 3.13$; and for $Ur = 1.10$, $\lambda_3 \simeq 0.81$ and $\lambda_4 \simeq 3.87$. The vertical asymmetry (λ_3), however, assumes remarkably high values already at relatively low nonlinearity. A similar result was also obtained in unidirectional condition using the KdV equation by Pelinovsky and Sergeeva (Pelinovsky and Sergeeva 2006).

We now investigate how the statistical properties of the simulated surfaces can change when two spectral peaks with different mean directions are considered. In Figs. 2 and 3, we show the skewness and kurtosis as a function of the angle ϑ^*. The existence of a second spectral peak results in an increase of the skewness and kurtosis as the angle between the spectra is increased. This enhancement, however, is almost nil when the systems are almost collinear ($\vartheta^* \leq 5°$) and becomes more relevant as the wave trains assume well-separated directions. The maximum magnitudes of the third- and fourth-order moments are obtained for $\vartheta^* \approx 30°$: (1) for $Ur = 0.39$, $\lambda_3 \simeq 0.47$ and $\lambda_4 \simeq 3.33$; (2) for $Ur = 0.62$, $\lambda_3 \simeq 0.63$ and $\lambda_4 \simeq 3.65$; (3) for $Ur = 1.10$, $\lambda_3 \simeq 0.85$ and $\lambda_4 \simeq 4.30$. It is important to remark that the energy remains constant, and hence this variation is only due to the nonlinear interaction of intersecting wave trains. When $\vartheta^* > 30°$, the increasing trend decreases. For low and moderate nonlinearity, the kurtosis approximately conserved the values found for $\vartheta^* \approx 30°$; the skewness, in contrast, slowly decreases as ϑ^* approaches 45°. For high degrees of nonlinearity, however, both skewness and kurtosis show a decreasing trend for large ϑ^*. Note that the angle at which the statistical moments of the surface elevation maximize is to some extent consistent with the critical angle at which two non-collinear solitons with wavelength equal to λ_p and amplitude equal to $H_s/2$

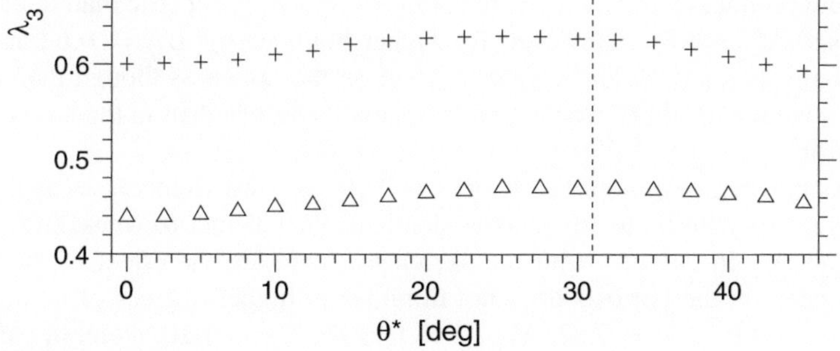

Fig. 2 Skewness (λ_3) as a function of the mean direction ($\vartheta_1 = -\vartheta_2 = \vartheta^*$) of the spectral peaks from numerical simulation: $Ur = 0.39$ (△); $Ur = 0.62$ (+); and $Ur = 1.10$ (o)

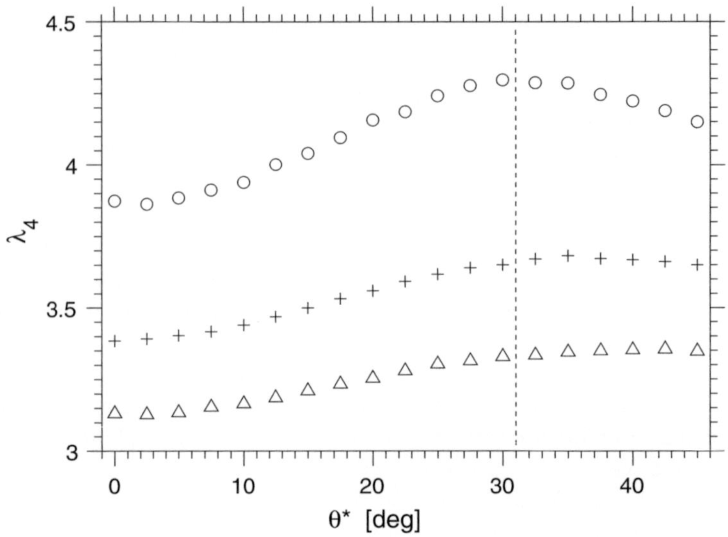

Fig. 3 Kurtosis (λ_4) as a function of the mean direction ($\vartheta_1 = -\vartheta_2 = \vartheta^*$) of the spectral peaks from numerical simulation: $Ur = 0.39$ (△); $Ur = 0.62$ (+); and $Ur = 1.10$ (o)

produce an intersection peak with maximum height and length (Peterson et al. 2003). For $\vartheta^* > 30°$, however, the condition of weak directionality, i.e., $\zeta = (k_y/k_x)^2 \ll 1$, is no longer respected: $\zeta > 0.34$ for $\vartheta^* > 30°$. Therefore, the physical meaning of this critical angle is uncertain, and further investigations using the Euler equations as well as laboratory experiments are needed to confirm this finding.

It is now instructive to look at a normalized form of the skewness and kurtosis. In this respect, we define two additional parameters, which describe how much the skewness and kurtosis in bimodal systems differ from the one in unimodal conditions: $\lambda_3^* = \lambda_3/\lambda_{3,\ \vartheta^*=0}$; $\lambda_4^* = \lambda_4/\lambda_{4,\ \vartheta^*=0}$, where $\lambda_{3,\ \vartheta^*=0}$ and $\lambda_{4,\ \vartheta^*=0}$ are the skewness and kurtosis in unimodal condition, respectively. In Figs. 4 and 5, λ_3^* and λ_4^* are presented as a function of ϑ^*. For low nonlinearity, $Ur = 0.39$, the bimodality of the system produces waves which are about 8 % more asymmetric than in unimodal conditions ($\lambda_3^* \simeq 1.08$). However, for higher nonlinearity, $Ur = 0.62$ and 1.10, the second spectral peak has a weaker effect on the skewness though the behavior is qualitative identical; λ_3^* reaches, in fact, smaller values than in the case of lower nonlinearity ($\lambda_{3\ max}^* \simeq 1.05$).

For a small degree of nonlinearity, $Ur = 0.39$, the kurtosis increases up to 7 %; at such nonlinearity, the bimodality has an identical effect of both skewness and kurtosis, i.e., $\lambda_4^* \approx \lambda_3^*$. In contrast with the skewness, however, the effect of the second spectral peak on the kurtosis does not diminish at higher degrees of nonlinearity: $\lambda_{4\ max}^* = 1.09$ for $Ur = 0.62$; $\lambda_{4\ max}^* \simeq 1.11$ for $Ur = 1.10$. Note, in particular, the relevant enhancement of kurtosis at high degrees of nonlinearity, at which very large amplitudes may be expected.

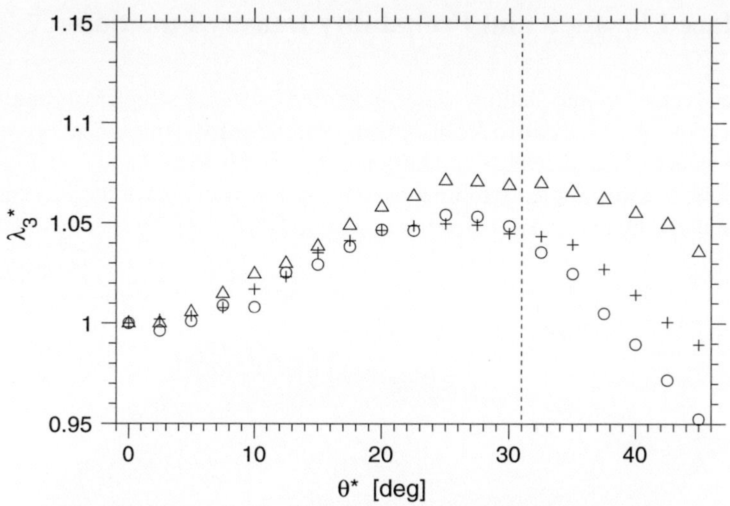

Fig. 4 Normalized skewness ($\lambda_3^* = \lambda_3/\lambda_3^{(\vartheta^*=0)}$) as a function of the mean direction ($\vartheta_1 = -\vartheta_2 = \vartheta^*$) of the spectral peaks from numerical simulation: $Ur = 0.39$ (Δ); $Ur = 0.62$ (+); and $Ur = 1.10$ (o)

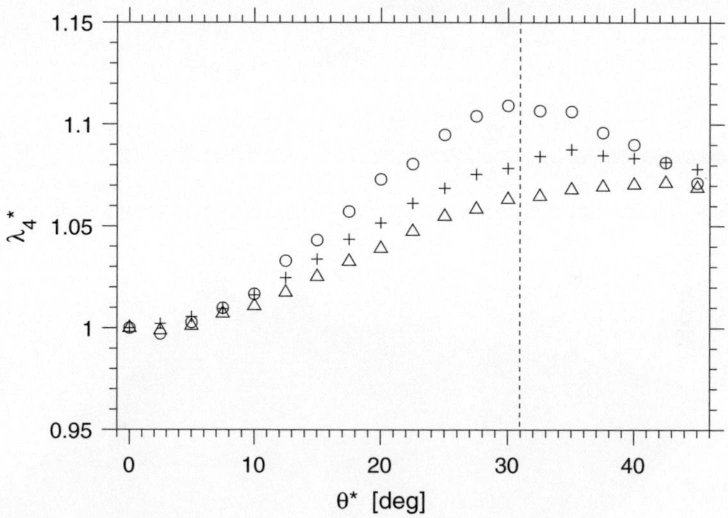

Fig. 5 Normalized kurtosis ($\lambda_4^* = \lambda_4/\lambda_4^{(\vartheta^*=0)}$) as a function of the mean direction ($\vartheta_1 = -\vartheta_2 = \vartheta^*$) of the spectral peaks from numerical simulation: $Ur = 0.39$ (Δ); $Ur = 0.62$ (+); and $Ur = 1.10$ (o)

5 Surface Elevation and Probability Density Function

As waves propagate into shallow water, a definite excess of steep crests and shallow troughs can be observed due to the increasing of nonlinearity; this is also expressed by the high values of the skewness and kurtosis observed in Figs. 2 and 3. In Figs. 6 and 7, we show, as an example, two-dimensional shallow water wave fields, which have been simulated by unimodal ($\vartheta^* = 0°$) and bimodal ($\vartheta^* = 30°$) spectral conditions.

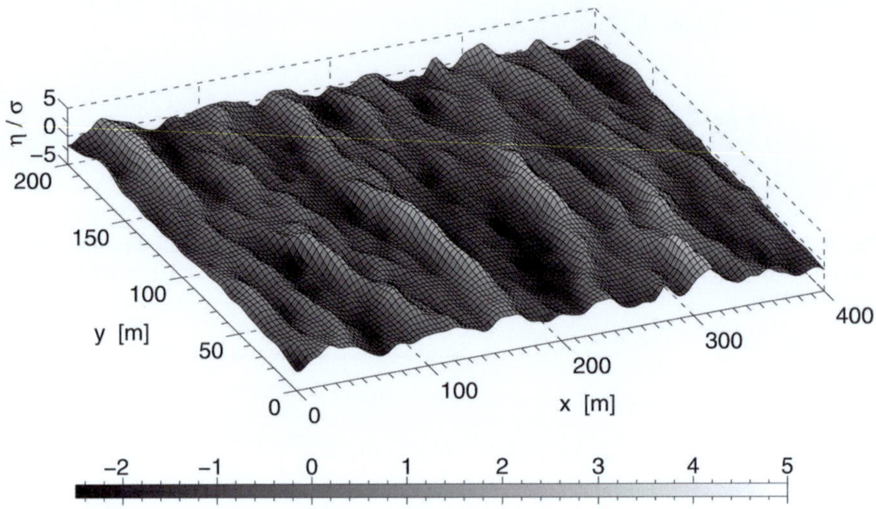

Fig. 6 Nonlinear sea surface (η/σ) from numerical simulations: $\vartheta^* = 0°$ (unimodal sea)

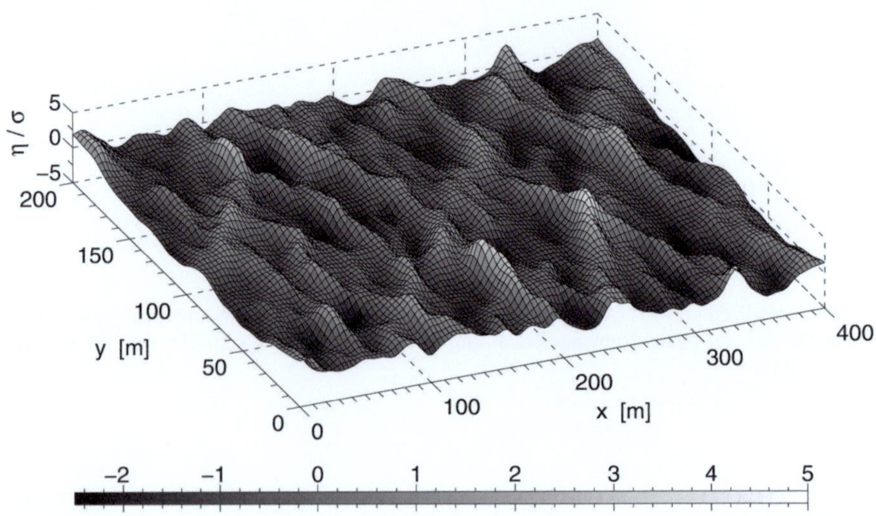

Fig. 7 Nonlinear sea surface (η/σ) from numerical simulations: $\vartheta^* = 30°$ (bimodal sea)

Using the KP equation, we observed that when two wave systems coexist, beside the sharpening of the wave crests and the flattening of the wave troughs, steep and large amplitude peaks arise in the intersecting region of two wave fronts similar to what observed in the case of the two-soliton solution of the KP equations (see, e.g., Miles 1977a, b). Considering the case of two non-collinear solitons, the geometrical properties of these peaks have already been described in details by many authors (Peterson et al. 2003; Segur and Finkel 1985; Soomere and Engelbrecht 2005). It is important to note that the interaction peak of two non-collinear solitons can, in principle, propagate unaltered for infinite distance. In a dispersive system, however, the amplitude of the interaction peak arises and decreases within a distance of a few wavelengths: two up to three times the dominant wavelength in the case of not well-separated directions (e.g., $\vartheta^* \leq 20°$); one dominant wavelength in case of well-separated directions (e.g., $\vartheta^* > 20°$).

It is now instructive to investigate how the interaction peaks modify the probability density function for the surface elevation. In Figs. 8, 9, and 10, we present, as an a example, the probability density function in unimodal ($\vartheta^* = 0°$) and bimodal ($\vartheta^* = 30°$) wave fields for different degrees of nonlinearity. The wave elevation is expressed in normalized form using the standard deviation σ of the surface displacements. Note that the statistical distribution is calculated by only using the surface elevation $\eta(x, y)$ at the final time step of each random realization, because at this stage, the surface can be assumed statistically stationary (see, e.g., Fig. 1). The distribution of the simulated elevation is compared with the normal probability density function, which describes linear (Gaussian) waves.

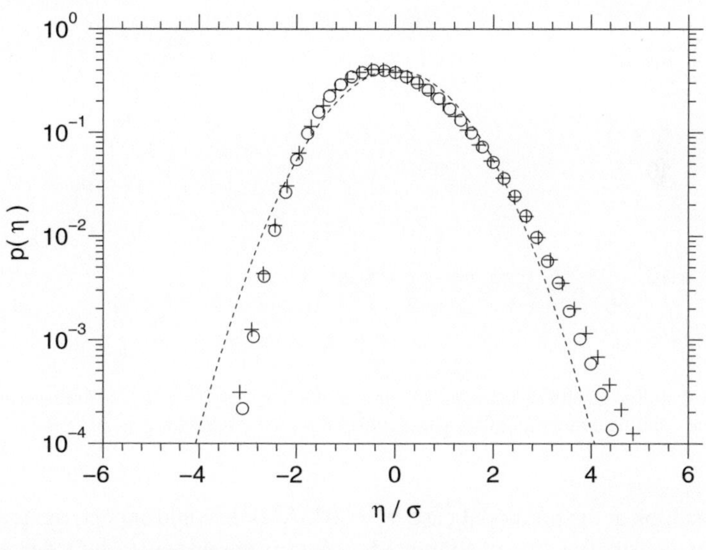

Fig. 8 Probability density function for a degree of nonlinearity $Ur = 0.39$: Gaussian distribution (*dashed line*); simulated unimodal sea (o); simulated bimodal sea with $\vartheta^* = 30°$ (+)

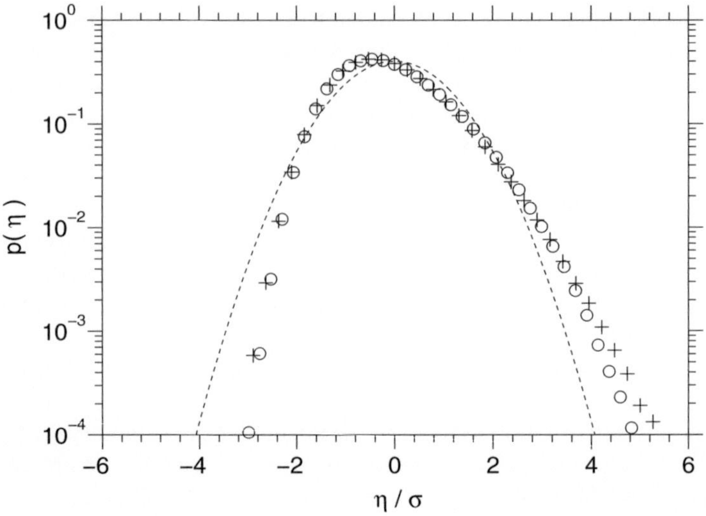

Fig. 9 Probability density function for a degree of nonlinearity $Ur = 0.62$: Gaussian distribution (*dashed line*); simulated unimodal sea (o); simulated bimodal sea with $\vartheta^* = 30°$ (+)

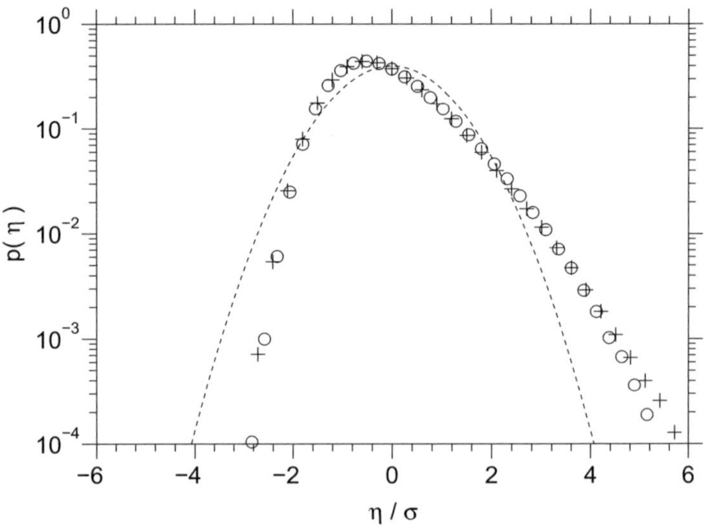

Fig. 10 Probability density function for a degree of nonlinearity $Ur = 1.10$: Gaussian distribution (*dashed line*); simulated unimodal sea (o); simulated bimodal sea with $\vartheta = 30°$ (+)

We first look at the unimodal case $\vartheta^* = 0°$. As one would expect, the form of the probability density function of simulated shallow water waves strongly deviates from the Gaussian statistics. The excess of sharp crests and shallow troughs results in a shift toward negative values of the peak of the distribution and a remarkable deviation of its tails. When crossing waves are considered, on the other hand, the peaks arising at the intersection of the wave fronts produce a modification of the upper tail of the distribution. However, a significant deviation from the probability density function of unimodal wave fields only occurs at low probability levels ($p(\eta) < 0.001$). The interaction between two non-collinear wave systems, moreover, seems not to have any particular effect on the form of the lower tail of the distribution, which remains almost unchanged.

6 Wave Height Distribution

For an empirical analysis, the wave height H is normally defined as the difference in surface elevation between the crest and trough of an individual wave (trough-to-crest wave height). Although it is straightforward to extract individual waves from time series (see, e.g., Goda 2000), it is not clear how to define them from a certain surface $\eta(x, y)$. In the following, therefore, we make use of the recorded time series to calculate the wave heights. In this respect, we assume that an individual wave is the portion of a wave record between two consecutive zero-downcrossings. In Fig. 11, the exceedance probability of the dimensionless wave height (the standard deviation

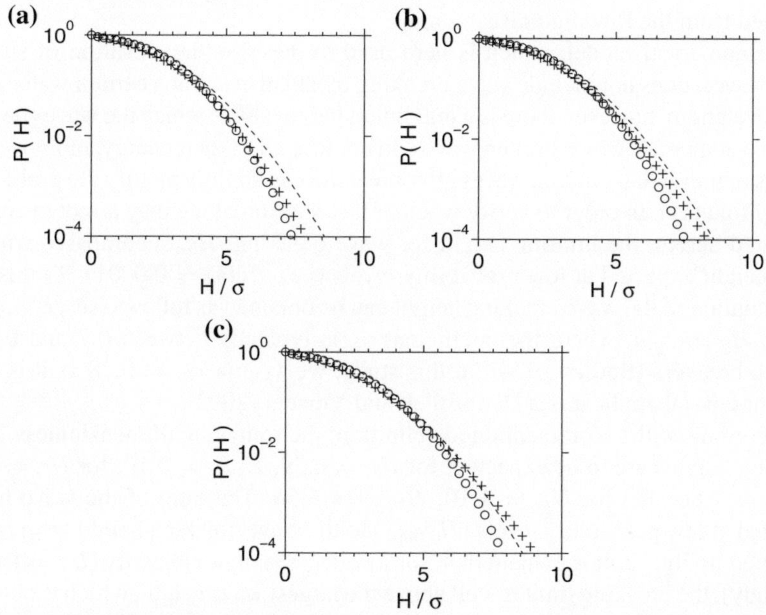

Fig. 11 Exceedance probability of wave height for different degrees of nonlinearity: **a** $Ur = 0.39$; **b** $Ur = 062$; **c** $Ur = 1.10$. The unimodal, $\vartheta = 0°$ (o), and bimodal, $\vartheta = 30°$ (+), cases are compared with the Rayleigh distribution (*dashed line*)

σ is used as normalizing factor) is presented for the unimodal and bimodal cases. For the latter, only the case with $\vartheta^* = 30°$ is shown, because the wave height is expected to maximize for this angular condition. The statistical distribution of the simulated wave heights is then compared with the Rayleigh density function, which approximates the wave height distribution of Gaussian random processes (Longuet-Higgins 1952).

We first consider the unimodal spectral conditions. The analysis shows that the simulated wave heights are systematically overestimated by the Rayleigh density function. Although this deviation may be attributed to the skew non-Gaussian nature of shallow water waves, it is more likely related to the finite bandwidth of the wave spectrum, and the difference in definition between the trough-to-crest wave height and the envelope-based representation, which is relevant for the Rayleigh distribution (Longuet-Higgins 1952; Tayfun 1981, 1980). When two spectral peaks coexist, however, the wave height increases. For a low degree of nonlinearity ($Ur = 0.39$), nevertheless, the bimodality of the wave spectrum does not lead to any significant deviation of the wave height distribution; as indicated in Fig. 11a, the tail of the distribution only slightly changes. When higher degrees of nonlinearity are taken into account, the interaction between non-collinear wave trains becomes more relevant and the wave height tends to be significantly higher than in unimodal conditions. As a result, the deviation of the tail of the distribution becomes more evident as the Ursell number is increased (Fig. 11b, c). For high nonlinearity ($Ur = 1.10$), furthermore, the tail of the distribution indicates that the wave height can overtake the value expected from the Rayleigh distribution.

The numerical model, which is here used to describe the elevation of shallow water waves, does not include wave breaking mechanisms. For a certain water depth and wavelength, however, there is a maximum height above which the waves become unstable and break. Since broken waves do not lose all of their energy, moreover, the heights of individual random waves after breaking can slightly modify the probability density function. In order to verify whether the wave breaking may affect the results presented herein, the limiting height for wave breaking, H_b, is compared with the wave height expected at low probability levels (i.e., $P(H) = 0.0001$). To this end, an estimation of the wave breaking height can be obtained as follows (see, e.g., Goda 2000): $H_b = \gamma_b h$, where the coefficient γ_b is typically between 0.7 and 0.9 for spilling breakers (Battjes 1974); in this study, we assume $\gamma_b = 0.78$ as it is often used for coastal applications Demirbilek and Vincent (2002).

According to the aforementioned definition, the following (dimensionless) wave breaking heights are to be expected: for $Ur = 0.39$, $H_b/\sigma = 9.10$; for $Ur = 0.62$, $H_b/\sigma = 7.80$; and for $Ur = 1.10$, $H_b/\sigma = 6.50$. The ratio of the wave height expected at low probability levels ($H_{0.0001}$) to the wave breaking height (H_b) is then presented in Fig. 12. It is evident that, for low degrees of nonlinearity ($Ur = 0.39$ in this study), the breaking limit is well above the largest wave height, which is obtained in crossing seas. Therefore, the wave breaking should not have any significant effects on the statistical properties of the surface elevation.

For a moderate nonlinearity ($Ur = 0.62$), the breaking limit approximately matches the wave height at the probability level $P(H) = 0.0001$ (see Fig. 12).

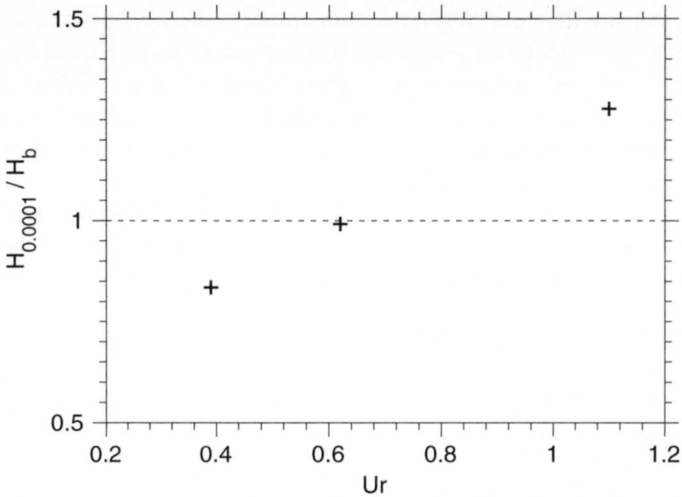

Fig. 12 Ratio of the simulated large wave height ($H_{0.0001}$) to wave breaking height (H_b) as a function of the Ursell number

However, since a very limited number of waves have been recorded to be higher than $H_{0.0001}$, we may expect that also for the case $Ur = 0.62$ the wave breaking does not affect the probability density function of wave elevation and wave height significantly. When $Ur > 0.62$, nevertheless, the breaking limit is rather small. Thus, wave heights at probability levels lower than 0.01 would break. As a result, the probability density function and the statistical moment would be significantly different than the ones obtained from the simulations.

7 Conclusions

The statistical properties of bimodal, shallow water wave fields have been discussed. A generalization of the Korteweg–de Vries equation in the (x, y) plane, the Kadomtsev–Petviashvili equation, has been used to describe two-dimensional waves in shallow water. The input spectra have been generated as a summation of two equivalent single-peaked spectra with different mean directions. We have used a JONSWAP spectral formulation to model the energy distribution in the frequency domain and a *cos–2s* function to model the directional domain. Considering several spectral cases with constant energy and different angles between the two peaks, many realizations have been simulated with different random phases. Different degrees of nonlinearity, as measured by the Ursell number, have been investigated.

Shallow water waves are characterized by an excess of high crests and shallow troughs, which produces a deviation from the Gaussian statistics. The simulated surface elevation, during single-peaked spectral conditions, shows a very skewed probability density function. When two non-collinear wave trains are considered in the wave field, their nonlinear interaction produces steep and high-amplitude peaks. These peaks yield a modification of the upper tail of the probability density function for the surface elevation $\eta(x, y)$, which significantly deviates from the distribution of wave elevation in the unimodal condition. The coexistence of two spectral peaks, therefore, enhances the nonlinearity of the wave field, which results in an increase of the skewness and kurtosis. Whereas this enhancement is negligible for nearly collinear waves, the skewness and kurtosis reach high values when the two spectral peaks have well-separated directions. In particular, it has been observed that for $\vartheta^* \simeq 30°$ their magnitudes maximize. Beyond the threshold $\vartheta^* \simeq 30°$, the increasing trend of the skewness and kurtosis vanishes; for high degrees of nonlinearity, furthermore, skewness and kurtosis decrease. Formally speaking, the KP equation results from an expansion in which the directionality is a small parameter; therefore, additional verification should be made with the Euler equations and laboratory experiments to confirm this finding.

It is important to mention that the model does not consider wave breaking. However, for a certain water depth and wavelength, waves can actually break. In this respect, the comparison between the wave height expected at low probability levels ($P(H) = 0.0001$) and the breaking limit has shown that wave breaking may only affect the statistical properties when large degree of nonlinearity ($Ur > 0.62$) and large angle ϑ^* are accounted for.

This work was carried out in the framework of the F.W.O. project G.0228.02 and G.0477.04., and the E.U. project SEAMOCS (contract MRTN–CT–2005–019374). The numerical simulations were performed using the K.U. Leuven's High Performance Computing (HPC) facilities.

References

Battjes JA (1974) Surf similarity. In: Proceedings of the 14th international conference on coastal engineering, Copenhagen, Denmark

Bitner-Gregersen E, Eknes M (2001) Ship accidents due to bad weather. Technical Report report 2001-1330, Det Norske Veritas

Demirbilek Z, Vincent L (2002) Water wave mechanics. In: Demirbilek Z, Vincent L (eds), Coastal engineering manual, part II, hydrodynamics, pp Chapter II–1, Engineer Manual 1110–2–1100. U.S. Army Corps of Engineers, Washington, DC

Donelan M, Magnusson AK (2005) The role of meteorological focusing in generating rogue wave conditions. In: Proceedings of the 14th Aha Huliko a Hawaiian winter workshop, University of Hawaii at Manoa, USA, 24–28 January 2005

Fornberg B, Whitham GB (1978) A numerical and theoretical study of certain nonlinear wave phenomena. Philos Trans R Soc Lond Ser A 289:373–404

Forristall GZ (2000) Wave crests distributions: observations and second-order theory. J Phys Ocean 30:1931–1943

Goda Y (2000) Random seas and design on marine structures. Advanced series on ocean engineering, vol. 15. World Scientific Publishing company, Singapore

Greenslade DJM (2001) A wave modelling study of the 1998 Sydney to Hobart yacht race. Aust Meteorol Mag 50:53–63

Hauser D, Kahma KK, Krogstad HE, Lehner S, Monbaliu J, Wyatt LW (eds) (2005) Measuring and analysing the directional spectrum of ocean waves. Cost Office, Brussels

Herbers THC, Orzech M, Elgar S, Guza RT (2007) Shoaling transformation of wave frequency-directional spectra. J Geophys Res **108**(C1). doi:10.1029/2001JC001304

Janssen TT, Herbers THC, Battjes JA (2006) Generalized evolution equations for nonlinear surface gravity waves over two-dimensional topography. J Fluid Mech 552:393–418

Johnson RS (1997) A modern introduction to the mathematical theory of water waves. Cambridge University Press, Cambridge

Kadomtsev BB, Petviashvili VI (1970) On the stability of solitary waves in weakly dispersive media. Sov. Phys. Dokl. 15:539–541

Komen GJ, Cavaleri L, Donelan M, Hasselmann K, Hasselmann H, Janssen PAEM (1994) Dynamics and modeling of ocean waves. Cambridge University Press, Cambridge

Lehner S, Günther H, Rosenthal W (2005) Extreme wave observations from radar data sets. In: Ocean waves measurements and analysis, fifth international symposium WAVES 2005, Madrid, Spain, 3–7 July 2005. paper 69

Longuet-Higgins MS (1952) On the statistical distribution of the heights of sea waves. J Mar Res 11:1245–1266

Miles JW (1977a) Diffraction of solitary waves. Z Ang Math 28:889–902

Miles JW (1977b) Note on solitary wave on a slowly varying channel. J. Fluid Mech. 80:149–152

Onorato M, Osborne AR, Serio M (2006) Modulation instability in crossing sea states: a possible mechanism for the formation of freak waves. Phys Rev Lett 96.014503

Osborne AR, Petti M (1994) Laboratory-generated, shallow-water surface waves: analysis using the periodic, inverse scattering transform. Phys Fluids 6(5):1727–1744

Pelinovsky E, Sergeeva A (2006) Numerical modeling of the kdv random wave field. Eur J Mech B/Fluids 25:425–434

Peterson P, Soomere T, Engelbrecht J, van Groesen E (2003) Soliton interaction as a possible model for extreme waves in shallow water. Nonlinear Proc Geophys 10:503–510

Segur H, Finkel A (1985) An analytical model of periodic waves in shallow water. Stud Appl Math 73:183–220

Shukla PK, Kaurakis I, Eliasson B, Marklund M, Stenflo L (2006) Instability and evolution of nonlinearly interacting water waves. Phys Rev Lett 97.094501

Soomere T, Engelbrecht J (2005) Extreme elevation and slopes of interacting solitons in shallow water. Wave Motion 41:179–192

Soomere T, Engelbrecht J (2006) Weakly two-dimensional interaction of solitons in shallow water. Eur J Mech B/Fluids 25:636–648

Tanaka M (2001) A method of studying nonlinear random field of surface gravity waves by direct numerical simulations. Fluid Dyn Res 28:41–60

Tayfun AM (1981) Distribution of crest–to–trough wave heights. J Waterw Port Coast Ocean Eng 107(3):149–158

Tayfun MA (1980) Narrow-band nonlinear sea waves. J Geophys Res 85(C3):1548–1552

Toffoli A, Lefèvre JM, Bitner-Gregersen E, Monbaliu J (2005) Towards the identification of warning criteria: analysis of a ship accident database. Appl Ocean Res 27:281–291

Toffoli A, Onorato M, Monbaliu J (2006) Wave statistics in unimodal and bimodal seas from a second-order model. Eur J Mech B/Fluids 25:649–661

Ursell F (1953) The long wave paradox in the theory of gravity waves. Proc Camb Philos Soc 49:685–694

Searching for Factors that Limit Observed Extreme Maximum Wave Height Distributions in the North Sea

Gerrit Burgers, Frits Koek, Hans de Vries and Martin Stam

Abstract The probability that individual waves are much larger than the significant wave height is studied in a large set of observations. It is investigated whether steepness and shallow water effects are limiting factors for extreme wave heights. The relation between observations and a model freak wave index is examined. Measurements from two locations in the North Sea are used, one with a depth of 80 m and one with a depth of 20 m. The data consist of significant wave height, wave period and maximum wave height of 20 min records. The total amount of the records covers several years. The freak wave model index from the European Centre for Medium-Range Weather Forecasts (ECMWF) wave model is collocated with the observations. The instrumental data show Rayleigh-like distributions for the ratio of maximum wave height to significant wave height. Our analysis is limited by uncertainties in the instrumental response in measuring maximum wave height. The data indicate that steepness is a limiting factor for extreme wave height. At the shallow water location, extreme waves are not more frequently observed than at the deep water location. The relation between the freak wave index of the ECMWF wave model and enhanced extreme wave probability is studied.

1 Introduction

During the All Saints Day storm of November 2006, a waverider buoy recorded extreme individual waves of 17 and 20 m, around twice the significant wave height at that time (see Fig. 1). For the same storm, the ECMWF (European Centre for Medium-Range Weather Forecasts) wave model WAM indicated an enhanced probability of extreme waves in the North Sea, see Fig. 2. For a water depth of 20 m, a

G. Burgers (✉)
Rijkswaterstaat Water, Verkeer en Leefomgeving, Zuiderwagenplein 2,
8224 AD Lelystad, The Netherlands
e-mail: Gerrit.Burgers@rws.nl

F. Koek · H. de Vries · M. Stam
Royal Netherlands Meteorological Institute, P.O. Box 201,
NL-3730 AE De Bilt, The Netherlands

E. Pelinovsky and C. Kharif (eds.), *Extreme Ocean Waves*,
DOI 10.1007/978-3-319-21575-4_6

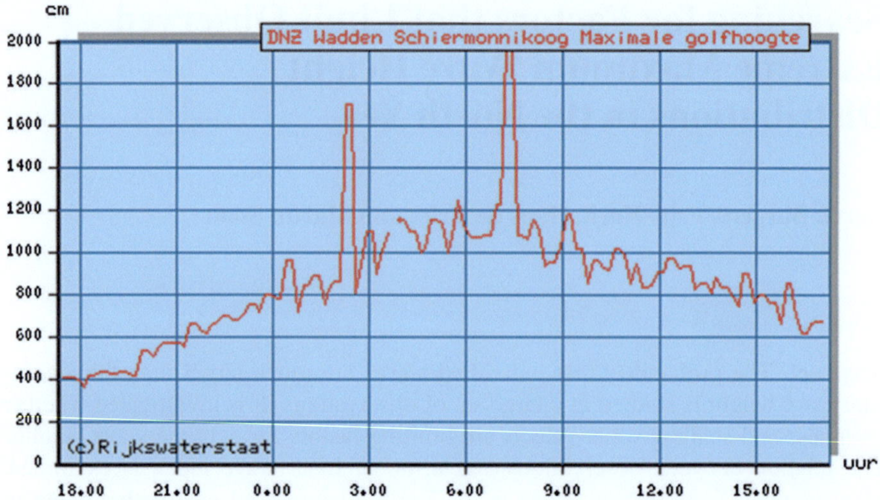

Fig. 1 Registration of maximum wave height at Schiermonnikoog Island (SMN) at (53°35N,6°10E) during the night of 1 November 2006

20 m wave would be quite exceptional. Whether the record is correct remains unclear: analysis of the buoy by the manufacturer showed that the measurement is not reliable in these conditions (Datawell 2006) in the same storm damage was reported at a platform more than 15 above sea level (Bojanowski 2007). But the buoy has been in place for years, so in principle we can determine multi-year return times for wave extremes.

In this paper, we study the following questions: how exceptional are extreme waves in long records? Can the WAM model identify conditions with enhanced extreme wave probability? More precisely, we focus on the ratio $r = H_{max}/H_s$, where H_s is the significant wave height, and H_{max} the maximum wave height in a 20 min record, and from the WAM model we use the BFI index (Janssen 2003). Our study differs from the one by Holliday et al. (2006) in that we do use the 20 min record summary information instead of individual wave records, and that we have used data that accumulate to a much longer time. However, because various instruments and algorithms have been used, the interpretation as one single dataset is problematic.

Standard linear wave theory (see, e.g. the textbook of Holthuijsen (2007) or the introduction by Berg and Rhome (2005)) gives rise to a Rayleigh distribution for wave height (see Appendix A). Freak waves are sometimes defined as waves that are higher than twice the significant wave height. According to the Rayleigh distribution, about one in about 3000 waves is a freak wave. For waves with a period of about 10 s, this is of the order of once every 8 h. In practice, of course, usage of the term freak wave is often restricted to cases where the absolute value is exceptionally high. Over the last few years, several mechanisms have been proposed that give rise to enhanced extreme wave distribution compared to standard Rayleigh theory (Janssen 2003; Mori and Janssen 2006). In Fig. 3, which is discussed in Appendix,

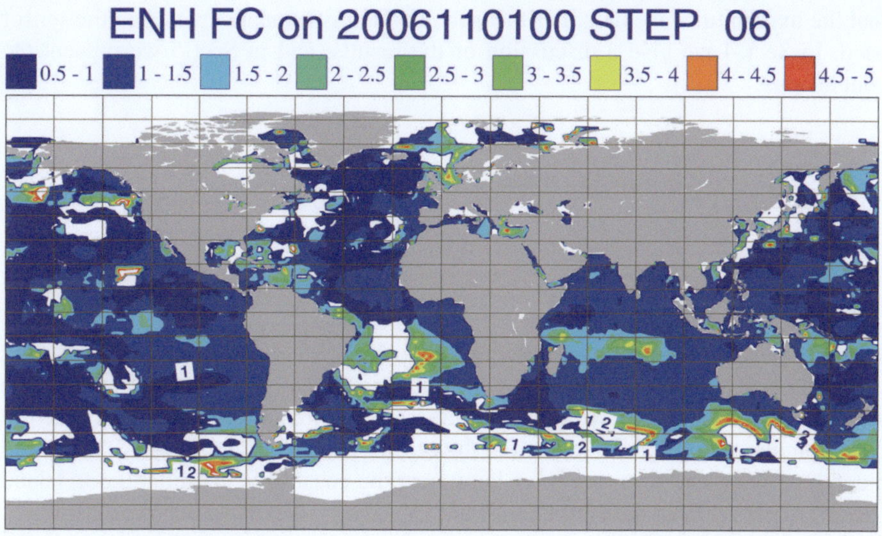

Fig. 2 WAM enhancement factor. This factor gives the ratio of the WAM model estimate to the standard linear model probability for waves with a height more than twice the significant wave height

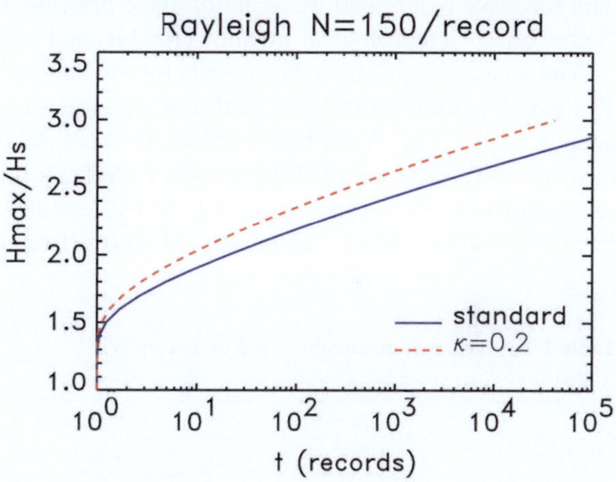

Fig. 3 Rayleigh and Janssen distribution of Hmax/Hs of the return time of the ratio of H_{max}/H_s of maximum wave height over significant wave height. Time is measured in 150-wave records. The *thick line* is the Rayleigh distribution, the *dashed line* distribution according to Janssen theory for the case that the kurtosis of the sea surface $\kappa = 0.2$, corresponding to $BFI = 0.33$

such an enhanced extreme wave distribution is compared to the standard Rayleigh distribution. The WAM model index is based on a non-linear effect that enhance the extreme wave height distribution. On the other hand, for very large r, when steepness becomes a limiting factor, non-linear effects will lead to a suppression with respect to the Rayleigh distribution. Moreover, in shallow water behaviour might be quite different. So it is interesting to examine the behaviour of the observed distribution.

In Sect. 2, we discuss the observational data set as well as the WAM model index. In Sect. 3, we present and analyse maximum wave height distributions, including a comparison between model results and observations. It is clear from the results that the maximum wave height as measured by an instrument depends on the sensor used. In Sect. 4 we give a discussion of the results, and in Sect. 5 we present the conclusions.

2 Data Sources

2.1 North Sea Data

The data consist of reports of the Meetnet Noordzee (MNZ), a network of measuring instruments at a number of platforms and buoys in the North Sea set up by the Dutch authorities in cooperation with platform operating companies. In this paper, we use data from the Auk platform located in the central North Sea at (56°24N,0°02E) and a depth of 80 m, and from the wave buoy near the coast of Schiermonnikoog Island (SMN) in the north of the Netherlands at (53°35N,6°10E) and a depth of 20 m.

A report consists of a set of wave parameters extracted from a 20 min record, the frequency of the reports is 3 h. In this paper, in addition to significant wave height and wave period, also the maximum wave height of the record is used. The MNZ data for wave height and wave period have been monitored for years by the Royal Netherlands Meteorological Institute (KNMI) and we know them to be reliable.

The measurements have been made by several types of instruments, see Table 1. The radars operate from fixed platforms and measure surface elevation with a frequency of 5.12 Hz. Wave buoys calculate wave data on the basis of acceleration measurements. The sample frequency is 1.28 Hz. A linear time-domain filter is used to reconstruct a wave height record, and to estimate the maximum wave height from these accelerations. The number of waves (from the estimated zero-upcrossing

Table 1 Overview of instruments used in this study

Label	Location	Type	Period
1	Auk	Waverider	19840326–19861010
2	Auk	Saab radar	19860413–19870714
A2	Auk	Saab radar	19930616–19990625
A3	Auk	Wavec	19901024–19991109
AUK1	Auk	Saab radar	20000204–20050928
AUK2	Auk	Wavec	20010718–20010925
AUK2	Auk	Directional Waverider	20020621–20030731
4	Schier	Wavec	19901024–19930331
W4	Schier	Wavec	19931101–19991109
SMN1	Schier	Wavec	19931109–20031216
SMN1	Schier	Directional Waverider	20020220–20070831

At the locations of Auk at 56°24N,0°02E and Schiermonnikoog Island (Schier) at 53°35N,6°10E period) in a record falls typically in the range 150–200. The results of Sect. 3 show that wave buoys give systematically lower estimates for the maximum wave height than the radar estimates, and that there are significant differences between the various combinations of buoys and filters that have been used over the years.

Quality control included rejection of duplicates and the rejection of gross errors. Example for the radar altimeter, some short periods with on average unrealistically high values were skipped. There are some features of the data set we cannot explain. For example, there seems to be a preference for 'nice' values of the ratio H_{max}/H_s such as 1.5 or 2. We have not been able to trace what part of the processing is responsible for this feature.

2.2 WAM Model BFI

The Benjamin Feir index *BFI* proposed by Janssen (2003) is a measure of the strength of the effect non-linear interactions on wave height distribution. Non-linear effects are stronger if (1) waves are steeper and (2) the wave spectrum is more narrow allowing for waves to travel longer together. For a narrowband spectrum the definition of the *BFI* is

$$BFI = \frac{\sqrt{2}km_o^{1/2}}{(\sigma/\omega)} \tag{1}$$

where k denotes the dominant wave number and ω the frequency of the spectrum, σ the spectral width and $m_o^{1/2} = 0.25H_s$ the amplitude of the spectrum. In the numerator $km_0^{1/2}$ is the steepness, the denominator is the narrowness of the spectrum.

For general spectra, the above expression for the spectral width is rather ambiguous and Janssen (2003) uses the following expression for the BFI:

$$BFI = \sqrt{2\pi}km_o^{1/2}Q_p, \tag{2}$$

with

$$Q_p = \frac{2}{m_0^2}\int d\omega\omega E^2(\omega) \tag{3}$$

where $E(\omega)$ is the spectral density.

The BFI was added as an output parameter to ECMWF's wave model in the fall of 2003. Since then all model forecasts have been archived and from these archives BFI and significant wave height have been extracted from 6 October 2003 until 31 December 2006 in 6-hourly steps for model grid points near platform Auk and Schiermonnikoog.

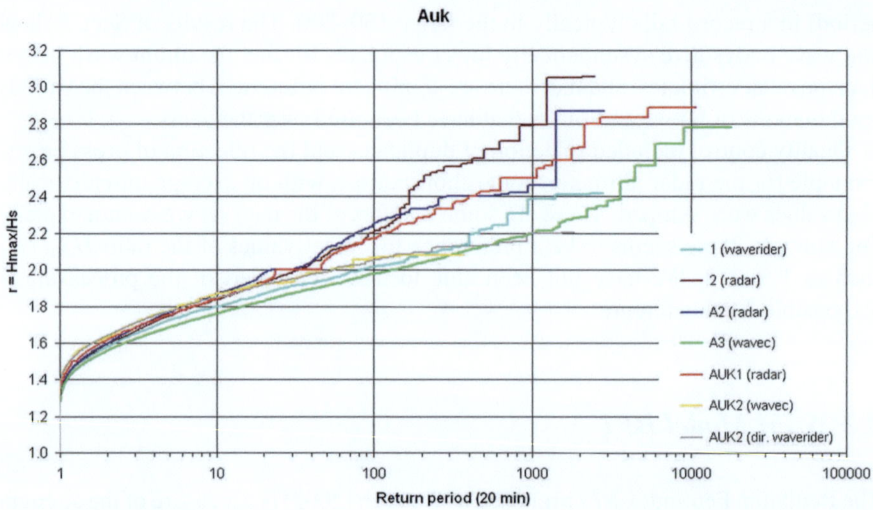

Fig. 4 Return times at the deep water station at Auk. Time is measured in 20 min records. The labels refer to different instruments and periods, see Table 1

3 Results

The first quantity we consider is the return period of $r = H_{max}/H_s$ for the deep water station at Auk. In Fig. 4, this quantity is plotted as a function of return time (in units of 20 min records), for various instruments and periods. It appears that the differences between the instruments and periods are large. The radar altimeter values are close to those which one would expect from a Rayleigh, the wave buoy data are generally lower.

For situations where the average steepness (H_s/λ) is large, steepness may be a limiting factor. First we check in Fig. 5, which gives a plot of H_{max} versus T if wave heights do not exceed the limiting steepness line of $H = \lambda/7$. This figure shows that it is not uncommon that the steepness approaches the limiting steepness.

If steepness is a limiting factor, then one would expect that for a given significant wave height, longer periods that go with less steep waves would lead to an enhanced probability of high values of r. Figure 6, where the average value of r for records with $H_s \approx 4$ m is plotted as a function of T, gives some evidence for this fact.

Now we turn to the shallow water results. In Fig. 7, the return periods for the shallow water station Schiermonnikoog (SMN) are shown. Comparing these results to the deep water data of Auk is hampered by the fact that different stations and periods are hard to compare. Even when matching periods and instruments there remain problems: the SMN1 Wavec is much lower than the AUK2 Wavec, while the SMN W4 is only slightly lower than the AUK A3. We checked that this also holds when we restricted the comparison to periods that both instruments yielded data (not shown). We note that period of the W4 versus A3 comparison is much

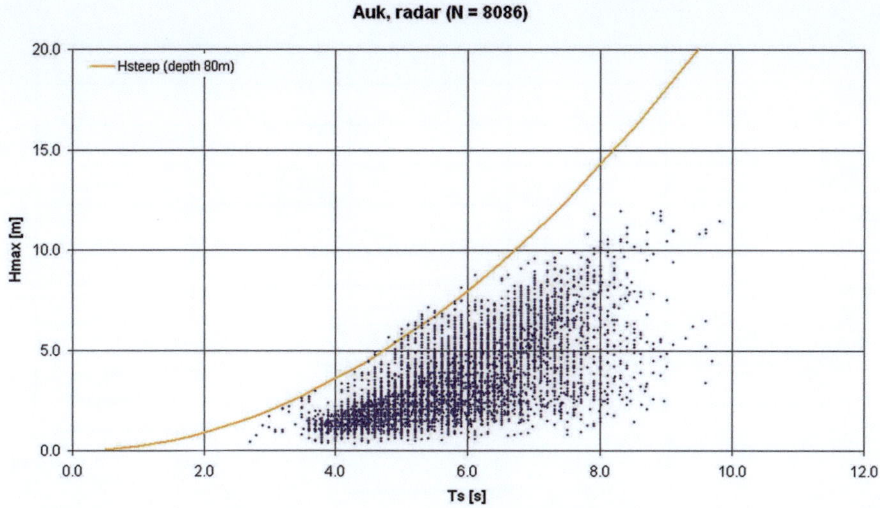

Fig. 5 Scatter plot of maximum wave height H_{max} versus mean period T in deep water for the radar measurements at Auk

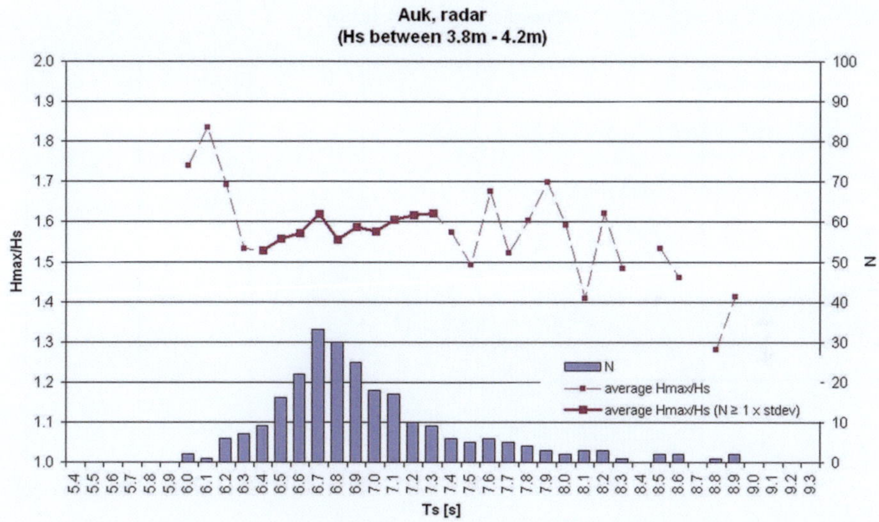

Fig. 6 The ratio $r = H_{max}/H_s$ as function of the mean period T, for radar measurements at Auk. The histogram indicates the number of data in each bin. The *solid line* connects the average value of r for bins with nine or more entries

longer (19931101–19991109) than the period of the AUK2 versus SMN1 comparison (20010718–20010925). What we can conclude is that there is no indication that outside the surf zone, values of r are higher in shallow water than in deep water.

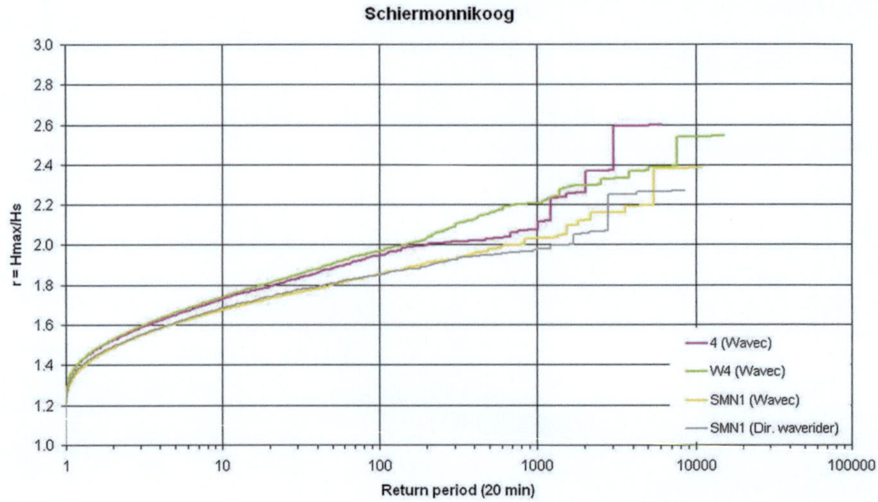

Fig. 7 Return times at the shallow water station Schiermonnikoog Island (SMN). Time is measured in 20 min records. The labels refer to different instruments and periods, see Table 1

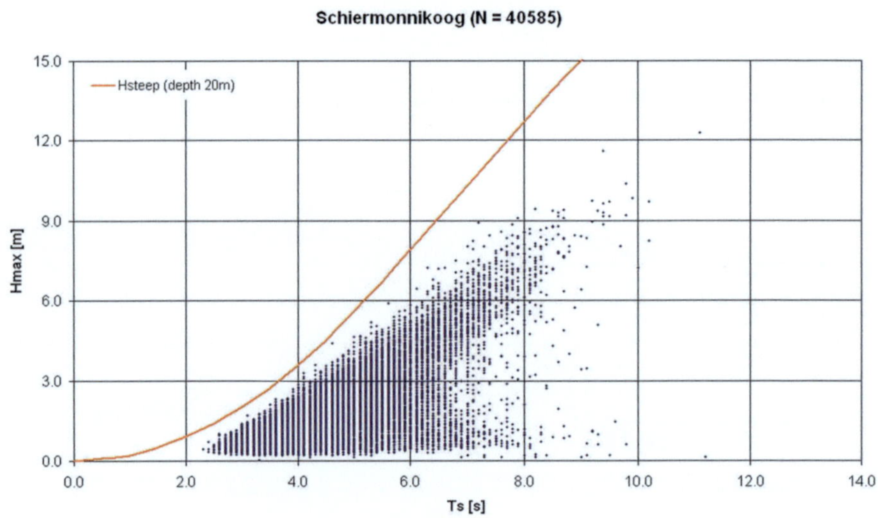

Fig. 8 Scatter plot of maximum wave height H_{max} versus mean period T at the shallow water station Schiermonnikoog Island (SMN)

For this shallow water location, the constant steepness line (yellow) in a H-T diagram has a different shape than for deep water. Figure 8, which gives a plot of H_{max} versus T for SMN, shows that for this station maximum wave heights are not as close to yellow line as for deep water waves. We checked in plots of H_s against steepness, see Figs. 9 and 10, that while the general picture that high waves are more

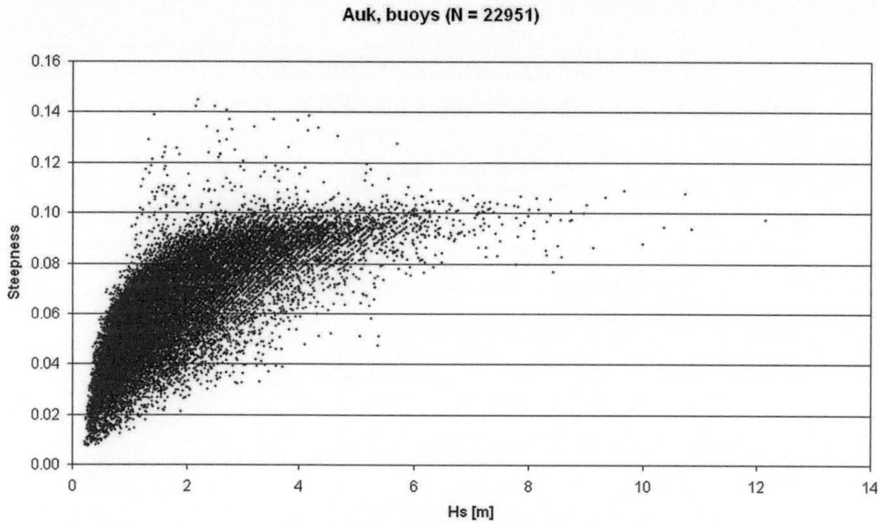

Fig. 9 Scatterplot of wave steepness versus H_s at the deep water station Auk

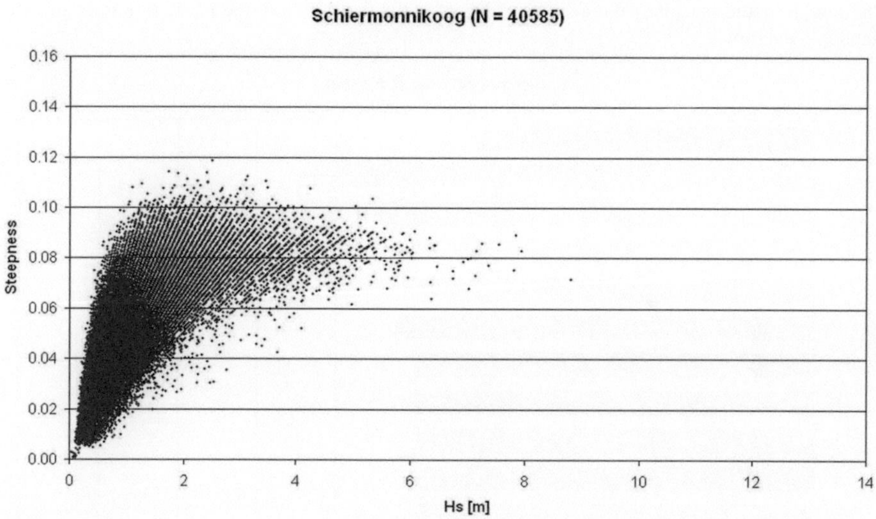

Fig. 10 Scatterplot of wave steepness versus H_s at the shallow water station Schiermonnikoog Island

often steep than low waves remains valid, for waves with $H_S > 2\,\text{m}$, the limiting steepness at the shallow water location decreases with significant wave height, while it stays roughly constant at the deep-water location.

Finally, we checked whether there is a relation between the BFI index of WAM of the ECMWF and the probability of high r values. To this end, average values

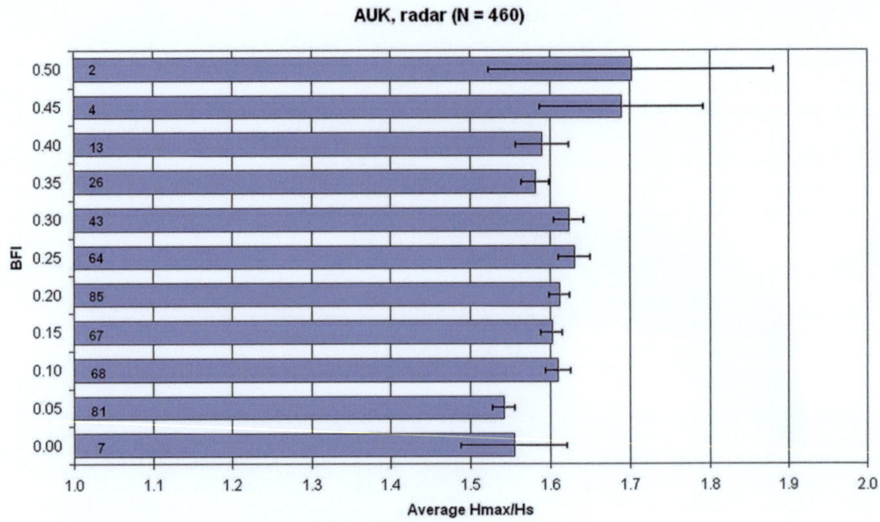

Fig. 11 Model BFI versus measured H_{max}/H_s in deep water at Auk. Observations have been binned according to model BFI, the number of observations is indicated by the black number in each bar. The length of the bar gives the average value of r for that bin, and the *black line* indicates the standard deviation

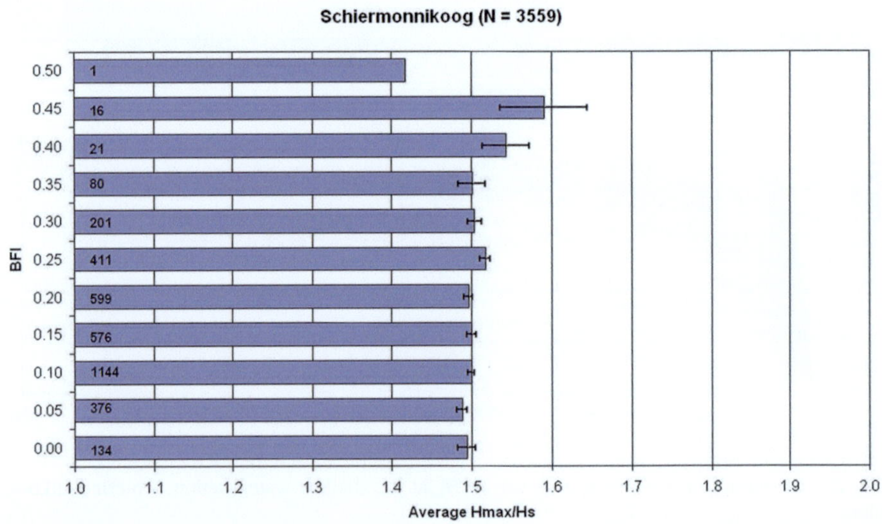

Fig. 12 Model BFI versus measured H_{max}/H_s in shallow water at SMN. Observations have been binned according to model BFI, the number of observations is indicated by the black number in each bar. The length of the bar gives the average value of r for that bin, and the *black line* indicates the standard deviation

of r have been collocated with the model BFI. The results are shown in Fig. 11 for Auk and Fig. 12 for SMN. During the period for which the model BFI was available, there were many more observations at the shallow water station than at the deep water station. The data for the deep water station at Auk do not exclude a relation between high r observations and high WAM model BFI indices. For the shallow water location, there is hardly a correlation between high r observations and high WAM model BFI indices.

4 Discussion

From the measurements it appears that there are systematic differences between the various instruments. Because the radar has a higher sampling frequency, and makes a more direct measurement than the accelerations measured by the buoys, we consider the radar measurements to be more reliable. Additional confidence in the radar results comes from the fact that they are close to those predicted by standard Rayleigh theory. Apparently, the frequency of around 1 Hz of accelerations by the buoys is not sufficient for capturing maximum wave heights well, making that the buoy results are systematically below the radar results. That differences between various periods are so large indicate that not only the instrument but also the processing algorithm has a large impact on the distribution of the measurements. This makes a comparison of the deep water station with the shallow water station far from straightforward. What remains is a number of series of several months of data. The most obvious result is that there is no change in behaviour for timescales ranging from hours to months. There is neither evidence for an enhanced tail because of non-linear enhancement effects, nor for a damped tail because of limiting steepness effects. Considering subclasses, steepness may have an effect: there is some indication that, given the significant wave height and the period, steeper waves are more likely in the case of long periods. In shallow water, wave energy can converge and give rise to high waves. But our results do not indicate a higher probability of extreme waves in shallow water than in deep water. If any, we observe the opposite effect: extreme waves are less likely in shallow water. We do find a clear difference in plots of steepness versus wave height between deep water and shallow water: in deep water there is a limiting steepness that does not depend on wave height, in shallow water this limit decreases with wave height, probably because of bottom friction effects.

For deep water, Janssen theory expects an increase of the mean value of r of the order of 0.1 if the BFI is varied from 0 to 0.5 (Peter Janssen, personal communication). Such an increase is compatible with our results for the deep water station at Auk in Fig. 11. Janssen theory does not expect a correlation between BFI and r in shallow water, because in shallow water conditions for four-wave interactions differ from deep water. This is confirmed by our shallow water results in Fig. 12.

5 Conclusion

The analysis has been hampered by the fact the instruments report approximations for maximum wave height, and that those approximations differ between different instruments and observing periods. Our main result is that the instrumental data are consistent with a Rayleigh-like extreme wave distributions up to return periods of many months. There is slight indication that for long waves steepness can be a limiting factor for maximum wave height. In shallow water, there is some evidence that extreme waves are less common than in deep water. The distributions of wave height versus steepness in shallow water and deep water are different, which may be related to the overall damping effect of bottom friction that causes a reduction of significant wave height in shallow water. As expected, in shallow water, there is no relation between the WAM model BFI and the probability of extreme waves. For deep water, such a relation cannot be ruled out.

Acknowledgments We wish to thank Peter Janssen for valuable and stimulating discussions.

Appendix: Rayleigh Distribution

According to linear wave theory, see e.g. Holthuijsen (2007), the probability $p(r)$ that an individual wave has a height $H = rH_s$ is given by

$$p(r) = \exp(-r^2) \tag{4}$$

From this expression it follows the probability $P(\bar{r})$ that in a series of N waves all waves have $r < \bar{r}$ is

$$P(\bar{r}) = (1 - p(\bar{r}))^N \tag{5}$$

The exact number of waves in a 20 min record is not determined. In the North Sea, periods are shorter than in the open ocean and vary from 5 to 10 s, in very severe storms the dominant period can be higher. So a typical number of waves in a 20 minute record is about 150. In Fig. 3 return maximum wave height as a function of the number of records is plotted as a solid line. The maximum wave height ratio r reaches a value of 2 for about 20 records (about 6 h) and increases slowly with the number of records. Even for 10^5 records, that is about 4 years, r is below 3.

For comparison, a distribution that follows from the theory of Mori and Janssen (2006) is shown as well (dotted line). The case shown corresponds to a BFI of 0.33. In Janssen theory, the BFI is directly related to κ, the kurtosis of the sea surface, by $\kappa = (\pi\sqrt{3})BFI^2$, so the case of Fig. 3 corresponds to $\kappa = 0.2$.

References

Berg R, Rhome J (2005) Expecting the unexpected wave: how the national weather service marine forecasts compare to observed seas. Mar Weather Log 49(2):4–7

Bojanowski A (2007) Monsterwellen bedrohen Schiffe im Nordatlantik (in German). http://www.spiegel.de/wissenschaft/natur/0,1518,470359,00.html

Datawell (2006) Versnellingswaardein: Detectie en reparatie van extreme versnellingswaarden (in Dutch). Datawell internal report, Datawell BV, Haarlem, The Netherlands

Holliday N, Yelland M, Pascal R, Swail V, Taylor P, Griffitchs C, Kent E (2006) Were extreme waves in the Rockall trough the largest ever recorded? Geophys Res Lett 33(L05):613. doi:10.1029/2005GL025238

Holthuijsen L (2007) Waves in oceanic and coastal waters. Cambridge University Press, Cambridge, 404 pp. doi:10.2277/0521860288

Janssen P (2003) Nonlinear four-wave interactions and freak waves. J Phys Oceanogr 33:863–884

Mori N, Janssen P (2006) On kurtosis and occurrence probability of freak waves. J Phys Oceanogr 36:1471–1483

Extremes and Decadal Variations in the Baltic Sea Wave Conditions

Tarmo Soomere

Abstract Average and extreme wave conditions, their seasonal cycle and decadal variations, and extreme wave storms in Baltic Sea Proper and in the largest sub-basins of this sea are studied based on long-term time instrumentally measured time series of wave properties at Almagrundet and the Darss Sill, visual wave observations from several coastal sites of the eastern Baltic Sea, wave statistics from the northern Baltic Proper and Gulf of Finland, long-term reconstructions of the wave climate and numerical modelling of an extreme wave storm. The wave climate is highly intermittent and occasionally contains very strong wave storms. Significant wave heights $H_S \geq 4$ m occur with a probability of about 1 % among all wave fields in the open Baltic Proper. Extreme wave conditions with $H_S \geq 7$ m occur approximately twice in a decade. The overall recorded maximum H_S is 8.2 m. The estimated maximum of H_S was about 9.5 m in cyclone Gudrun in January 2005. No clear trend exists in the wave properties in the Baltic Sea. The 99 %-iles of the significant wave height exhibit a complicated spatial pattern of changes and have significantly decreased between the islands of Öland and Gotland and to the south of these islands.

1 Introduction

The complexity of physics and dynamics of the Baltic Sea extend far beyond the typical features of many other water bodies of comparable size (e.g. Feistel et al. 2008; Leppäranta and Myrberg 2009). The combination of a relatively small size of this water body and vulnerability of its ecosystem makes this region extremely susceptible with respect to climate changes and shifts. Its complex geometry, high variability of wind patterns and extensive archipelago areas with specific wave propagation properties (Tuomi et al. 2014) give rise to large spatio-temporal variability in the wave properties. The presence of relatively shallow areas and often occurring convergent wind patterns may lead to occasional wave energy concentration in some areas (Soomere 2003, 2005; Soomere et al. 2008). This feature requires a high

T. Soomere (✉)
Institute of Cybernetics at Tallinn University of Technology, Tallinn, Estonia
e-mail: soomere@cs.ioc.ee

E. Pelinovsky and C. Kharif (eds.), *Extreme Ocean Waves*,
DOI 10.1007/978-3-319-21575-4_7

spatial resolution of wave simulations and also a careful choice of wave measurement sites. The seasonal ice cover of large parts of the sea considerably affects the wave patterns during winter and early spring. As wave measurement devices are removed well before the ice season (Kahma et al. 2003; Tuomi et al. 2011), the wave data have long gaps and several commonly used characteristics (e.g. annual mean wave height or period) may become meaningless (Tuomi et al. 2011). The situation is additionally complicated by specific features of wave generation by offshore winds over irregular coastline (Kahma 1981; Kahma and Calkoen 1992) or under so-called slanting fetch (when wind blows obliquely across the coastline; Pettersson et al. 2010).

Numerous changes in the forcing conditions and in the reaction of the water masses of the Baltic Sea have been reported during the later decade (BACC Author Team 2008, BACC II Author Team 2015). The apparently increasing storminess in the Baltic Sea during the second half of the 20th century (Alexandersson et al. 1998) has caused extensive erosion of depositional coasts (Orviku et al. 2003). The changes in the average wave climate of the entire Baltic Sea and adjacent seas have been found marginal, at least, until the mid-1990s (WASA Group 1995; Mietus and von Storch 1997).

Very rough seas that occurred twice in December 1999, all-time highest significant wave height 8.2 m in the Baltic Proper in December 2004 (Tuomi et al. 2011) and all-time highest single wave in the Gulf of Finland in November 2012 (9.4 m, www.fmi.fi), reinforced the discussion as whether the wave conditions in the Baltic Sea have become rougher compared with the situation a few decades ago. The exceptional storm Erwin/Gudrun (January 2005) highlighted inadequate awareness of extreme wave properties (Soomere et al. 2008) and of the height and spatial extent of extreme water levels (Suursaar et al. 2006). Recent numerical simulations (Soomere and Räämet 2011a, b; Tuomi et al. 2011), analysis of long-term directional wave measurements in the Arkona Basin (Soomere et al. 2012) and reconstructions of the nearshore wave climate back to the mid-1940s based on visual wave observations (Soomere 2013) have shed much more light on spatio-temporal variations in the Baltic Sea wave climate.

Recognition of the wave climate changes, in particular, changes in extremes, presumes a thorough knowledge of the typical and extreme wave conditions. The global wave data set KNMI/ERA-40 Wave Atlas (09.1957–08.2002, Sterl and Caires 2005) allows the production of a reliable wave climatology for open ocean conditions, based on 6-hourly means of wave properties over an average of $1.5° \times 1.5°$ areas. This resolution is too sparse for the Baltic Sea conditions. As typical for semi-enclosed shallow basins, wave properties are additionally modified here through wave-bottom interaction (refraction, shoaling, breaking or reflection) and diffraction behind obstacles.

The Baltic Sea has probably the longest history in the world of almost one and half centuries of systematic visual observations of wave properties from fixed coastal locations (Rosenhagen and Tinz 2013). Similar observations from lightships started about 90 years ago in Danish waters (Hünicke et al. 2015) and slightly later in Swedish waters (Wahl 1974). Systematic observations of wave properties from many coastal sites were launched in the eastern Baltic Sea since the mid-1940s, and such observations are performed at a few sites until today (Soomere 2013). The outcome of these historical observations combined with similar data from ships and results of various early hindcasts has been formulated in several generations of textbooks (Davidan et al. 1978, 1985; Lopatukhin et al. 2006b) and wave atlases for the Baltic Sea (Rzheplinsky 1965; Russian Shipping Registry 1974; DWD 2006; Lopatukhin et al. 2006a) and its sub-basins (Druet et al. 1972; Rzheplinsky and Brekhovskikh 1967; Schmager 1979; Sparre 1982).

The properties of wind waves primarily depend on the wind speed and duration, and effective fetch length. The pattern of predominant winds (Mietus 1998; Soomere and Keevallik 2001) and the geometry of the Baltic Sea suggest that the highest and longest waves occur either at the entrance of the Gulf of Finland, off the coasts of Saaremaa, Hiiumaa, and Latvia, or along the Polish coasts (Schmager et al. 2008). Wave data from the northern parts of the Baltic Sea Proper thus adequately represent both the average and the roughest wave situations in the region.

The wave properties in the Baltic Sea can be modelled with the use of local models, because the waves from the rest of the World Ocean practically do not affect this water body. Numerical reconstruction of the Baltic Sea wave fields is still a complicated task, and the results usually contain extensive uncertainties (Cieślikiewicz and Paplińska-Swerpel 2008; Kriezi and Broman 2008; Räämet et al. 2009). The largest source of uncertainties is the wind information (Nikolkina et al. 2014). Its quality has considerably increased within the last decade, but this information still suffers from substantial temporal inhomogeneity (Tuomi et al. 2011), spatial variations in its quality (Räämet et al. 2009; Soomere and Räämet 2011b), and occasional mismatch of modelled and actual air flow directions (Keevallik and Soomere 2010).

Several numerical wave studies are performed for the southern part of the Baltic Sea (e.g. Gayer et al. 1995; Paplińska 1999; Blomgren et al. 2001; Siewert et al. 2015). Early estimates of wave statistics for the entire Baltic Proper have been performed using the second-generation spectral wave model HYPAS and wind data from a few years (1999–2000; Jönsson et al. 2003, 2005; Danielsson et al. 2007). Although third-generation wave models such as WAM (e.g. Komen et al. 1994) and SWAN have been implemented for the northern Baltic Sea at the turn of the millennium (e.g. Tuomi et al. 1999; Soomere 2001), wave statistics based on such models was available only for limited areas until 2005 (Soomere 2003, 2005). Extensive simulations based on different wind information were performed only starting from about 2005 (Schmager et al. 2008; Soomere and Räämet 2011a, b; Tuomi et al. 2011). These simulations together with the increasing pool of instrumental measurements (Tuomi et al. 2011; Soomere et al. 2012) and reconstructions of wave properties from historical visual observations (Zaitseva-Pärnaste 2013) made it possible to identify not only the basic properties of the wave climate of the Baltic Sea (Hünicke et al. 2015) but also to distinguish a remarkable pattern of its spatial and decadal variations (Soomere and Räämet 2011b, 2014; Suursaar 2013).

This chapter presents a description of the basic properties of average and extreme wave conditions and depicts their spatio-temporal variations in the Baltic Proper based on available long-term wave measurements (both instrumental and visual), numerical reconstructions of wave climate and wave properties in a specific event.

The wave patterns in the Gulf of Finland (an elongated basin with a length of about 400 km and a maximum width of about 135 km) and Arkona Basin are frequently connected with those in the Proper, and are addressed to some extent as well.

The analysis is mostly based on (i) instrumental measurements in 1978–2003 at Almagrundet (located near the western coast of the northern Baltic Proper), in 1990–2011 at the Darss Sill in the Arkona Basin, since 1996 in the northern Baltic Proper and since 2001 in the Gulf of Finland, (ii) visual observations from 12 sites along the eastern coast of the Baltic Sea from the vicinity of Kaliningrad to the neighbourhood of Saint Petersburg and (iii) numerical reconstructions of the entire Baltic Sea wave fields using geostrophic winds and the outcome of HIRLAM model. To a limited extent, data from waveriders in the northern Baltic Proper and at Bogskär are used. The results predominantly represent Type A statistics in terms of the classification of Kahma et al. (2003): almost no corrections have been made to compensate for missing values, for the uneven distribution of data (except for the use of daily mean wave heights for certain parameters) or for ice cover. Modelled data are used in the estimates of extreme waves in windstorm Gudrun in January 2005.

2 Long-Term Wave Measurements and Hindcasts in the Baltic Sea

2.1 Instrumental Data Sets

Contemporary instrumental wave measurements were launched in the northern Baltic Sea in the framework of wave power studies at the end of the 1970s near the lighthouse of Almagrundet and south of Öland. A waverider buoy was simultaneously deployed near Hoburg, south of Gotland. The measurements were mostly performed during a few years (Mårtensson and Bergdahl 1987).

The data from Almagrundet (1978–2003, 59°09′N, 19°08′ E, Fig. 1, Broman et al. 2006) form the longest instrumentally measured wave time series in the region. Almagrundet is a 14 m deep shoaling area about 10 nautical miles south-east of Sandhamn in the Stockholm archipelago. It is sheltered from a part of predominant winds. The fetch length for winds from the south-west, west and north-west is quite limited. The above-discussed anisotropy of the Baltic Sea wave fields has caused some discussion about whether the data correctly represent the open-sea wave conditions for some wind directions (Kahma et al. 2003).

An upward-looking echo sounder from Simrad was placed at a depth of about 30 m in 1978 (Mårtensson and Bergdahl 1987) and was active until mid-September 1995. An analogous device from WHM was installed in a neighbouring location at a depth of 29 m in 1992 and produced usable data in 1993–2003 (Broman et al. 2006). The position of the water surface was sampled during 640 s each hour. Wave components with periods of less than 1.5 s as well as the data probably reflecting wave interference and breaking waves and possibly very steep waves were discarded (Mårtensson and Bergdahl 1987).

Single waves were identified based on the classical zero-downcrossing method (IAHR 1989). An estimate of the significant wave height $H_{1/3}$ (the average height of 1/3 of the highest waves) was found somewhat untypically, from the 10th highest wave H_{10} in a record of N waves under the assumption that wave heights are Rayleigh distributed:

$$H_{1/3} = \frac{H_{10}}{\sqrt{\frac{1}{2}\ln\frac{N}{10}}}. \tag{1}$$

The set of 95 458 measurements using the Simrad device in 1978–1995 reliably describes the wave properties (Broman et al. 2006). Later 46 671 recordings using the WHM device in 1993–2003 have certain quality problems: the data contain a number of modest, but still evidently unrealistic peaks and the values of wave period are unreliable. Broman et al. (2006) recommend considering the WHM data for 1993–2003 as merely indicative.

A non-directional waverider was operated in 1983–1986 near Bogskär at 59°28.0′N, 20°21.0′ E (Kahma et al. 2003). The wave properties were measured hourly during total 14 630 h, or about 2 years of uninterrupted measurements. The data set is concentrated in the autumn season and thus represents the wave climate during relatively windy months.

A directional waverider was deployed in the northern Baltic Proper at a depth of about 100 m (Fig. 1, 59°15′N, 21°00′E) in September 1996 and operated since then during the ice-free seasons (Kahma et al. 2003). This device as well as contemporary spectral wave models estimate the significant wave height as $H_S = 4\sqrt{m_0} \approx H_{1/3}$, where m_0 is the zero-order moment of the wave spectrum (the total variance of the water surface displacement, e.g. Komen et al. 1994). This data set is the most representative of the northern Baltic Sea wave fields; however, only its few sections have been analysed in the literature.

Directional wave measurements in the Gulf of Finland in 1990–1991, 1994 and from November 2001 (59°57.9′N, 25°14.1′E, water depth about 60 m, Fig. 1) during the ice-free seasons have considerably increased the awareness of wave conditions in semi-enclosed sub-basins of the Baltic Sea (Kahma and Pettersson 1993; Pettersson 2001; Kahma et al. 2003). Similar wave measurements have been performed almost continuously since 29 January 1991 at a 20-m-deep site on the Darss Sill (Soomere et al. 2012; 54°41.9′N, 12°42.0′E). This data set together with similar measurements since 2002 to the north-west of Cape Arkona (54°52.9′N, 13°51.5′E) forms the most valuable source of the wave information in the SW Baltic. Some elements of the wave climate in the southern and south-eastern Baltic Sea have been presented in (Paplińska 1999; Cieślikiewicz and Paplińska-Swerpel 2008; Siewert et al. 2015).

The number of contemporary wave measurement locations has increased in the Baltic Sea since 2006 (Pettersson et al. 2007). The main results are described on annual basis on the HELCOM website http://helcom.fi/baltic-sea-trends/environment-fact-sheets/hydrography/wave-climate-in-the-baltic-sea/. Instrumentally measured wave

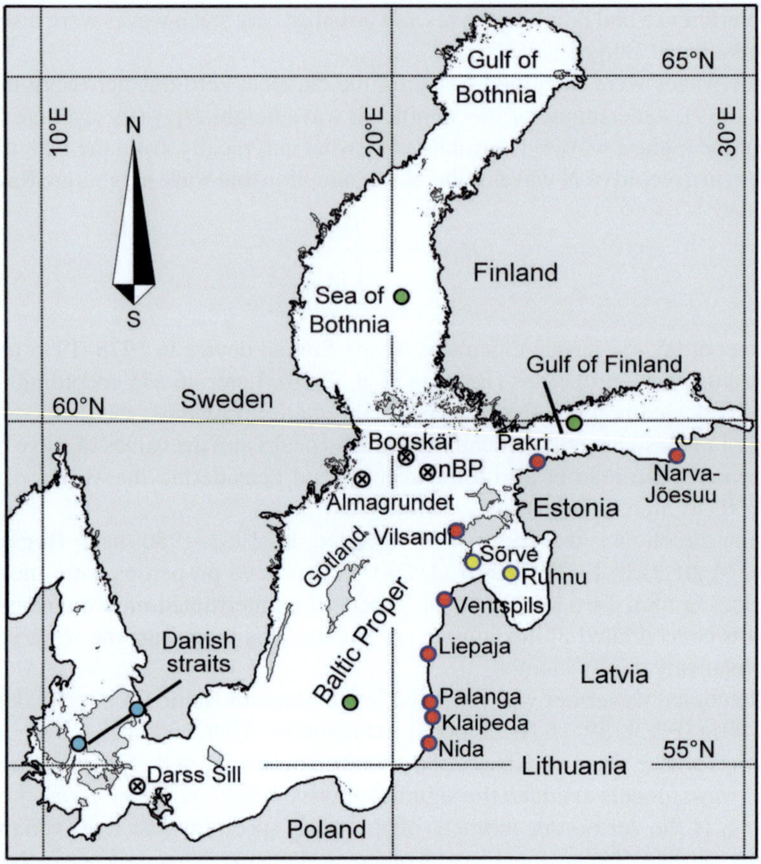

Fig. 1 Location scheme of the Baltic Sea, showing the sites of instrumental measurements discussed in detail in this chapter (*crossed circles*), sites of directional waveriders in the southern Baltic Proper, Gulf of Finland and Sea of Bothnia (*green circles*) and visual wave observation sites at the eastern coast of the Baltic Sea (*red circles*) and in the Gulf of Riga (*yellow circles*)

data from the coastal areas of Estonia, Latvia and Lithuania have been mostly obtained using pressure-based sensors (Soomere 2005) or ADCP-s (Suursaar 2013), and cover only shorter sections of a few months. Satellite altimeter data for wave properties have been used in a very few studies (Cieślikiewicz and Paplińska-Swerpel 2008; Tuomi et al. 2011).

2.2 Visual Observations

A reasonable source of the open-sea wave information in the past formed visual observations (Davidan et al. 1985; Hogben et al. 1986). The visually observed wave height generally matches the significant wave height well (Gulev and Hasse 1998, 1999). The visually estimated wave periods are, on average, a few tenths of a second shorter than the peak period (Gulev and Hasse 1998, 1999). Wave climate changes estimated from data observed from merchant ships are consistent with those shown by the instrumental records (Gulev and Hasse 1999; Gulev et al. 2003).

Visual observations from the coast are less frequently used for wave climate studies. Such data frequently represent only wave properties in the immediate vicinity of the observation point (Orlenko et al. 1984; Soomere 2005). They pose intrinsic quality and interpretation problems, contain a large fraction of subjectivity (Zaitseva-Pärnaste et al. 2009), have a poor temporal resolution, have many gaps caused by inappropriate weather conditions or by the presence of ice, may give a distorted impression of extreme wave conditions, etc. The interval between subsequent observations is often much longer than the typical saturation time of rough seas in the northern Baltic Proper (about 8 h, Soomere 2001) or the duration of wave storms (that seldom exceeds 10 h, Broman et al. 2006; Lopatukhin et al. 2006b). These data sets have, however, exceptional temporal coverage for the Baltic Sea. Regular wave observations have been performed using a unified procedure at many locations since the mid-1940s during up to almost 70 years (Zaitseva-Pärnaste et al. 2011; Pindsoo et al. 2012) and are thus one of the few sources for detecting the long-term alterations of the wave climate. These observations have been carried out usually 3 times per day using perspectometers (binoculars with a specific scaling), buoys or bottom-fixed structures to better characterise the wave properties. The data, however, well represent the general features of the Baltic Sea wave fields: relatively low overall wave activity, short wave periods and substantial seasonal variation of wave conditions (Soomere and Räämet 2011b).

The observers scanned at least 4-m-deep areas about 200–400 m from the waterline. To evaluate the wave height, the observer noted the five highest waves during a 5-minute time interval and filed the highest single wave $H_{\max}$ and the mean height H_{mean} of these waves. The visually observed wave height, obtained using this or similar procedure, tends to overestimate the wave heights as the observer often picks up the largest waves and not necessarily in a single location. There have been many efforts to link the visually observed wave heights with the instrumentally measured ones (Massel 2013). As the typical wave periods in the coastal zone of the Baltic Sea are 3–4 s (Broman et al. 2006; Zaitseva-Pärnaste et al. 2011), $H_{\max}$ is approximately equal to the average height of 2.5–3 % of the highest waves and the mean wave height H_{mean} is approximately equal to the average height of 5–7 % of the highest waves (Zaitseva-Pärnaste 2013). Consequently, $H_{\max}$ roughly represents the 97.5 %-ile of single wave heights and H_{mean} exceeds the significant wave height by 15–20 %. This conjecture matches the outcome of a comparative analysis of the visually observed data from lightships and instrumentally measured data. In the south-western Baltic Sea, $H_S \approx 0.94 H_{mean}$, and in the Bay of Bothnia, $H_S \approx 0.81 H_{mean}$ (Wahl 1974).

The mean height H_{mean} has usually the largest coverage in the visually observed data sets (Zaitseva-Pärnaste et al. 2011; Pindsoo et al. 2012; Eelsalu et al. 2014). As the routine of observations has been slightly changed over decades (Eelsalu et al. 2014), during some time, only $H_{\max}$ is available in the observation diaries. In these occasions, $H_{\max}$ is used to evaluate the average wave properties over longer time intervals. As the average difference between H_{mean} and $H_{\max}$ is about 6 % (Soomere and Zaitseva 2007), doing so apparently has a fairly minor influence on climatological values of wave heights.

The wave direction was visually identified with a resolution of 45° (so-called eight rhumb system) as the direction from which the waves approached. The wave period was found as an arithmetic mean from three consecutive observations of passing time of 10 waves each time. These waves were not necessarily the highest ones. There exist different opinions about the interrelations of visually observed and instrumentally measured wave periods (Guedes Soares 1986; Massel 2013). The mismatches can be often reduced to different definitions of the wave period in different observation routines (Massel 2013). Davidan et al. (1985) found that for periods <7 s (which is the case in the Baltic Sea) the visually observed wave periods (using the above-described routine) almost exactly matched the average wave periods from the zero-crossing analysis.

The visual observation conditions vary considerably along the eastern Baltic Sea coast. For example, at Pakri the observer was located on the top of a 20-m high cliff and the water depth of the area over which the waves were observed was 8–11 m (Zaitseva-Pärnaste et al. 2009). Contrariwise, at Vilsandi, the observation site was chosen from two options according to the approaching wave directions and the water depth in the observation areas was only about 4 m (Soomere and Zaitseva 2007).

All coastal sites in Fig. 1 only conditionally represent the open-sea wave conditions. The largest distortions are evidently due to the sheltering effect of the mainland and the relatively shallow water depth. It is still likely that long-term variations and trends in the offshore wave properties are evident in the coastal observations (Soomere and Räämet 2011b). In northern locations, only 1–2 observations per day were possible in autumn and winter. Most of the gaps in the data sets occur from January to March apparently owing to the presence of sea ice. To eliminate the bias caused by a varying number of observations per day, the analysis has mostly been performed using the set of daily mean wave heights or measurements at a single observation time (Soomere and Zaitseva 2007; Zaitseva-Pärnaste et al. 2009, 2011; Pindsoo et al. 2012).

2.3 Long-Term Wave Hindcasts

The relatively small size of the Baltic Sea, frequent large-scale homogeneity in the wind fields and the short reaction and saturation time and memory of wave fields

(Soomere 2003) allow to use greatly simplified wave hindcast schemes (Soomere 2005), high-quality wind data from a few points (Blomgren et al. 2001) or parametric wave models (Suursaar and Kullas 2009a, b; Suursaar 2010, 2013) to reproduce the local wave statistics. The use of such models for the identification of extreme wave conditions is limited as they basically rely on the properties of the local wind field. Several attempts have been made to perform long-term numerical reconstructions of the entire Baltic Sea wave fields, but the relevant publications are scarce and often concentrated on the used methods (Cieślikiewicz and Paplińska-Swerpel 2008; Kriezi and Broman 2008; Alari 2013). The main features of the wave climate and its possible changes in this water body were established using simulations for 1958–2002 based on the output of National Centres for Environmental Prediction and for Atmospheric Research (NCEP/NCAR) wind reconstructions (Augustin 2005) and depicted in (Schmager et al. 2008; Weisse and von Storch 2010).

A more detailed description of the Baltic Sea wave climate was produced for 1970–2007 using the WAM wave model (Komen et al. 1994) Cycle 4 forced by adjusted geostrophic winds (Räämet and Soomere 2010; Soomere and Räämet 2011b). The bathymetry for the model run was based on the so-called Warnemünde data set of water depths for the Baltic Sea (Seifert et al. 2001). The calculation was carried out over a regular rectangular grid with a resolution of about 3 nautical miles. The grid covers the area from 09°36′E to 30°18′E and from 53°57′N to 65°51′N and contains 239 × 208 points (11 545 sea points). It was cut off in the narrowest part of the Danish straits, and the model was run independently from the North Sea. The entire grid was optimised for wave calculations, e.g. deeply indented bays were omitted (Soomere 2003) and very large bottom gradients at a few locations were smoothed. The run was performed in shallow water mode with depth refraction but without depth-induced wave breaking. The presence of sea ice was ignored. Doing so may substantially modify the average and extreme wave properties in the northern sub-basins of the Baltic Sea (Tuomi et al. 2011). The energy spectrum contained 24 equally spaced directions at each sea point. The range of wave frequencies was extended to properly resolve the wave growth under relatively low winds and short fetch. The model accounted for 42 wave components with frequencies ranging from 0.042 Hz to about 2 Hz and arranged in a geometrical progression with an increment of 1.1.

The wind data were extracted from the Swedish Meteorological and Hydrological Institute (SMHI) geostrophic wind database. The original geostrophic wind components were presented as gridded information with a spatial resolution of 1 × 1°. To derive an approximation of the 10 m level wind, the geostrophic wind vector was first rotated by 15° counterclockwise, and its length was multiplied by a factor of 0.6 to mirror the effect of surface roughness. This scheme explicitly ignores many details of the vertical structure of realistic winds (Bumke and Hasse 1989), but it is still quite popular in studies of the Baltic Sea dynamics (Myrberg et al. 2010). The resulting values were first externally interpolated to a grid with a step of about 6 nautical miles (123 × 107 points) and finally into the resolution of the WAM model internally in this model. The wind input time step was 6 h before September 1977 and 3 h after that. The calculations tend to underestimate the long-term average wave heights and

99 %-iles of the significant wave height by about 10 % (Tuomi et al. 2011) but almost exactly match the measured 95 %-iles of the significant wave height in the Arkona Basin (Soomere et al. 2012).

3 Wave Climate

3.1 Statistics of Wave Heights and Periods

The above-described sources of the information about the Baltic Sea wave fields make it possible to reliably identify the basic long-term wave properties (average and extreme wave height, typical periods, occurrence distributions of different heights

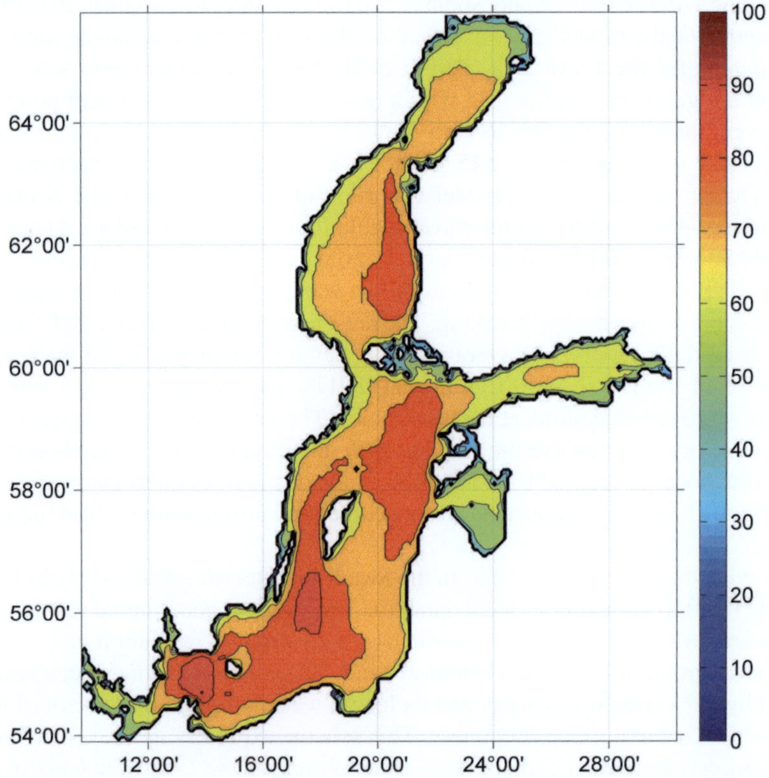

Fig. 2 Numerically simulated average significant wave height (cm; isolines plotted after each 10 cm) in the Baltic Sea in 1970–2007 based on adjusted geostrophic winds from the Swedish Meteorological and Hydrological Institute (Räämet and Soomere 2010). A local maximum in the Arkona Basin is evidently caused by overestimation of the 10-m wind speeds from the geostrophic wind data for this region

and periods, and height–period combinations) and their spatial variations in this water body. The wave climate of the Baltic Sea is relatively mild. As mentioned above, simulations of Räämet and Soomere (2010) (Fig. 2) apparently underestimate the long-term significant wave height (around 0.9 m in the Baltic Proper) by 10–15 %. The significant wave height in the open part of the Baltic Proper according to most of reconstructions slightly exceeds 1 m and is somewhat less between Gotland and the Swedish mainland (Kahma et al. 2003; Broman et al. 2006; Schmager et al. 2008; Tuomi et al. 2011). The differences between the various estimates of the long-term wave heights are about ± 15 % and mostly stem from differences in the underlying wind fields (Nikolkina et al. 2014).

The long-term average wave heights are 0.6–0.8 m in the open parts of larger subbasins of the Baltic Sea such as the Gulf of Finland (Soomere et al. 2010; Suursaar 2013) or Arkona Basin (Soomere et al. 2012), around 0.5 m in the open part of the Gulf of Riga (Eelsalu et al. 2014) and well below 0.5 m in semi-sheltered bays such as Tallinn Bay (Soomere 2005; Kelpšaitė et al. 2009). These values are by 10–20 % lower in the nearshore regions (Suursaar and Kullas 2009a, b; Suursaar 2010; Soomere 2013) and considerably lower in these nearshore areas of larger subbasins (such as the western Gulf of Riga) that are sheltered with respect to predominant wind directions (Eelsalu et al. 2014).

The spatial patterns of average wave heights are slightly different in different hindcasts. They contain either an elongated maximum (Augustin 2005, Tuomi et al. 2011) or several local maxima in the eastern Baltic Proper (Jönsson et al. 2003). The calculations of Räämet and Soomere (2010) based on geostrophic winds suggest that another maximum may exist to the south of Gotland. This disparity may partially reflect large inter-annual and decadal variability in wind patterns over the area (Soomere and Räämet 2014) but more likely it indicates inconsistency of forcing wind fields (Nikolkina et al. 2014).

The probability distributions of the occurrence of different wave heights at open-sea measurement sites (Almagrundet, Bogskär, northern Baltic Proper, Arkona Basin) resemble a Rayleigh distribution (Fig. 3) with typical values of the shape parameter of 1.5–1.8 (Soomere et al. 2011). The median wave heights are about 20 % and the most frequent wave heights (usually in the range of 0.5–0.75 m) up to 30 % lower than the long-term average wave height (Kahma et al. 2003; Soomere et al. 2008, 2012). This distribution for visually observed data sets has a distinguished peak for very low wave heights and calms, and resembles analogous distributions for wave heights in semi-sheltered bays of the Baltic Sea (Soomere 2005). The excess proportion of calms in the data sets of visual observations (often >30 %, Soomere and Zaitseva 2007; Zaitseva-Pärnaste et al. 2009, 2011; Pindsoo et al. 2012) evidently is due to the absence of observable waves in many cases of easterly winds. Removing a fraction of calms from these sets therefore is roughly equivalent to ignoring the observations that inadequately reflect the open-sea wave fields in such wind conditions. For example, if the number of calms is reduced to 6 % from the total number of recordings (which is the level typical for the northern Baltic Proper, Fig. 2), the average wave height for wave systems propagating onshore at Vilsandi is 0.74 m (Soomere and Zaitseva 2007).

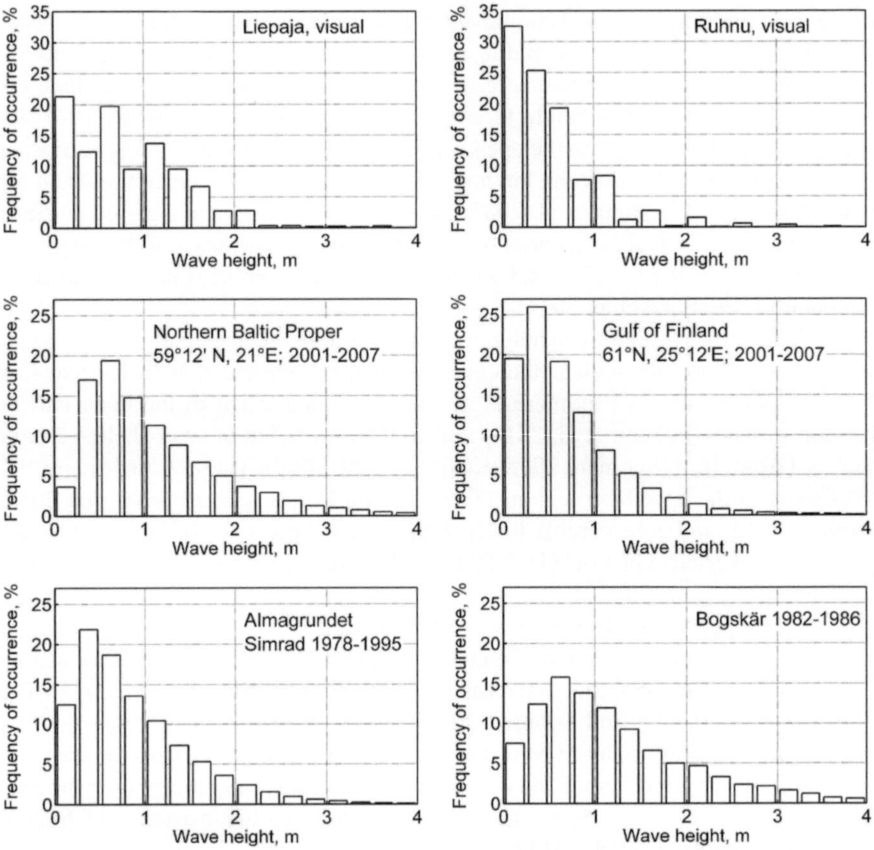

Fig. 3 Frequency of occurrence of wave heights at Almagrundet 1978–1995 (Broman et al. 2006), Bogskär 1982–1986 (Kahma et al. 2003), according to the hindcast for 2001–2007 in the northern Baltic Proper and in the Gulf of Finland (Tuomi et al. 2011), and according to visual observations at Liepaja and Ruhnu (Pindsoo et al. 2012; Eelsalu et al. 2014)

Most frequently waves with periods of 4–6 s dominate in the middle of the Baltic Proper, whereas in the more sheltered and coastal regions, waves with periods of 3–4 s predominate (Fig. 4). Periods up to 7–8 s are also common on the open sea. This difference in periods apparently comes from a relatively large number of short-fetched waves at sheltered measurement sites. A large proportion of the combinations of wave heights and periods roughly correspond to saturated wave fields with a Pierson–Moskowitz (PM) spectrum (Soomere et al. 2008; Räämet et al. 2010; Soomere et al. 2011). Such wave systems generally occur in the Baltic Sea at wind speeds up to about 8 m s^{-1} (Schmager et al. 2008). The properties of the roughest seas, however, match better a JONSWAP spectrum. These wave fields correspond to fetch-limited seas and are characterised by shorter periods (equivalently, they are steeper) than wave fields with a PM spectrum. Fetch-limited seas are typical in more

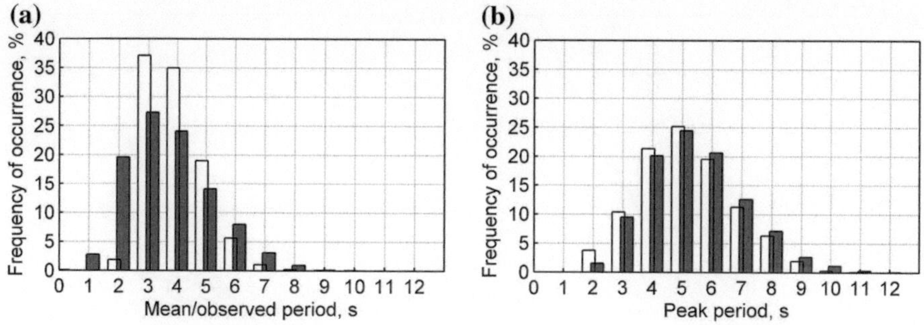

Fig. 4 Frequency of occurrence of wave periods: **a** Almagrundet 1978–1995 (*white bars*, Broman et al. 2006) and Vilsandi 1954–2005 (*filled bars*, Soomere and Zaitseva 2007), **b** Bogskär 1982–1986 (*white bars*) and the northern Baltic Proper 1996–2000 (*filled bars*, Kahma et al. 2003)

sheltered areas such as the Darss Sill (Soomere et al. 2011). The proportion of intense swells is very limited in all parts of the Baltic Sea.

The joint distributions of wave heights and periods (Fig. 5) suggest that the proportion of relatively steep seas is quite large in the Baltic Sea. Periods of 2–3 s usually correspond to wave heights well below 1 m, whereas waves with periods of 4–5 s have a typical height of about 1 m. Periods 6–7 s commonly correspond to wave heights of about 1.5–2 m. In coastal areas, dominating periods are 7–8 s only when wave heights are about 3 m or higher. Even longer waves are infrequent. Mean periods $T_m > 8$ s (peak periods $T_p \geq 10$ s) dominate either in very rough seas (wave heights >4 m) or in remote low swell conditions when the wave heights are well below 1 m. For example, at Almagrundet, the mean period never exceeded 9.5 s in very rough seas and was about 10 s in one case of rough seas with $H_{1/3} \sim 4$ m. Even in the final stage of the January 1984 storm when $H_{1/3} \sim 7$ m, the mean period was below 10 s. The waverider in the northern Baltic Proper registered the peak period of about 12 s about twice a year and at Bogskär roughly once in 2 years (Kahma et al. 2003).

3.2 Extreme Conditions

The Baltic Sea wave climate is highly intermittent. The sea occasionally hosts furious wave storms in certain seasons when the conditions are favourable for the generation of high waves. The highest waves in this water body are, however, much smaller than in the open ocean. Rough seas with the wave heights over 4 m occurred with a probability of 0.42 % in 1978–1995 at Almagrundet, of 1 % at Bogskär, and of 1.4 % in the northern Baltic Proper. Such seas usually occur several times a year, each time during a few hours. The area in which the significant wave height may exceed 4 m within 1 % of time (equivalently, the 99 %-ile of the wave height exceeds 4 m) is almost fully located to the east of the geometrical centreline of the Baltic Proper

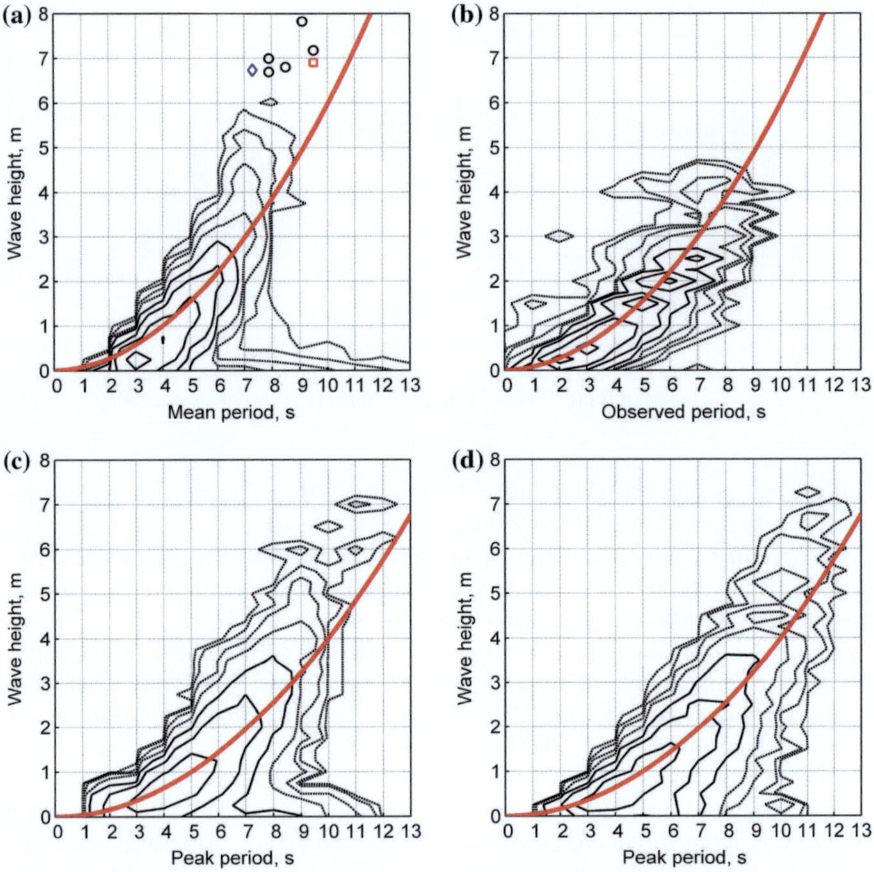

Fig. 5 Joint distribution of wave heights and periods: **a** Almagrundet 1978–1995 (Broman et al. 2006), **b** Vilsandi 1954–1994 (Soomere and Zaitseva 2007), **c** Bogskär 1982–1986, **d** northern Baltic Proper (Kahma et al. 2003). The wave height step is 0.25 m, and the period step is 1 s. The range of periods is shown on the horizontal axis: 2 s stands for $1.5 \le T_p < 2.5$ s, 3 s stands for $2.5 \le T_p < 3.5$ s, etc. Isolines for the probability of occurrence of 0.0033 %, 0.01 %, 0.033 %, 0.1 % (*dashed lines*), 0.33 %, 1 %, 3.3 %, and 10 % (*solid lines*) are plotted. Wave conditions with $H_{1/3} > 6.5$ m at Almagrundet are shown as follows: *circles*—the January 1984 storm, diamond—a storm in January 1988, square—a storm in August 1989. The *bold line* in panels **c** and **d** indicates the relationship $H_S = g\sqrt{\alpha/5}/\pi^2 \times T_p^2 \approx 0.04T_p^2$ between the significant wave height and peak period for saturated wave conditions with a Pierson–Moskowitz spectrum corresponding to the classical value $\alpha = 0.0081$. Similar lines in panels **a** and **b** assume that the mean period is about 80 % of the peak period

(Tuomi et al. 2011). Wave heights exceed 6 m with a probability of 0.1 % (8.8 h yr^{-1}) almost in the same area.

Seas in which $H_S > 7$ m are extremely rough in the Baltic Sea basin. Waves of this height cannot be observed from coastal observation stations because of the limited depth at all sites. This threshold was not reached at Bogskär in 1982–1986. The significant wave height >7 m has been recorded only five times in the northern Baltic Proper since 1996: twice in December 1999, on 22 December 2004, on 9 January 2005 during windstorm Gudrun (Soomere et al. 2008) and in December 2011 (Pettersson et al. 2012). The peak periods during these events slightly exceeded 12 s. Similar wave heights have been recorded twice in the southern Baltic at 55°55′N, 18°47′E (7.4 m on 14 October 2009 and 7.2 m in February 2011, Pettersson et al. 2010, 2012).

The largest instrumentally measured significant wave height $H_S = 8.2$ m occurred in the northern Baltic Proper on 22 December 2004. The highest single wave reached 14 m (Tuomi et al. 2011). It took thus 20 years to overshoot the previous maximum. It was set by a ferocious storm at Almagrundet on 13–14 January 1984 when $H_{1/3}$ calculated using Eq. (1) reached 7.82 m and the highest single wave was 12.75 m high[1] (Broman et al. 2006). An alternative estimate of the significant wave height in this storm from wave spectrum was $H_S = 7.28$ m. The wave periods remained fairly modest ($T_m = 9.1$ s, $T_p = 10.7$ s, Broman et al. 2006). This was the only case during which $H_{1/3} \geq 7$ m was registered at Almagrundet. Instrumentally measured extreme wave heights are generally lower in the southern Baltic Sea. This is evidently caused by the lack of wave measurement devices in this part of the Baltic Sea that, according to hindcasts, should host as severe extreme waves as the northern Baltic Proper. As mentioned above, a significant wave height >7 m has been measured only twice in the southern Baltic Sea.

3.3 The Ratio of Extreme and Average Wave Heights

In many coastal engineering applications, it is implicitly assumed that the ratio of the extreme and average wave heights is approximately constant. This assumption is used, for example, in express estimates of the closure depth of (almost) equilibrium beach profiles. This depth indicates the water depth until which storm waves substantially and regularly affect the shape of the coastal profile (Kraus 1992; Dean 1991). It is a widely used concept in coastal engineering and a fundamental variable in modelling of coastal evolution and morphology. It basically depends on the roughest wave conditions that persist for a reasonable time at a given site (Hallermeier 1981).

---

[1] The Almagrundet data set from 1993–2003 contains several contradicting extreme wave records. A severe storm in March 1997 that affected nearly the whole Baltic Proper caused $H_{1/3} = 7.83$ m. As H_S estimated from the wave spectrum was 5.7 m and the highest single wave reached 10.24 m, this value of $H_{1/3}$ evidently overestimates the wave conditions. An extremely high single wave (12.79 m) was recorded on 25 December 1996 when $H_{1/3} = 6.37$ m but the significant wave height, estimated from the wave spectrum, was only 3.8 m. The listed values are apparently doubtful although they do represent quite severe wave fields (Broman et al. 2006), and the value of $H_{1/3}$ in December 1996 is consistent with the data from the waverider in the northern Baltic Proper. More reliable are the data from 1978–1995. The wave height reached $H_{1/3} = 6.9$ m in a relatively short but violent storm in August 1989 and $H_{1/3} = 6.73$ m in another severe storm on 30 January 1988. The significant wave height on the open sea apparently exceeded 7 m during these events. No reliable data are available for a severe storm in January 1993.

A convenient measure of the severest part of the wave climate in this respect is the significant wave height $H_{0.137}$ that is exceeded during 12 h in a year, that is, with a probability of 0.137 % (Hallermeier 1981). The simplest approximation for the closure depth h_c as a linear function is $h_c \cong q_1 H_{0.137}$. The commonly used values are $q_1 = 1.5$ (Birkemeier 1985) and $q_1 = 1.57$ (Hallermeier 1981). Houston 1996 extended this relationship towards the use of the long-term significant wave height H_{mean} as $h_c \cong q_1 H_{0.137} \cong q_2 H_{mean}$ and evaluated that $q_2 = 6.75$. It is, however, not necessarily correct for areas with a specific wave climate as it implicitly assumes the relationship $H_{0.137} \cong 4.5 H_{mean}$. This relationship has been established for wave fields with a Pierson–Moskowitz spectrum that is common for the observed wave statistics along the US coasts (Houston 1996).

The proportions of the mean and extreme wave heights are different in semi-enclosed seas like the Baltic Sea where intense swells are almost absent and windseas carry a large part of the wave energy. As discussed above, a specific feature of the Baltic Sea wave climate is that the average wave conditions are relatively mild but very rough seas may episodically occur in long-lasting severe storms (Soomere 2005; Broman et al. 2006; Soomere et al. 2012). Waves in such storms are much higher than expected from the mean wave conditions. This feature leads to a marked difference between the factor q_2 used in $h_c \cong q_2 H_{mean}$ for the open ocean coasts and its analogue for the Baltic Sea. The typical ratio of these two measures for the coasts of the open Baltic Proper is $H_{0.137}/H_{mean} \cong 5.5$ (Soomere et al. 2013). This result suggests that an appropriate express formula for the closure depth in the Baltic Sea conditions is $h_c^B \cong 1.5 H_{0.137} \cong 8.25 H_{mean}$, while the above expression with $q_2 = 6.75$ is valid for short relatively sheltered sections of the Baltic Sea shore located in bayheads.

3.4 Sub-basins of the Baltic Sea

The average and, in particular, the maximum wave heights in the semi-sheltered sub-basins of the Baltic Sea are much smaller than in the Baltic Proper. The 'memory' of wave fields is relatively short, and the changes in the wind field are fast reflected in the wave pattern. As a consequence, the wave fields in smaller sub-basins (such as Tallinn Bay or Narva Bay) largely mimic the changes in the open-sea winds (Soomere 2005).

The largest wave heights among the major sub-basins of the Baltic Sea occur in the Sea of Bothnia where the numerically simulated maxima of H_S exceed 7 m (Tuomi et al. 2011) and may reach 7.6 m (Soomere and Räämet 2011b). The maximum significant wave height 6.5 m has been measured in this water body on 09 December 2011 (Pettersson et al. 2012).

Based on data from 1990–1991 to 1994, the maximum H_S occurring once in 100 years in the Gulf of Finland was estimated to be 3.8 m and the corresponding single wave height 7.1 m. Wave conditions with $H_S > 4$ m were thought to occur extremely seldom (Alenius et al. 1998; Pettersson 2001). The peak periods in rough seas (with $H_S \sim 4$ m) were 8–9 s (Kahma and Pettersson 1993). Recent data show that considerably rougher seas may occur in this gulf. In November 2001, seas with $H_S = 5.2$ m and $T_p \approx 11$ s occurred (Pettersson and Boman 2002). The same significant wave height (Pettersson et al. 2013) and the all-time highest wave of 9.4 m (www.fmi.fi) were measured on 30 November 2012 during an easterly storm. Wave fields with $T_p \geq 10$s, however, usually correspond to a penetration of long-period swell of moderate height into the gulf. Only a few observations reveal such long periods: $T_p \approx 11$ s occurred only three times in 1990–1994 and during a short time in another very strong storm in November 2001 (Pettersson 2001). The average wave directions are often concentrated along the gulf axis (Pettersson et al. 2010) although the wind directions are more evenly spread (Soomere and Keevallik 2003). This phenomenon is attached to the slanting fetch conditions in which the wind direction is oblique to the coastline. Shorter waves are usually aligned with the wind, while somewhat longer and higher waves (that often dominate the wave field) propagate along the gulf axis (Holthuijsen 1983; Kahma and Pettersson 1994; Pettersson et al. 2010).

The frequency of occurrence of waves with $H_S > 4$ m is very small for all sub-basins except for the Sea of Bothnia (Tuomi et al. 2011), and waves with $H_S >$ 2 m can be considered as very severe for many semi-sheltered areas such as the Darss Sill (Soomere and Kurkina 2011; Soomere et al. 2012). The maximum H_S in the Darss Sill area was 4.47 m (wave period 6.2 s) on 03 November 1995 during a strong north-easterly storm (Soomere and Kurkina 2011). Numerical simulations indicate that H_S up to 6.2–6.7 m may occur in the region of the Darss Sill in the existing wave climate (Soomere et al. 2012). Other sub-basins (incl. the Gulf of Riga) are not covered by regular wave measurements. Still, during violent storms from unfavourable directions, even sheltered bays may experience very strong waves (Davidan et al. 1985). For instance, $H_S > 4$ m apparently occurred in the interior of Tallinn Bay on 15 November 2001 (Soomere 2005).

4 Extremes During Windstorm Gudrun

Earlier estimates of extreme wave conditions with the use of the WAM model forced by homogeneous wind patterns suggested that the significant wave height generally does not exceed 8–8.5 m in the Baltic Proper (Soomere 2001). This estimate was confirmed by Lopatukhin et al. (2006a). Later simulations indicated that the significant wave height may reach 9.5–10 m in the north-eastern Baltic Proper at the entrance of the Gulf of Finland, to the north-west of the Latvian coast and in the south-eastern part of the sea in the Gulf of Gdańsk (Schmager et al. 2008; Soomere et al. 2008; Tuomi et al. 2011). The properties of waves in a specific region and storm event substantially depend on the match of the geometry of the particular sea area and the wind pattern in the storm (Augustin 2005; Soomere et al. 2008; Schmager et al. 2008).

4.1 The Storm and Waves

The above estimates for extreme wave conditions turned out to be inadequate when windstorm Gudrun, an extratropical cyclone, also known as Erwin in Ireland, the United Kingdom and Central Europe, attacked northern Europe on 7–9 January 2005. It reached the power of a hurricane, according to the Saffir–Simpson hurricane scale (Simpson and Riehl 1981), in the North Sea region. In the Baltic Sea, it remained slightly below the hurricane level; yet, it was one of the strongest storms in Denmark, Sweden, Latvia and Estonia for at least 40 years. It caused widespread property damage, exceptionally high coastal floods along the Western Estonian coast and in the Gulf of Finland, and loss of 18 lives (Suursaar et al. 2006; Bengtsson and Nilsson 2007). The coastal wind data suffered from failures of meteorological equipment during Gudrun (Suursaar et al. 2006). Forecast winds from the German Weather Forecast Service (DWD, Deutscher Wetterdienst), the Danish Meteorological Institute (DMI) and the Finnish Institute of Marine Research (FIMR) suggest that the maximum 10-min average wind speed on the open sea (Fig. 6) was 28–29 m s^{-1}. Forecasts released on 6–7 January predicted the windstorm maximum to hit the entrance of the Gulf of Finland. The significant wave height was forecast to exceed 10 m in the location of the waverider in the northern Baltic Proper, to reach 11–12 m at the latitudes of the Gulf of Finland, and to be >6 m in the central part of this gulf (Soomere et al. 2008). Such wave conditions would have been considerably rougher than during any other storm in the northern Baltic Sea in the history of contemporary shipping (K. Kahma, personal communication on 8 January 2005). The area with the largest wind speeds crossed the Baltic Sea somewhat more southwards than originally forecast (Soomere et al. 2008), and the wave conditions were not so rough.

The waverider in the northern Baltic Proper (Fig. 1) adequately reflects extreme wave conditions in the case of south-western winds. The Gudrun's strongest winds were from west to west-south-west and occurred between Gotland and Saaremaa. The waverider therefore was located much to the north of the maximum of the wave storm. Even with these non-ideal conditions, the significant wave height reached $H_S = 7.16$ m at 03:00 and 07:00 GMT on 9 January and was close to 7 m during about 12 h. The peak period T_p exceeded 10 s for nearly 24 h and was about 11–12 s at the wavestorm maximum.

Very long (T_p up to 12 s) and high ($H_S > 4$ m) waves also occurred in the Gulf of Finland during Gudrun. The significant wave height was close to 4 m in the early morning of 9 January and exceeded 3 m during the rest of this day according to the directional waverider (Fig. 1). The peak periods were over 10 s during almost the whole day and reached 11–12 s at noon. The wave height was about 4 m in the morning of 9 January in the vicinity of the island of Naissaar in 14-m-deep water at 59°37.1′N, 24°29.1′E and reached 4.5 m at the storm maximum. The peak periods were ~12 s during about 10 h (Soomere et al. 2008).

The occurrence of long and high waves in the interior of the Gulf of Finland is an important feature of this storm. The maximum wind speed in the northernmost part of the Baltic Proper and at the entrance of the gulf apparently was about 20–

$24\,\mathrm{m\,s^{-1}}$ (Fig. 6) and well below $20\,\mathrm{m\,s^{-1}}$ during a large part of the storm in the gulf (Suursaar et al. 2006). Storms with a wind speed of about $20\,\mathrm{m\,s^{-1}}$ may excite peak periods about 12 s only if the fetch length is $\geq$600 km and the wind duration is $\geq$18 h (Rosenthal 1986). Although growth curves of Kahma and Calkoen (1992) suggest that somewhat shorter duration ($\sim$15 h) and fetch ($\sim$350 km) are sufficient for the generation of such seas, it is still probable that some other factors eventually contributed to the observed wave system in the Gulf of Finland. For example, topographic refraction caused by the coastal slopes of the entrance of the gulf may gradually redirect a part of waves propagating from the southern parts of the Baltic Proper.

4.2 Modelled Wave Fields

The wave conditions in the Baltic Sea during windstorm Gudrun were routinely forecast by operational centres of Deutscher Wetterdienst (DWD), Danish Meteoro-

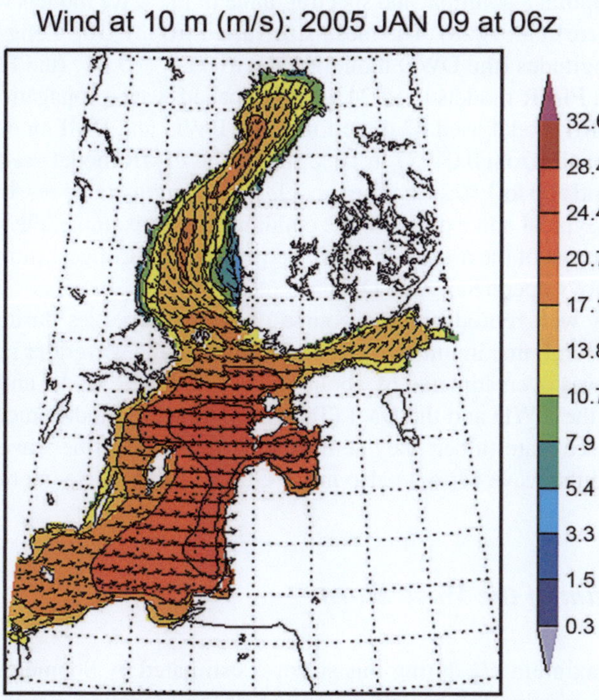

Fig. 6 Modelled wind speed ($\mathrm{m\,s^{-1}}$) and direction (*arrows*) 10 m above water surface at 06:00 GMT on 9 January in the DMI 54-hour forecast valid at 00:00 GMT on 9 January. Courtesy of the Danish Meteorological Institute

Table 1 Relative errors of operational wave models and the estimated overall maximum of the significant wave height in the Baltic Sea during windstorm Gudrun

Model	Overall maximum at the location of the waverider in the northern Baltic Proper	Over-prediction (m)	Relative error (%)	Modelled overall maximum of H_S (m)	Estimated overall maximum of H_S (m)
FIMR	7.6	0.44	5.8	10.2	9.6
DMI	8.96	1.80	20	11.7	9.4
DWD	8.17	1.01	12.4	10.95	9.59

logical Institute (DMI), and Finnish Institute of Marine Research (FIMR). They all run the wave model WAM cycle 4 (Komen et al. 1994) on a regular rectangular grid in shallow water mode without data assimilation. The models used hourly to three-hourly forecast winds at the standard height of 10 m above the surface level from different atmospheric models. The land–sea masks, bathymetry, computational grid, spatial and temporal resolution, and spectral range of the wave models were different as well (Soomere et al. 2008). The mesh size varied from $1/10°$ along latitudes and $1/6°$ along longitudes (the DWD model) down to $0.08 \times 0.08°$ (the FIMR model). The DWD and FIMR models used 24 equally spaced wave propagation directions, whereas the DMI model used 12 directions. The DWD and DMI models employed 25 frequency bands from 0.04177 in 10 % steps. The FIMR model used an extended range of 35 bands up to 1.073 Hz. The models have demonstrated reasonable performance in both typical and extreme wave conditions. For example, the mean relative error of the forecast of the maximum wave height in the 5 strongest storms was about 15 % for 13 buoys operated by the DMI.

The models well reproduced the course of wave properties during windstorm Gudrun. The overall maximum of H_S at the location of the waverider in the northern Baltic Proper was overestimated by about 6 % by the FIMR model and 12–20 % by the models of the DWD and the DMI (Table 1). The wave models mostly followed the measured sea state (albeit they somewhat overpredicted the wave heights and underpredicted the wave periods) also in the Gulf of Finland (Soomere et al. 2008).

4.3 Maximum of the Wave Storm

The overall maximum H_S during this storm is estimated by Soomere et al. (2008) by means of correcting the overall maximum of the modelled H_S with the use of the relative errors of the models calculated from observed data (Table 1). Doing so presumes that the wave models adequately represent the spatial patterns of wave properties and that the relative errors of the models are roughly the same over the

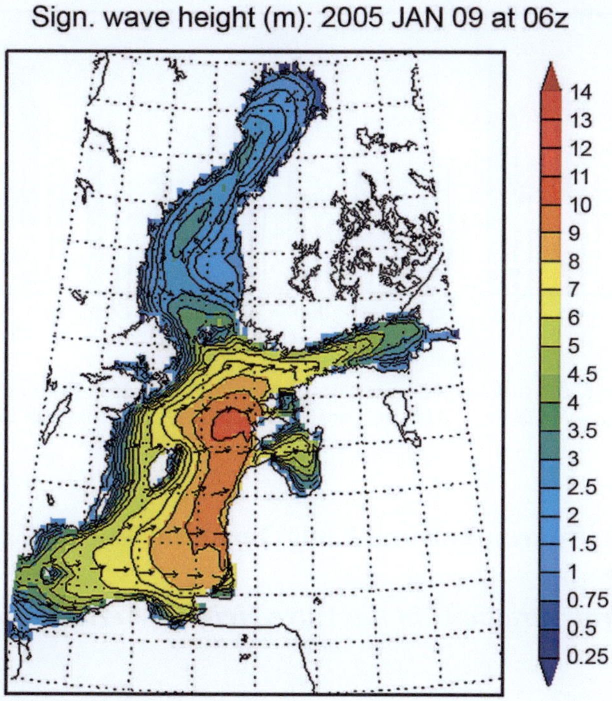

Fig. 7 Modelled significant wave heights (m) and wave propagation directions (*arrows*) at 06:00 GMT on 9 January in the DMI 54-h forecast valid at 00:00 GMT on 9 January. Courtesy of the Danish Meteorological Institute

entire area of intense waves. Since a large part of properties of the wave fields during Gudrun were located within the 'corridors' formed by outputs of the three models, a reasonable estimate of this maximum eventually lies between the values defined by these models.

The overall maximum $H_S \approx 9.5$ m during windstorm Gudrun evidently occurred about 200 km south-eastwards of the location of the waverider in the northern Baltic Proper, off the coast of Saaremaa (about 57°N, 20.4°E, Fig. 7). This estimate was later confirmed by Tuomi et al. (2011). Such wave conditions are much rougher than those expected to happen once in a century (Lopatukhin et al. 2006a). Waves were also remarkably long: peak periods up to 13 s were forecast (and eventually occurred) in the eastern part of the sea (Soomere et al. 2008). The described procedure can be applied to the Gulf of Finland only conditionally. The significant wave height evidently reached 5 m in the gulf but most probably did not exceed the historical maximum $H_S = 5.2$ m.

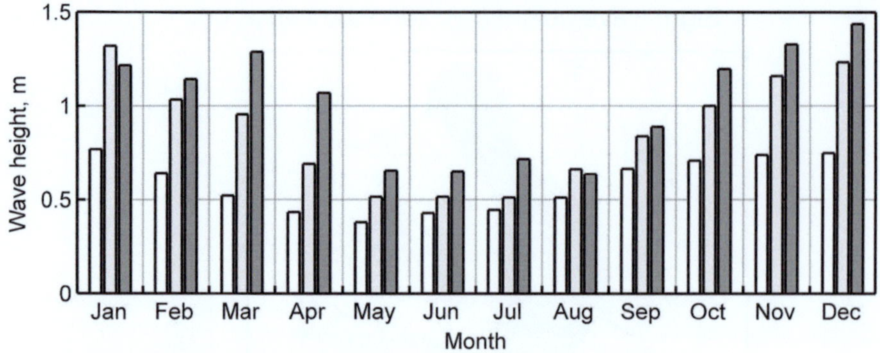

Fig. 8 Annual variation in the monthly mean wave height at Vilsandi 1954–2005 (*white bars*, based on the daily mean wave height) and at Almagrundet 1978–95 (*light grey bars*) and 1993–2002 (*dark grey bars*)

5 Seasonal, Interannual and Long-Term Variations

5.1 Temporal Variations in the Wave Heights, Periods and Directions

The extensive seasonal variation in the wind speed over the entire Baltic Sea basin (Mietus 1998) naturally causes substantial differences (by a factor of two in coastal areas and up to three times in the offshore regions) in wave heights at monthly scales (Schmager et al. 2008; Soomere and Räämet 2011b). This variation is impressive, for example, at Almagrundet, from about 0.5 m during summer to 1.3–1.4 m in winter (Fig. 8). It is much less pronounced in the data sets of visual wave observations from the coastal stations and has very limited amplitude at locations such as Sõrve (Eelsalu et al. 2014) that are sheltered from the predominant strong wind directions.

The highest monthly mean wave height occurs from October to January. Some locations reveal another minor wave height maximum, for example, in March at Almagrundet (Broman et al. 2006). It may be connected with easterly winds during late winter and early spring at the latitudes of the Gulf of Finland (Mietus 1998; Soomere and Keevallik 2003). These winds almost do not impact the wave fields in the rest of the measurement sites. The calmest period is the late spring and summer months from April to July–August.

The most intriguing question is whether any long-term changes in the extreme wave heights or their spatial patterns can be identified in the Baltic Proper. The existing visual observations reveal no long-term ($\sim$70 yr) trend for wave heights, but the wave fields exhibit extensive trends over a few decades and substantial decadal variability. The overall course of wave activity (Fig. 9) is quasiperiodic. The interval between subsequent periods of high or low wave activity is two to three decades.

Historical visual observations suggest that there was a relatively rapid decrease in the (annual mean) wave heights in the 'pre-instrumental' era from the mid-1940s until the end of the 1970s (Soomere 2013). The further course of wave heights was different in different parts of the sea. The wave height rapidly increased in the northern Baltic Proper at a rate of 1.3–2.8 % per annum from the mid-1980s until the mid-1990s (Broman et al. 2006; Soomere and Zaitseva 2007). This increase was consistent (albeit faster) with the analogous trends for the south-western Baltic Sea and for the North Atlantic (Kushnir et al. 1997; Gulev and Hasse 1999; Vikebo et al. 2003; Weisse and Günther 2007) and also with the increase in the storminess (Alexandersson et al. 1998) and the wind speed over the northern Baltic Sea (Broman et al. 2006). The increase was followed by a drastic decrease since 1997. The relevant data from Almagrundet were even estimated as doubtful by Broman et al. (2006), because the annual mean wind speed continued to increase and intensification of beach processes was reported along the downwind side of the coasts (Orviku et al. 2003). The extensive similarity of changes at Almagrundet and Vilsandi suggests that both data sets reflect certain real changes (albeit possibly overestimate their magnitude). The timing of these variations matches an almost twofold increase in the number of low- pressure observations below 980 hPa at Härnösand over the 1990s (Bärring and von Storch 2004).

The wave height at the south-eastern Baltic Sea (Lithuanian) coast showed the opposite behaviour. A rapid decrease occurred until about 1996 and a rapid increase since then (Zaitseva-Pärnaste et al. 2011). The data from Ventspils and particularly from Liepaja (Pindsoo et al. 2012) confirm that the wave height variations were fairly different in different parts of the Baltic Proper. Such changes do not necessarily become evident in all sub-basins of the Baltic Sea, for example, almost no change occurred in the annual mean significant wave height at the Darss Sill (Soomere et al.

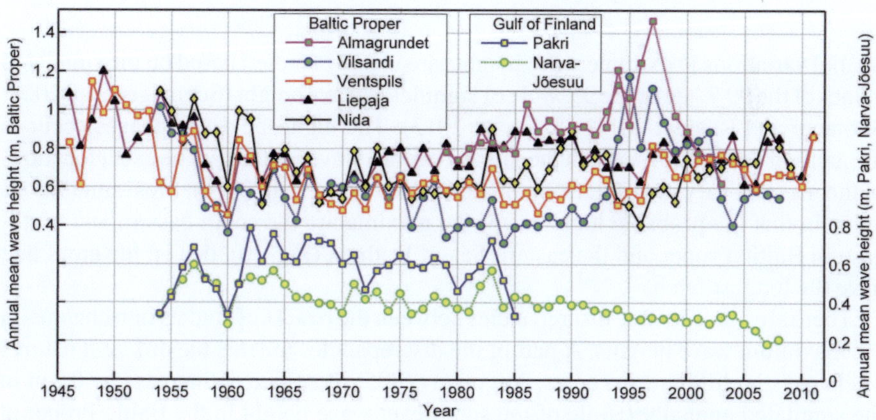

Fig. 9 Annual mean visually observed wave height at seven coastal observation sites and instrumentally measured wave height at Almagrundet (Hünicke et al. 2015). Almagrundet data from 1978 reflect only windy months November and December, and data for 1998 are missing (Broman et al. 2006). The wave heights in the mid-1990s are probably overestimated at Vilsandi

2011). Fetch-based models using one-point coastal wind have demonstrated that the wave intensity changed quasi-periodically and revealed no statistically significant trend in Estonian coastal waters (Suursaar and Kullas 2009a, b; Suursaar 2010, 2013).

Although the mean wind speed does not necessarily exactly match the average wave height, it is intuitively clear that a larger wind speed generally causes greater wave activity. The drastic changes to the mean wave height on the background of the gradual increase in the mean wind speed (Broman et al. 2006; Räämet et al. 2009) suggest that the local wave generation conditions may have substantially changed within relatively short time intervals. Quite large variations in the average wave periods (from about 2.3 s in the mid-1970s up to 2.65 s around 1990) were hindcast for selected sites (Suursaar and Kullas 2009a, b). This change, reconstructed using fetch-based models, suggests that the wind direction may have changed.

The most prominent changes in the Baltic Sea wave fields became evident as a remarkable rotation (up to 90° from the north-west to the south-west) of the most frequently observed wave directions at Narva-Jõesuu during the last half century (Räämet et al. 2010). This change mirrors substantial changes in the local wind direction (Jaagus 2009; Jaagus and Kull 2011). Similar changes (of much smaller amplitude) observed at the Lithuanian coast were interpreted as a possible reason for changes in the distribution of erosion and accumulation areas (Kelpšaitė et al. 2011). Also, changes in the wave direction can explain certain specific features in the patterns of wave-driven sediment transport (Viška and Soomere 2012; Soomere et al. 2015).

5.2 Variations in the Wave Heights in Very Rough Seas in the Baltic Proper

Spatial variations in extreme wave heights are usually studied based on the simulated values of the 99 %-ile or the 95 %-ile of significant wave height (Schmager et al. 2008; Soomere and Räämet 2011b; Suursaar 2013). The highest extreme waves occur in the areas of the Baltic Sea with the highest overall wave intensity. The exact locations of the maxima vary in different simulations (Nikolkina et al. 2014). Most simulations agree in that the probable locations of the maxima are the south-eastern and north-eastern Baltic Proper and the eastern Sea of Bothnia (Fig. 10), that is, the areas that have the longest fetch.

There are considerable discrepancies between the results of studies into changes in the maximum wave heights. A part of the discrepancies mirrors the different natures of changes in different sea areas. Augustin (2005) identified an increase by 0.3 m in the simulated annual 99 %-ile of the significant wave height in the Baltic Proper at 58°N, 20°E. This was mostly caused by an increase in the frequency of severe wave events (BACC Author Team 2008).

Simulations over the period 1970–2007 using geostrophic winds (Soomere and Räämet 2011b) confirmed the presence of an increasing trend at this location but also brought evidence about a complex spatio-temporal pattern of changes (with scales down to about 100 km) to the Baltic Sea wave fields (Fig. 10). The pattern of changes in the extreme wave heights is almost identical to the similar pattern for the average wave heights. The most drastic decrease in the hindcast wave intensity occurred in an area between the islands of Öland and Gotland, and to the south of these islands down to the Polish coast. A probable reason for this decrease is a major rotation of the geostrophic air flow over the southern Baltic Sea from the year 1988 (Soomere and Räämet 2014; Soomere et al. 2015). The model also suggests a considerable increase in the wave activity near the coast of Latvia, between the Åland Archipelago and Sweden, and in the sea area between the Sea of Bothnia and the Bay of Bothnia. Such an increase is consistent with the outcome of fetch-based wave models based on one-point forcing (Suursaar and Kullas 2009a, b; Zaitseva-Pärnaste et al. 2009), which indicate a pronounced increase in the 90 %-ile and 99 %-ile near the Western Estonian Archipelago.

5.3 *Mismatch of Trends for the Averages and Extremes in the Gulf of Finland*

An interesting feature is the possible mismatch of the changes in the average and in the extreme wave heights (Soomere and Healy 2008) that become most contrast in the Gulf of Finland (Soomere et al. 2010; Soomere and Räämet 2011b). The

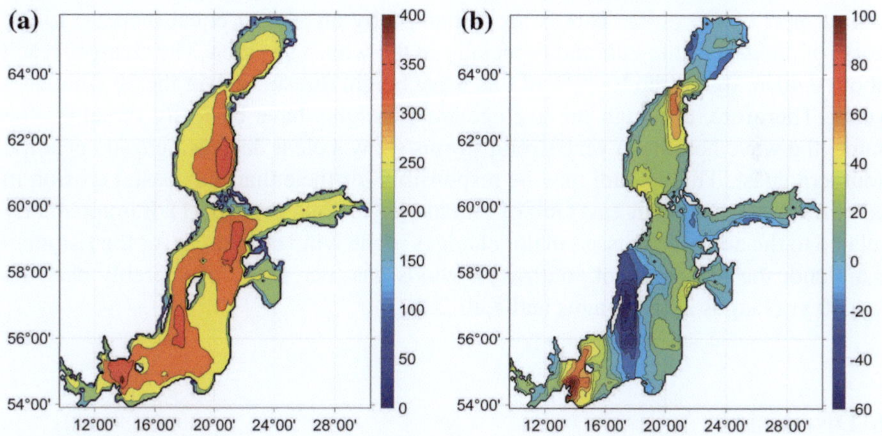

Fig. 10 **a** Numerically simulated 99 %-ile of significant wave height (cm; isolines plotted after each 50 cm) and (**b**) its linear trend in the Baltic Sea in 1970–2007 (cm; isolines plotted after each 10 cm) in the Baltic Sea in 1970–2007 based on adjusted geostrophic winds from the Swedish Meteorological and Hydrological Institute. Similar to Fig. 2, a local maximum in the Arkona Basin in both panels is evidently caused by an overestimation of the 10-m wind speeds from the geostrophic wind data for this region (Soomere and Räämet 2011b)

gulf is oriented obliquely with respect to predominant wind directions, and marine meteorological conditions are characterised here by remarkable wind anisotropy (Soomere and Keevallik 2003). The wind field in the gulf contains strong eastern and western winds blowing along the axis of the gulf. These winds are specific to the Gulf of Finland. They do not become evident in other parts of the Baltic Sea and are much weaker in the eastern part of the gulf. As discussed above, both average and maximum wave heights in the gulf are about 60 % of those in the Baltic Proper, whereas the wave periods in typical conditions are almost the same as in the Baltic Proper (Soomere et al. 2011).

The above-discussed changes in storminess in the northern Europe does not necessarily become evident in the interior of the Gulf of Finland, where the ageostrophic component of the surface level wind is at times substantial (cf. Keevallik and Soomere 2010). In contrast to the gradual increase in the mean wind speed over most of the Baltic Proper (Pryor and Barthelmie 2003; Broman et al. 2006), there is a very slow decrease (about $0.01\,\mathrm{m\,s^{-1}\,yr^{-1}}$) in the annual mean wind speed at Kalbådagrund (Soomere et al. 2010). Therefore, drastic long-term variations in the average wave properties are unlikely in this gulf. Consistently with this picture, numerical simulations indicate only minor changes in the annual mean wave height in the entire gulf, including its entrance area (Soomere et al. 2010).

Interestingly, Suursaar and Kullas (2009b) noted a decreasing trend in 99 %-iles near the north Estonian coast and a weak gradually increasing trend in the average wave height. Simulations using the WAM model show that, differentl from the average wave height, the maximum wave heights exhibit a clear pattern of changes since the 1970s (Fig. 11). There has been a substantial decrease (by about 10 %) in this threshold in question near the southern coast of the gulf (especially in the narrowest central part of the gulf). This is accompanied by an almost equal increase to the north of the axis of the gulf and especially in the widest sea area. The changes reach about 0.40 m, that is, up to 20 % of this wave height threshold over the 38 simulated years. Therefore, although the average wave heights have basically remained the same, the wave heights in very strong storms show a clear decreasing trend near the southern coast. This feature may be responsible for the enhanced coastal erosion in certain areas in the north-east Gulf of Finland (Ryabchuk et al. 2011). It is apparently related to the above-discussed major changes in the wind direction over the Estonian mainland: the frequency of south-west winds has increased considerably over the last 40 yr (Jaagus 2009; Jaagus and Kull 2011).

6 Discussion

The presented data indicate that substantial changes in several core properties of wave fields such as the average wave intensity (in terms of the annual mean significant wave height), very rough wave conditions (understood as the wave height occurring with a probability of 1–5 %) and wave propagation directions have occurred in certain regions of the Baltic Sea since the middle of the 20th century. Interestingly, there is

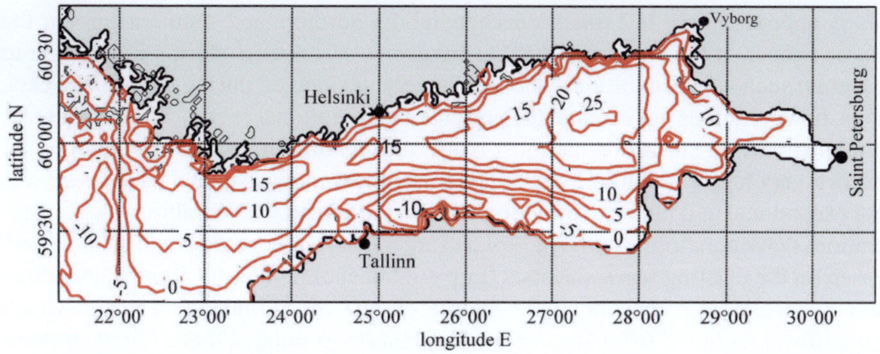

Fig. 11 Spatial distribution of the long-term changes (cm) 99 %-ile of significant wave height in the Gulf of Finland (Soomere and Räämet 2011b)

a very minor change in the spatially averaged wave height and almost no temporal changes in the wave periods. This means that remarkable local changes and associated substantial increase or decrease in the severity of wave-driven hazards may be often overlooked.

An increase in the height of the hindcast extreme waves in the Arkona Basin based on simulations forced by geostrophic winds is not unexpected because the wind speed in this area has grown markedly over the last decades (Pryor and Barthelmie 2003, 2010). A decrease of the same magnitude in the neighbouring sea area to the north-east of Bornholm is counter-intuitive but substantiated well by the abrupt change in the air flow direction over this region (Soomere and Räämet 2014; Soomere et al. 2015).

The annual mean wave height considerably increased in the 1980s and was exceptionally high in the mid-1990s, but quickly decreased starting from about 1997 in the northern Baltic Proper. However, no overall increase in the average wave height has occurred in the northern Baltic Proper within the second half of the 20th century. Such a complicated pattern of spatio-temporal changes in various properties of wave fields is not completely unexpected as many studies have shown that the magnitudes of the trends in wave properties can greatly vary in different sea areas (e.g. Weisse and von Storch 2010; Martucci et al. 2010). It is, however, notable that a steep decrease in wave heights may occur in open-sea areas adjacent to those hosting an equally steep increase in wave heights.

Ironically, most of the existing long-term wave observation and measurement sites are located in areas where the simulations have revealed almost no long-term changes in wave properties. The situation is even more complicated because of the mismatch of the long-term behaviour of the mean wave height with the gradual increase in the mean wind speed. A partial explanation is offered by the observation that the wind direction has changed but this highly interesting feature still needs further investigation.

There is now increasing evidence that extreme wave conditions with $H_S \geq 7\,\mathrm{m}$ (first observed in January 1984 in the northern Baltic Sea) occur more or less regularly approximately 1–2 times a decade in both northern and southern parts of the Baltic Proper. It is very likely that the frequency of extreme storms which able to generate such wave conditions has been largely unchanged during the last 30 years. This, however, does not exclude the presence of exceptional events. The strong reaction of the water surface is the most interesting feature of windstorm Gudrun that excited very high and long waves, although the maximum sustained wind speed was not exceptional and the wind direction was not particularly favourable for wave generation. Wave conditions with $H_S \sim 9.5$ m are much rougher than could be expected, based on the existing wave statistics (Lopatukhin et al. 2006a, b). Remarkably, long and high waves also appeared in the interior of the Gulf of Finland, in an area which generally is sheltered from long waves. It might be speculated that a future storm of the same strength and duration, but corresponding to more favourable wave generation conditions (e.g. a strong and large cyclone travelling to the north-north-east), may create even higher waves. Since only a few cyclones do so (Suursaar et al. 2006), such a 'perfect storm' is not likely to occur. However, if it did happen, it probably would excite even rougher wave conditions at the entrance of the Gulf of Finland and off the south-western coast of Finland than Gudrun did near Saaremaa. The possibility of such rough seas within the existing climatological conditions is of paramount importance for navigational safety and design of offshore structures.

The future climate changes are likely to modify factors controlling the volume of the water body, the mean temperature, salt water inflow conditions, the overall transport scheme of waters, the distribution of upwelling and downwelling patterns, the location of areas of the largest wave intensity and wave-induced mixing, and therefore the vertical and horizontal distributions of salinity, temperature, and other decisive constituents of the local ecosystem. In particular, the increased sea surface temperature leads to the reduction of ice cover in the northern parts of the sea. The potential increase in wind stress at sea surface during relatively windy winter months may lead to further changes in the wave climate, in particular, to enhancing the extremes in wave heights and sea levels. Timely detection of such changes is a major challenge for scientists. Launching of adaptation measures is an accompanying challenge of decision-makers.

Acknowledgments The underlying research was supported by institutional financing IUT33-3 of the Estonian Ministry of Education and Research, grant 9125 by the Estonian Science Foundation and through support of the ERDF to the Centre of Excellence in Non-linear Studies CENS.

References

Alari V (2013) Multi-scale wind wave modeling in the Baltic Sea. Ph.D thesis, Tallinn University of Technology

Augustin J (2005) Das Seegangsklima der Ostsee zwischen 1958–2002 auf Grundlage numerischer Daten [Sea state climate of the Baltic Sea 1958–2002 based on numerical data]. Diploma Thesis. Institute for Coastal Research, GKSS Research Center Geesthacht, Germany (In German)

Alenius P, Myrberg K, Nekrasov A (1998) Physical oceanography of the Gulf of Finland: a review. Boreal Environ Res 3:97–125

Alexandersson H, Schmith T, Iden K, Tuomenvirta H (1998) Long-term variations of the storm climate over NW Europe. Glob Atmos Ocean Syst 6:97–120

BACC Author Team (2008) Assessment of climate change for the Baltic Sea basin. Springer, Berlin, 473 pp

BACC II Author Team (2015) Second assessment of climate change for the Baltic Sea basin. Springer, Cham, 501 pp

Bärring L, von Storch H (2004) Scandinavian storminess since about 1800. Geophys Res Lett 31:L20202

Bengtsson A, Nilsson C (2007) Extreme value modelling of storm damage in Swedish forests. Nat Hazards Earth Syst Sci 7:515–521

Birkemeier WA (1985) Field data on seaward limit of profile change. J Waterw Port Coast Ocean Eng-ASCE 111:598–602

Blomgren S, Larsson M, Hanson H (2001) Numerical modelling of the wave climate in the southern Baltic Sea. J Coast Res 17:342–352

Broman B, Hammarklint T, Rannat K, Soomere T, Valdmann A (2006) Trends and extremes of wave fields in the north-eastern part of the Baltic Proper. Oceanologia 48:165–184

Bumke K, Hasse L (1989) An analysis scheme for determination of true surface winds at sea from ship synoptic wind and pressure observations. Bound Lay Meteorol 47:295–308

Cieślikiewicz W, Paplińska-Swerpel B (2008) A 44-year hindcast of wind wave fields over the Baltic Sea. Coast Eng 55:894–905

Danielsson A, Jönsson A, Rahm L (2007) Resuspension patterns in the Baltic proper. J Sea Res 57:257–269

Davidan IN, Lopatoukhin LI, Rozhkov VA (1978) Windsea as a probabilistic hydrodynamic process. Gidrometeoizdat, Leningrad (In Russian)

Davidan IN, Lopatoukhin LI, Rozhkov VA (1985) Wind waves in the world oceans. Gidrometeoizdat, Leningrad (In Russian)

Dean RG (1991) Equilibrium beach profiles: characteristics and applications. J Coast Res 7:53–84

Druet C, Massel S, Zeidler R (1972) Statistical characteristics of wind waves in the Baltic coastal zone in the Gulf of Gdańsk and open Baltic Sea. Rozprawy Hydrotechniczne, Zeszyt 30, 49–84 (In Polish, cited after Schmager et al. 2008)

DWD (Deutscher Wetterdienst), (2006) Seegangsklimatologie der Ostsee (1952 bis 2002). Deutscher Wetterdienst, Hamburg

Eelsalu M, Org M, Soomere T (2014) Visually observed wave climate in the Gulf of Riga. In: 6th IEEE/OES Baltic Symposium "Measuring and modeling of multi-scale interactions in the marine environment," 26–29 May 2014, Tallinn Estonia. IEEE Conference Publications

Feistel R, Nausch G, Wasmund N (eds) (2008) State and evolution of the Baltic Sea 1952–2005. Wiley, Hoboken

Gayer G, Günther H, Winkel N (1995) Wave climatology and extreme value analysis for the Baltic Sea area off the Warnemünde harbour entrance. Deutsche Hydrographische Zeitschrift 47:109–130

Guedes Soares C (1986) Assessment of the uncertainty in visual observations of wave height. Ocean Eng 13:37–56

Gulev SK, Hasse L (1998) North Atlantic wind waves and wind stress fields from voluntary observing ship data. J Phys Oceanogr 28:1107–1130

Gulev SK, Hasse L (1999) Changes of wind waves in the North Atlantic over the last 30 years. Int J Climatol 19:1091–1117

Gulev SK, Grigorieva V, Sterl A, Woolf D (2003) Assessment of the reliability of wave observations from voluntary observing ships: insights from the validation of a global wind wave climatology based on voluntary observing ship data. J Geophys Res Oceans 108:3236

Hallermeier RJ (1981) A profile zonation for seasonal sand beaches from wave climate. Coast Eng 4:253–277

Holthuijsen LH (1983) Observations of the directional distribution of ocean wave energy in fetch limited conditions. J Phys Oceanogr 13:191–207

Hogben N, Da Cunha NMC, Oliver GF (1986) Global wave statistics. Unwin Brothers, London

Houston JR (1996) Simplified Dean's method for beach-fill design. J Waterw Port C ASCE 122:143–146

Hünicke B, Zorita E, Soomere T, Madsen KS, Johansson M, Suursaar Ü (2015) Recent change sea level and wind waves. In: The BACC II author team. Second assessment of climate change for the Baltic Sea basin. Springer, Berlin

IAHR working group on wave generation and analysis (1989) List of sea-state parameters. J Waterw Port Coast Ocean Eng 115:793–808

Jaagus J (2009) Long-term changes in frequencies of wind directions on the western coast of Estonia. In: Kont A, Tõnisson H (eds) Climate change impact on Estonian coasts, Publication 11/2009. Tallinn University, Tallinn, Estonia, Institute of Ecology, pp 11–24 (In Estonian)

Jaagus J, Kull A (2011) Changes in surface wind directions in Estonia during 1966–2008 and their relationships with large-scale atmospheric circulation. Est J Earth Sci 60:220–231

Johansson M, Boman H, Kahma KK, Launiainen J (2001) Trends in sea level variability in the Baltic Sea. Boreal Environ Res 6:159–179

Jönsson A, Broman B, Rahm L (2003) Variations in the Baltic Sea wave fields. Ocean Eng 30:107–126

Jönsson A, Danielsson A, Rahm L (2005) Bottom type distribution based on wave friction velocity in the Baltic Sea. Cont Shelf Res 25:419–435

Kahma K (1981) A study of the growth of the wave spectrum with fetch. J Phys Oceanogr 11:1503–1515

Kahma K, Calkoen CJ (1992) Reconciling discrepancies in the observed growth of wind-generated waves. J Phys Oceanogr 22:1389–1405

Kahma K, Pettersson H (1993) Wave statistics from the Gulf of Finland. Internal report 1/1993. Finnish Institute of Marine Research, Helsinki (In Finnish with English summary)

Kahma K, Pettersson H (1994) Wave growth in a narrow fetch geometry. Glob Atmos Ocean Syst 2:253–263

Kahma K, Pettersson H, Tuomi L (2003) Scatter diagram wave statistics from the northern Baltic Sea. MERI—Rep Ser Finnish Inst Mar Res 49:15–32

Keevallik S, Soomere T (2010) Towards quantifying variations in wind parameters across the Gulf of Finland. Est J Earth Sci 59:288–297

Kelpšaitė L, Parnell KE, Soomere T (2009) Energy pollution: the relative influence of wind-wave and vessel-wake energy in Tallinn Bay, the Baltic Sea. J Coast Res Spec Issue 56:812–816

Kelpšaitė L, Dailidienė I, Soomere T (2011) Changes in wave dynamics at the south-eastern coast of the Baltic Proper during 1993–2008. Boreal Environ Res 16(Suppl A):220–232

Komen GJ, Cavaleri L, Donelan M, Hasselmann K, Hasselmann S, Janssen PAEM (1994) Dynamics and modelling of ocean waves. Cambridge University Press, Cambridge

Kraus NC (1992) Engineering approaches to cross-shore sediment processes. In: Lamberti A (ed) Proceedings short course on design and reliability of coastal structures attached to 23rd international conference on coastal engineering, Venice, pp 175–209

Kriezi EE, Broman B (2008) Past and future wave climate in the Baltic Sea produced by the SWAN model with forcing from the regional climate model RCA of the Rossby Centre. In: IEEE/OES US/EU-Baltic international symposium, 27–29 May 2008, Tallinn, Estonia. IEEE, pp 360–366

Kushnir Y, Cardone VJ, Greenwood JG, Cane MA (1997) The recent increase of North Atlantic wave heights. J Clim 10:2107–2113

Leppäranta M, Myrberg K (2009) Physical oceanography of the Baltic Sea. Springer, Berlin, 378 pp

Lopatukhin LI, Bukhanovsky AV, Ivanov SV, Tshernyshova ES (2006a) Handbook of wind and wave regimes in the Baltic Sea, North Sea, Black Sea, Azov Sea and the Mediterranean. Russian Shipping Registry. Sankt-Petersburg, 450 pp (In Russian)

Lopatukhin LI, Mironov ME, Pomeranets KS, Trapeznikov ES, Tshernysheva ES (2006b) Estimates of extreme wind and wave conditions in the eastern part of the Gulf of Finland. Proc VNIIG 245:145–155

Mårtensson N, Bergdahl L (1987) On the wave climate of the southern Baltic. Report Series A:15. Department of Hydraulics, Chalmers University of Technology, Göteborg

Martucci G, Carniel S, Chiggiato J, Sclavo M, Lionello P, Galati MB (2010) Statistical trend analysis and extreme distribution of significant wave height from 1958 to 1999–an application to the Italian seas. Ocean Sci 6:525–538

Massel SR (2013) Ocean surface waves: their physics and prediction, 2nd edn. World Scientific, New Jersey 692 pp

Mietus M (co-ordinator) (1998) The climate of the Baltic Sea Basin, Marine meteorology and related oceanographic activities. Report No. 41. World Meteorological Organisation, Geneva

Mietus M, von Storch H (1997) Reconstruction of the wave climate in the Proper Baltic Basin, April 1947-March 1988, GKSS Report 97/E/28, Geesthacht

Myrberg K, Ryabchenko V, Isaev A, Vankevich R, Andrejev O, Bendtsen J, Erichsen A, Funkquist L, Inkala A, Neelov I, Rasmus K, Rodriguez Medina M, Raudsepp U, Passenko J, Söderkvist J, Sokolov A, Kuosa H, Anderson TR, Lehmann A, Skogen MD (2010) Validation of three-dimensional hydrodynamic models of the Gulf of Finland. Boreal Environ Res 15:453–479

Nikolkina I, Soomere T, Räämet A (2014) Multidecadal ensemble hindcast of wave fields in the Baltic Sea. In: 6th IEEE/OES Baltic Symposium "Measuring and modeling of multi-scale interactions in the marine environment", 26–29 May, Tallinn Estonia. IEEE Conference Publications

Orlenko LR, Lopatukhin LI, Portnov GL (eds) (1984) Studies of the hydrometeorological regime of Tallinn Bay. Gidrometeoizdat, Leningrad (In Russian)

Orviku K, Jaagus J, Kont A, Ratas U, Rivis R (2003) Increasing activity of coastal processes associated with climate change in Estonia. J Coast Res 19:364–375

Paplińska B (1999) Wave analysis at Lubiatowo and in the Pomeranian Bay based on measurements from 1997/1998–comparison with modelled data (WAM4 model). Oceanologia 41:241–254

Pettersson H (2001) Directional wave statistics from the Gulf of Finland 1990–1994. MERI—Rep Ser Finnish Inst Mar Res 44:1–37 (In Finnish with English summary)

Pettersson H, Kahma KK, Tuomi L (2010) Predicting wave directions in a narrow bay. J Phys Oceanogr 40(1):155–169

Pettersson H, Boman H (2002) High waves and sea level during the November storm. Annual report 2001, Finnish Institute of Marine Research, Helsinki, p 7

Pettersson H, Hammarklint T, Schräder D (2007) Wave climate in the Baltic sea 2006. HELCOM Baltic Sea Environment Fact Sheets 2012

Pettersson H, Lindow H, Schräder D (2012) Wave climate in the Baltic Sea 2011. HELCOM Baltic sea Environment Fact Sheets 2012. http://helcom.fi/Documents/Baltic%20sea%20trends/Environment%20fact%20sheets/Wave%20climate%20in%20the%20Baltic%20Sea%202011.pdf

Pettersson H, Lindow H, Brüning T (2013) Wave climate in the Baltic Sea 2012. HELCOM Baltic sea Environment Fact Sheets 2012. http://helcom.fi/baltic-sea-trends/environment-fact-sheets/hydrography/wave-climate-in-the-baltic-sea/

Pindsoo K, Soomere T, Zujev M (2012) Decadal and long-term variations in the wave climate at the Latvian coast of the Baltic Proper. In: Proceedings of the IEEE/OES Baltic 2012 international symposium "Ocean: past, present and future. Climate change research, ocean observation & advanced technologies for regional sustainability," 8–11 May 2012, Klaipėda, Lithuania. IEEE. doi:10.1109/BALTIC.2012.6249160

Pryor SC, Barthelmie RJ (2003) Long-term trends in near-surface flow over the Baltic. Int J Climatol 23:271–289

Pryor SC, Barthelmie RJ (2010) Climate change impacts on wind energy: a review. Renew Sust Energy Rev 14:430–437

Räämet A, Soomere T (2010) The wave climate and its seasonal variability in the northeastern Baltic Sea. Est J Earth Sci 59:100–113

Räämet A, Soomere T (2011) Spatial variations in the wave climate change in the Baltic Sea. J Coast Res Spec Issue 64:240–244

Räämet A, Soomere T, Zaitseva-Pärnaste I (2010) Variations in extreme wave heights and wave directions in the north-eastern Baltic Sea. Proc Est Acad Sci 59:182–192

Räämet A, Suursaar Ü, Kullas T, Soomere T (2009) Reconsidering uncertainties of wave conditions in the coastal areas of the northern Baltic Sea. J Coast Res Spec Issue 56:257–261

Rosenhagen G, Tinz B (2013) New historical data of the southern Baltic Sea coasts. In: Reckermann M, Köppen S (eds), 7th study conference on BALTEX, 10–14 June 2013, Borgholm, Island of Öland, Sweden. International BALTEX Secretariat, Publication No. 53, 84

Rosenthal W (1986) Wind waves and swell. In: Sündermann J (ed) Oceanography, vol 3C, Landolt-Börnstein: Numerical data and functional relationships in science and technology–New series. Springer, Berlin, pp 17–36

Russian Shipping Registry (1974) Wind and waves in oceans and seas. Part II. Data about wind and wave regime in (semi-enclosed) seas. Volume 1—The Baltic Sea (In Russian)

Rzheplinsky GV, Brekhovskikh YuP (1967) Wave atlas for Gulf of Finland. Gidrometeoizdat, Leningrad (In Russian)

Rzheplinsky GV (ed) (1965) Wave and wind atlas for the Baltic Sea. Tallinn (In Russian)

Ryabchuk D, Kolesov A, Chubarenko B, Spiridonov M, Kurennoy D, Soomere T (2011) Coastal erosion processes in the eastern Gulf of Finland and their links with geological and hydrometeorological factors. Boreal Environ Res 16(Suppl A):117–137

Schmager WG (1979) Atlas zur Ermittlung der Wellenhöhe in der südlichen Ostsee. Seehydrographischer Dienst der DDR. XIV, Rostock, 115 pp (In German)

Schmager G, Fröhle P, Schrader D, Weisse R, Müller-Navarra S (2008) Sea state, tides. In: Feistel R, Nausch G, Wasmund N (eds) State and evolution of the Baltic sea 1952–2005. Wiley, Hoboken, pp 143–198

Seifert T, Tauber F, Kayser B (2001) A high resolution spherical grid topography of the Baltic Sea, 2nd ed. Baltic Sea Science Congress, Stockholm 25–29 November 2001, Poster #147. www.io-warnemuende.de/iowtopo

Siewert M, Schlamkow C, Saathoff F (2015) Spatial analyses of 52 years of modelled sea state data for the Western Baltic Sea and their potential applicability for offshore and nearshore construction purposes. Ocean Eng 96:284–294

Simpson RH, Riehl H (1981) The hurricane and its impact. Lousiana State University Press, Baton Rouge

Soomere T (2001) Wave regimes and anomalies off north-western Saaremaa Island. Proc Est. Acad Sci Eng 7:157–173

Soomere T (2003) Anisotropy of wind and wave regimes in the Baltic proper. J Sea Res 49:305–316

Soomere T (2005) Wind wave statistics in Tallinn Bay. Boreal Environ Res 10:103–118

Soomere T (2013) Extending the observed Baltic Sea wave climate back to the 1940s. J Coast Res Spec Issue 65:1969–1974

Soomere T, Healy T (2008) Escalating extremes over descending trends of the northern Baltic Sea wave fields. In: Wallendorf L, Ewing L, Jones C, Jaffe B (eds) Solutions to coastal disasters 2008, pp 129–138. American Society of Civil Engineers

Soomere T, Keevallik S (2001) Anisotropy of moderate and strong winds in the Baltic Proper. Proc Est Acad Sci Eng 7:35–49

Soomere T, Keevallik S (2003) Directional and extreme wind properties in the Gulf of Finland. Proc Est Acad Sci Eng 9:73–90

Soomere T, Kurkina O (2011) Statistics of extreme wave conditions in the south-western Baltic Sea. Fund Appl Hydrophys 4(4):43–57 (In Russian)

Soomere T, Räämet A (2011a) Long-term spatial variations in the Baltic Sea wave fields. Ocean Sci 7:141–150

Soomere T, Räämet A (2011b) Spatial patterns of the wave climate in the Baltic Proper and the Gulf of Finland. Oceanologia 53(1-TI), 335–371

Soomere T, Räämet A (2014) Decadal changes in the Baltic Sea wave heights. J Mar Syst 129:86–95

Soomere T, Zaitseva I (2007) Estimates of wave climate in the northern Baltic Proper derived from visual wave observations at Vilsandi. Proc Est Acad Sci Eng 13:48–64

Soomere T, Behrens A, Tuomi L, Nielsen JW (2008) Wave conditions in the Baltic Proper and in the Gulf of Finland during windstorm Erwin/Gudrun. Nat Hazards Earth Syst Sci 8:37–46

Soomere T, Zaitseva-Pärnaste I, Räämet A, Kurennoy D (2010) Spatio-temporal variations of wave fields in the Gulf of Finland. Fund Appl Hydrophys 4(10):90–101 (In Russian)

Soomere T, Zaitseva-Pärnaste I, Räämet A (2011) Variations in wave conditions in Estonian coastal waters from weekly to decadal scales. Boreal Environ Res 16(Suppl A):175–190

Soomere T, Weisse R, Behrens A (2012) Wave climatology in the Arkona Basin, the Baltic Sea. Ocean Sci 8:287–300

Soomere T, Viška M, Eelsalu M (2013) Spatial variations of wave loads and closure depths along the coast of the eastern Baltic Sea. Est J Eng 19:93–109

Soomere T, Bishop SR, Viška M, Räämet A (2015) An abrupt change in winds that may radically affect the coasts and deep sections of the Baltic Sea. Clim Res 62:163–171

Sparre A (1982) The climate of Denmark. Summaries of observations from light vessels II: waves and currents at the surface. Danish Meteorological Institute Climatological Papers 9

Sterl A, Caires S (2005) Climatology, variability and extrema of ocean waves—the web-based KNMI/ERA-40 wave atlas. Int J Climatol 25:963–977

Suursaar Ü (2010) Waves, currents and sea level variations along the Letipea–Sillamäe coastal section of the southern Gulf of Finland. Oceanologia 52:391–416

Suursaar Ü (2013) Locally calibrated wave hindcasts in the Estonian coastal sea in 1966–2011. Estonian J Earth Sci 62:42–56

Suursaar Ü, Kullas T (2009a) Decadal variations in wave heights off Cape Kelba, Saaremaa Island, and their relationships with changes in wind climate. Oceanologia 51:39–61

Suursaar Ü, Kullas T (2009b) Decadal changes in wave climate and sea level regime: the main causes of the recent intensification of coastal geomorphic processes along the coasts of Western Estonia? In: Coastal processes. WIT Trans Ecol Environ 126:105–116

Suursaar Ü, Kullas K, Otsmann M, Saaremäe I, Kuik J, Merilain M (2006) Cyclone Gudrun in January 2005 and modelling its hydrodynamic consequences in the Estonian coastal waters. Boreal Environ Res 11:143–159

Tuomi L, Pettersson H, Kahma K (1999) Preliminary results from the WAM wave model forced by the mesoscale EUR-HIRLAM atmospheric model. MERI—Rep Ser Finnish Inst Mar Res 40:19–23

Tuomi L, Kahma KK, Pettersson H (2011) Wave hindcast statistics in the seasonally ice-covered Baltic Sea. Boreal Environ Res 16:451–472

Tuomi L, Pettersson H, Fortelius C, Tikka K, Björkqvist J-V, Kahma KK (2014) Wave modelling in archipelagos. Coast Eng 83:205–220

Vikebo F, Furevik T, Furnes G, Kvamsto NG, Reistad M (2003) Wave height variations in the North Sea and on the Norwegian Continental Shelf 1881–1999. Cont Shelf Res 23:251–263

Viška M, Soomere T (2012) Hindcast of sediment flow along the Curonian spit under different wave climates. In: Proceedings of the IEEE/OES Baltic 2012 international symposium "Ocean: past, present and future. climate change research, ocean observation & advanced technologies for regional sustainability," 8–11 May, Klaipėda, Lithuania. IEEE Conference Publications. doi:10.1109/BALTIC.2012.6249195

Zaitseva-Pärnaste I (2013) Wave climate and its decadal changes in the Baltic Sea derived from visual observations. Ph.D thesis, Tallinn University of Technology

Zaitseva-Pärnaste I, Suursaar Ü, Kullas T, Lapimaa S, Soomere T (2009) Seasonal and long-term variations of wave conditions in the northern Baltic Sea. J Coast Res Spec Issue 56:277–281

Zaitseva-Pärnaste I, Soomere T, Tribštok O (2011) Spatial variations in the wave climate change in the eastern part of the Baltic Sea. J Coast Res Spec Issue 64:195–199

Wahl G (1974) Wave statistics from Swedish coastal waters. Proceedings of the international symposium on dynamics of marine vehicles and structures in waves. Institute of Mechanical Engineering, London, pp 33–40

Weisse R, Günther H (2007) Wave climate and long-term changes for the Southern North Sea obtained from a high-resolution hindcast 1958–2002. Ocean Dyn 57: 161–172

Weisse R, von Storch H (2010) Marine climate and climate change. Storms, wind waves and storm surges, Springer, Berlin, Heidelberg, 220 pp

WASA Group (1995) The WASA project: changing storm and wave climate in the northeast Atlantic and adjacent seas? In: Proceedings of the fourth international workshop on wave hindcasting and forecasting, Banff, Canada, 16–20 October, pp 31–44. Also: GKSS Report 96/E/61

Runup of Long Irregular Waves on Plane Beach

Ira Didenkulova, Efim Pelinovsky and Anna Sergeeva

Abstract Runup of irregular waves, modeled as superposition of Furrier harmonics with random phases, is studied in frames of nonlinear shallow water theory. The possibility of appearance of freak waves on a beach is analyzed. The distribution functions of runup characteristics are computed. An incident wave represents an irregular sea state with Gaussian spectrum. The asymptotic of probability functions in the range of large amplitudes for estimation of freak wave formation in the shore is studied. It is shown that the average runup height of waves with wide spectrum is higher than of waves with narrow spectrum.

1 Introduction

Descriptions of unusually high waves appearing on the sea surface for a short time (freak, rogue, or killer waves) have been considered as a part of marine folklore for a long time. A number of instrumental registrations have appeared recently making the community to pay more attention to this problem and to reconsider known observations of freak waves: some of them are collected in the papers by Torum and Gudmestad (1990), Olagnon and Athanassoulis (2001), Kharif and Pelinovsky (2003) and Rosenthal (2003). Mechanisms of freak wave generation are described in Dysthe et al. (2008), Kharif et al. (2009), Didenkulova and Pelinovsky (2011) and Slunyaev et al. (2011). Such unusual waves are observed also in the coastal zone and the probability of their appearance is rather high. One of the first works (Sand et al. 1990) already presents data of freak wave observations in the shallow part of the North Sea (on the depth of 20 m). Didenkulova, I. (2011) analyzes the data of sea level elevation in the coastal zone of the Baltic Sea (2.7 m depth) and

I. Didenkulova (✉) · E. Pelinovsky · A. Sergeeva
Nizhny Novgorod State Technical University n.a. R.E. Alekseev, Nizhny Novgorod, Russia
e-mail: dii@hydro.appl.sci-nnov.ru

I. Didenkulova · E. Pelinovsky · A. Sergeeva
Institute of Applied Physics, Nizhny Novgorod, Russia

I. Didenkulova
Marine Systems Institute, Tallinn University of Technology, Tallinn, Estonia

E. Pelinovsky and C. Kharif (eds.), *Extreme Ocean Waves*,
DOI 10.1007/978-3-319-21575-4_8

Fig. 1 Freak wave attacks the breakwater in Kalk Bay, South Africa on August 26, 2005

demonstrates the existence of two different families of freak waves. Chien et al. (2002) report about 140 freak wave events in the coastal zone of Taiwan in the past 50 years (1949–1999) that caused loss of 500 people and destruction of 35 ships. According to (Didenkulova et al. 2006a) two-third of the freak wave events occurred in 2005 were observed onshore. A freak wave attacked the breakwater in Kalk Bay (South Africa) on August 26, 2005 and washed off the breakwater people, some of them were injured (Fig. 1). Two months later on October 16, 2005, two freak waves induced panic at Maracas Beach (Trinidad Island, Lesser Antilles), when a series of towering waves, many more than 25 feet high (maximal height of 8 m), flooded the beach, carried sea-bathers, venders, and lifeguards, running for their lives. The new catalogue of freak waves (Nikolkina and Didenkulova 2011, 2012) has demonstrated that the majority of freak wave accidents occurs in the shallow water and at the coast.

Thus, analysis of freak waves on a coast is an important task for practice. Here, we will investigate distribution functions of the runup height and velocity on a beach, assuming that distribution functions in the coastal zone are known and waves do not break. The analytical shallow water theory, described in Spielfogel (1976), Pedersen and Gjevik (1983), Synolakis (1987), Pelinovsky and Mazova (1992), Carrier et al. (2003), Tinti and Tonini (2005), Kânoğlu and Synolakis (2006), Didenkulova et al. (2006b), Didenkulova et al. (2007a, b), Didenkulova (2009) (Tadepalli and Synolakis, 1994) is used as theoretical model. The paper is organized as follows. The theoretical model of the long wave runup is described in Sect. 2. The runup of irregular waves on a plane beach is discussed in Sect. 3. Main results are summarized in Sect. 4.

2 Theoretical Model of the Long Wave Runup

The dynamics of a wave climbing the beach can be described in the framework of the shallow water equations. The simplified geometry of the coastal zone is shown in Fig. 2.

The wave comes onshore from the left. Sketchy, the incident wave is presented as a single crest, but then we will consider the incident wave as a continuous function representing random crests and troughs. The basic equations for water waves in shallow water are ($\eta(x,t)$ is the vertical displacement of the sea level, $u(x,t)$ is the depth averaged velocity of the water flow)

$$\frac{\partial \eta}{\partial t} + \frac{\partial}{\partial x}\left[\left(h\left(x\right) + \eta\right) u\right] = 0,$$
$$\frac{\partial u}{\partial t} + u\frac{\partial u}{\partial x} + g\frac{\partial \eta}{\partial x} = 0, \tag{1}$$

where $h(x) = -\alpha x$. In this case, the nonlinear shallow water Eq. (1) can be solved with the use of Riemann invariants and the Legendre (hodograph) transformation (Carrier and Greenspan 1958). Let us introduce the Riemann invariants

$$I_{\pm} = u \pm 2\sqrt{g\left(h + \eta\right)} + g\alpha t \tag{2}$$

and rewrite system (1) in the following form

$$\frac{\partial I_{\pm}}{\partial t} + c_{\pm}\frac{\partial I_{\pm}}{\partial x} = 0, \tag{3}$$

where characteristic speeds are

$$c_{\pm} = \frac{3}{4}I_{\pm} + \frac{1}{4}I_{\mp} - g\alpha t. \tag{4}$$

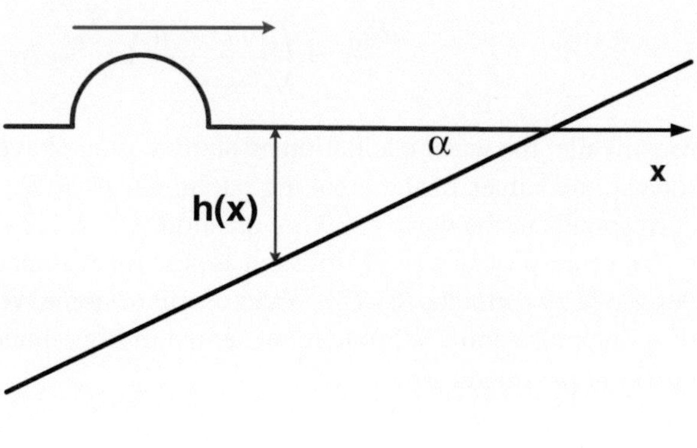

Fig. 2 Definition sketch for the wave runup problem

The system (3)–(4) is still nonlinear, as characteristic speeds $c_\pm$ contain time t; however, it can be reduced to linear by excluding the coordinate x. After introducing new variables

$$\lambda = \frac{I_+ + I_-}{2} = u + g\alpha t, \quad \sigma = \frac{I_+ - I_-}{2} = 2\sqrt{g(h+\eta)}, \tag{5}$$

we obtain the linear wave equation to describe the long wave runup process

$$\frac{\partial^2 \Phi}{\partial \lambda^2} - \frac{\partial^2 \Phi}{\partial \sigma^2} - \frac{1}{\sigma}\frac{\partial \Phi}{\partial \sigma} = 0, \tag{6}$$

and all physical variables can be expressed through the function $\Phi(\lambda, \sigma)$

$$\eta = \frac{1}{2g}\left(\frac{\partial \Phi}{\partial \lambda} - u^2\right), \quad u = \frac{1}{\sigma}\frac{\partial \Phi}{\partial \sigma}, \tag{7}$$

$$t = \frac{1}{\alpha g}\left(\lambda - \frac{1}{\sigma}\frac{\partial \Phi}{\partial \sigma}\right), \quad x = \frac{1}{2\alpha g}\left(\frac{\partial \Phi}{\partial \lambda} - u^2 - \frac{\sigma^2}{2}\right). \tag{8}$$

The physical sense of the variable σ is the total water depth, and $\sigma = 0$ corresponds to the moving shoreline. Various calculations of the wave field and runup characteristics using the Carrier-Greenspan transformation can be found in Spielfogel (1976), Pedersen and Gjevik (1983), Synolakis (1987), Pelinovsky and Mazova (1992), Carrier et al. (2003), Tinti and Tonini (2005), Kânoğlu and Synolakis (2006), Didenkulova et al. (2006b), Didenkulova et al. (2007a, b), Didenkulova (2009) (Tadepalli and Synolakis, 1994). A surprising result concluded from linear equation (6) is that the extreme runup characteristics (runup and rundown amplitudes, runup velocities) can be calculated in the framework of linear shallow water theory if the incident wave approaches to the beach from the open sea. Particularly, the runup amplitude R_{sin} of incident sine wave with amplitude A, wavelength λ, and frequency ω given at the point $x = L$ with the depth h is

$$\frac{R_{sin}}{A} = \left(\frac{16\pi^2\omega^2 h}{g\alpha^2}\right)^{1/4} = 2\pi\sqrt{\frac{2L}{\lambda}}. \tag{9}$$

Meanwhile, the water oscillation on shore will not have simple sine shape; see Fig. 3 for various values of the breaking parameter $Br = R_{sin}\omega^2/g\alpha^2$ (condition $Br = 1$ corresponds to the wave breaking on shore).

The runup of waves of different types, for instance solitary waves, can be also described by formulas (6)–(8). Water oscillations and velocities onshore for the runup of a sine pulse and a soliton are presented in Figs. 4 and 5 for different values of the breaking parameter Br.

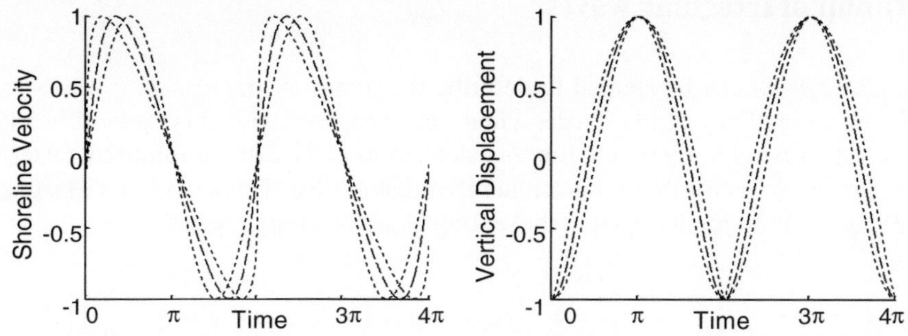

Fig. 3 Velocity and vertical displacement of the moving shoreline for incoming sine wave; the breaking parameter $Br = 0$ (*dotted line*), 0.5 (*dashed line*), and 1 (*solid line*); time is normalized by wave frequency ω^{-1}, vertical displacement by R_{sin}, and shoreline velocity by $\omega R_{sin}/\alpha$

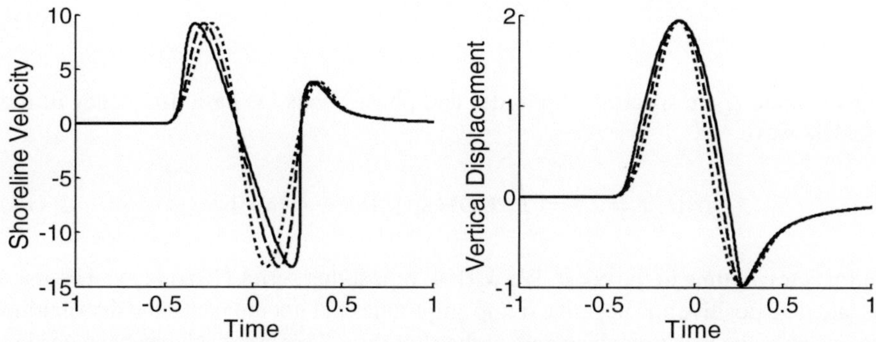

Fig. 4 Velocity and vertical displacement of the moving shoreline for incoming $\sin^4(\omega t)$ pulse; the breaking parameter $Br = 0$ (*dotted line*), 0.5 (*dashed line*) and 1 (*solid line*); time is normalized by wave frequency ω^{-1}, vertical displacement by R_{sin}, and shoreline velocity by $\omega R_{sin}/\alpha$

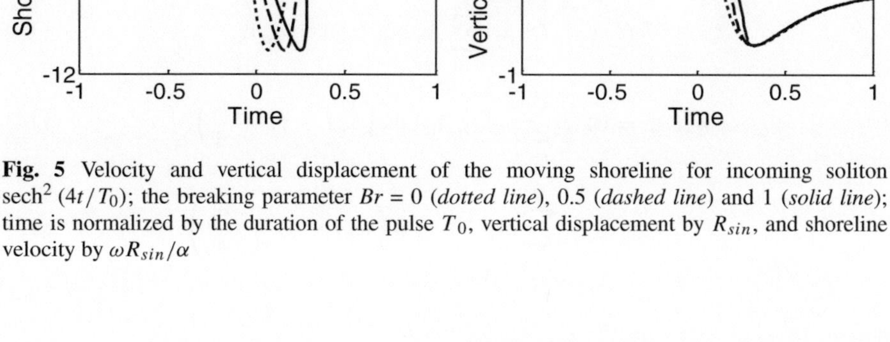

Fig. 5 Velocity and vertical displacement of the moving shoreline for incoming soliton $\text{sech}^2\,(4t/T_0)$; the breaking parameter $Br = 0$ (*dotted line*), 0.5 (*dashed line*) and 1 (*solid line*); time is normalized by the duration of the pulse T_0, vertical displacement by R_{sin}, and shoreline velocity by $\omega R_{sin}/\alpha$

3 Runup of irregular waves

Formulas (6)–(8) can be applied to describe the runup of irregular long waves as well. Due to implicity of the Carrier–Greenspan transformation it is rather difficult to calculate wave characteristics. But for calculations of the extreme runup characteristics, the linear approach can be applied (Synolakis 1991; Pelinovsky and Mazova 1992), and in this case we need to find extremes of the Fourier series

$$\eta\,(t, x = 0) = \int \left(\frac{16\pi^2 h}{g\alpha^2}\right)^{1/4} \omega^{1/2} A(\omega) \exp\left[i\left(\omega\,(t-\tau) + \phi\,(\omega) + \frac{\pi}{4}\right)\right] d\omega, \tag{10}$$

$$u\,(t, x = 0) = \frac{1}{\alpha} \int \left(\frac{16\pi^2 h}{g\alpha^2}\right)^{1/4} \omega^{3/2} A(\omega) \exp\left[i\left(\omega\,(t-\tau) + \phi\,(\omega) + \frac{3\pi}{4}\right)\right] d\omega, \tag{11}$$

where A and ϕ are spectral amplitudes and phases, ω is the basic frequency of the incident wave

$$\eta\,(t, x = L) = \int A(\omega) \exp\left[i\,(\omega t + \phi\,(\omega))\right] d\omega, \tag{12}$$

and τ is travel time to the coast. We wish to repeat that series (10)–(11) can be used to calculate positive and negative runup amplitudes but not moments and distribution functions of the water displacement at the shoreline. This approach has been used in Didenkulova et al. (2007a) to study the runup of nonsinusoidal waves.

Now we will consider the transformation of irregular waves when they climb a beach and estimate distribution functions of the water displacement at the shoreline assuming distribution functions of the water displacement at the coastal zone to be known and waves do not break.

The ensemble of realizations with random phases ϕ is taken for a numerical simulation of irregular waves. For this purpose, we quantize Fourier series (10)–(12) and use real functions, whereupon equations for incoming wave, displacement, and velocity of the shoreline in nondimensional variables can be rewritten as

$$\bar{\eta}\,(t, x = L) = \sum_{n=1}^{N} A_n \cos\,(\omega_n t + \phi_n), \tag{13}$$

$$\bar{\eta}\,(t, x = 0) = \sum_{n=1}^{N} \sqrt{\omega_n} A_n \cos\left(\omega_n t + \phi_n + \frac{\pi}{4}\right), \tag{14}$$

$$\bar{u}\,(t, x = 0) = \sum_{n=1}^{N} \omega_n^{3/2} A_n \cos\left(\omega_n t + \phi_n + \frac{3\pi}{4}\right), \tag{15}$$

where $A_n = \sqrt{2S(\omega_n)\Delta\omega}$ is calculated through the frequency spectrum of incoming wave $S(\omega)$, sampling rate $\Delta\omega$=$2\pi/T$, the size of time calculated domain T, and ω_n= $n\Delta\omega$. Random spectral phases ϕ_n are distributed uniformly at the interval (0, 2π).

First, let us consider random wave field with Gaussian statistics, where the frequency spectrum of incoming wave $S(\omega)$ is

$$S(\omega) = Q \exp\left[-\frac{(\omega-\omega_0)^2}{2l^2}\right], \tag{16}$$

with the central frequency ω_0 and the spectrum width l. Constant Q in (16) can be found from the condition

$$\sigma^2 = 2\int_0^\infty S(\omega)d\omega, \tag{17}$$

then

$$Q = \frac{\sigma^2}{\sqrt{2\pi}l\,\mathrm{erfc}(-\omega_0/\sqrt{2}l)}, \tag{18}$$

where

$$erfc(z) = \frac{2}{\sqrt{\pi}}\int_z^\infty \exp(-t^2)dt \tag{19}$$

is a complementary error function.

In this case, frequency spectra for the shoreline displacement $S_r(\omega)$ and the shoreline velocity $S_u(\omega)$ are

$$S_r(\omega) = \frac{4\pi L\omega}{c} Q \exp\left[-\frac{(\omega-\omega_0)^2}{2l^2}\right], \tag{20}$$

$$S_u(\omega) = \frac{4\pi L\omega^3}{c\alpha^2} Q \exp\left[-\frac{(\omega-\omega_0)^2}{2l^2}\right]. \tag{21}$$

All these spectra in nondimensional variables for $l = 0.5$ are shown on Fig. 6. It is obvious that spectra for the shoreline displacement $S_r(\omega)$ and the shoreline velocity $S_u(\omega)$ are asymmetric and shifted to the high-frequency area.

Distribution functions for maximal amplitudes (positive and negative) of the wave field, defined as maximum (minimum) between two zero points, are important for applications. Detailed calculations of the distribution functions of the runup amplitudes are given in (Sergeeva and Didenkulova 2005). The Fourier series of N= 512 harmonics and sampling rate $\Delta\omega$ = 0.01 are used. Spectrum width l is changed

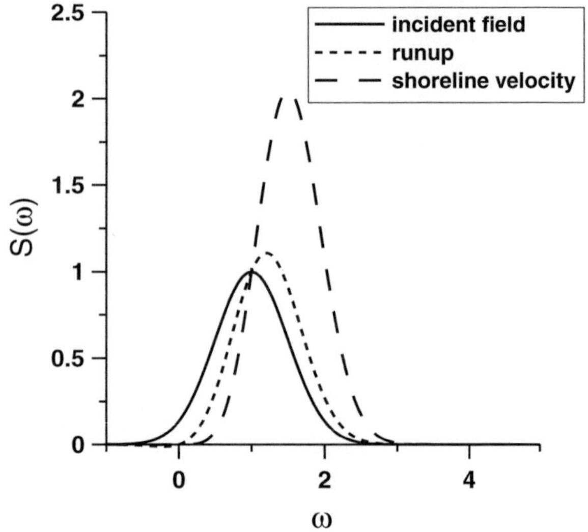

Fig. 6 Incident field, runup and shoreline velocity spectra for $l = 0.5$

from 0.1 to 0.7. All statistical characteristics are obtained with the use of ensemble averaging over 500 realizations.

The occurrence probability of the wave with amplitude A for a Gaussian narrow-band process can be described by Rayleigh distribution (Massel 1996)

$$P(A) = \exp\left(-2A^2\right), \tag{22}$$

where A is wave amplitude normalized on significant amplitude A_s, which is defined as $A_s \approx 2\sigma$. For the numerical estimation of positive (negative) amplitude distribution, the statistical "frequency" F (ratio of a number of waves m with fixed amplitude a to a general number of waves)

$$F = \frac{m}{N}, \tag{23}$$

and statistical distribution function of amplitude (occurrence frequency of waves with amplitude A larger than a)

$$P(a) = F(A > a). \tag{24}$$

are calculated. For the narrowband incident wave field ($l = 0.1$), the distribution functions of the runup characteristics are described by the Rayleigh distribution, as it is expected due to linearity expressions for extreme characteristics. If the spectrum of incident wave is wider ($l = 0.7$), the asymmetry of displacement and velocity spectra increases, but nevertheless distribution functions of the maximal shoreline displacement (Figs. 7 and 8) and the maximal shoreline velocity (Fig. 9) differ

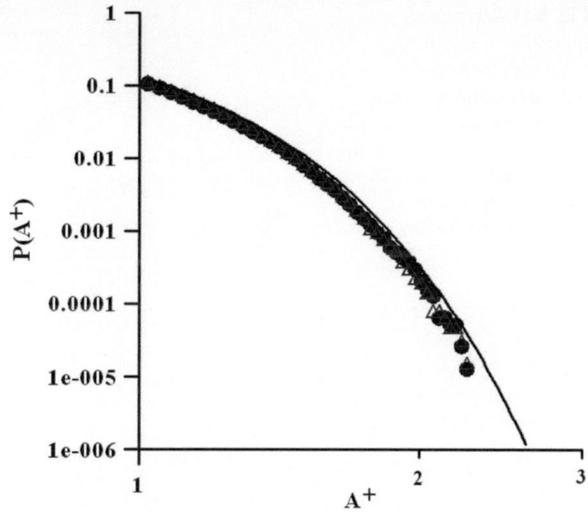

Fig. 7 Distribution functions of maximal positive amplitudes for incident wave (*triangles*) and shoreline displacement (*circles*) with the incident wave spectrum width $l = 0.7$; solid line corresponds to the Rayleigh distribution

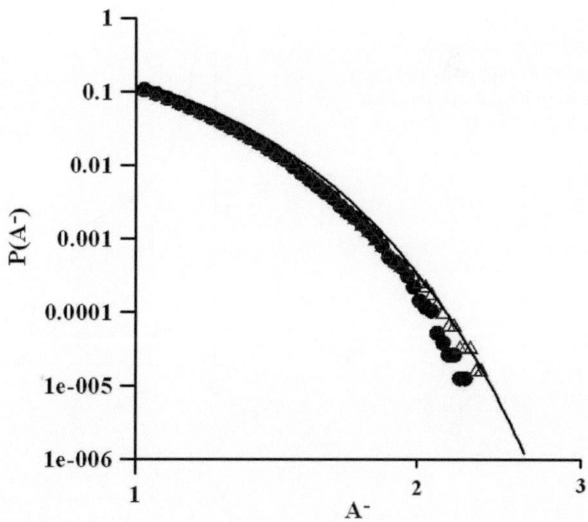

Fig. 8 Distribution functions of maximal negative amplitudes for incident wave (*triangles*) and shoreline displacement (*circles*) with the incident wave spectrum width $l = 0.7$; solid line corresponds to the Rayleigh distribution

from the Rayleigh low weakly. This effect has also been confirmed experimentally (Denissenko et al. 2013).

Knowing spectral and probability distributions of the wave field runup characteristics on a beach can be calculated. Thus the significant runup height of the wave on a beach is

$$R_s = \sqrt{\frac{4\pi\omega_0 L}{c}} A_s F\left(\frac{\omega_0}{l}\right) = 2\pi\sqrt{\frac{2L}{\lambda}} A_s F\left(\frac{\omega_0}{l}\right), \tag{25}$$

where function $F(z)$ describes influence of the incident wave spectrum width

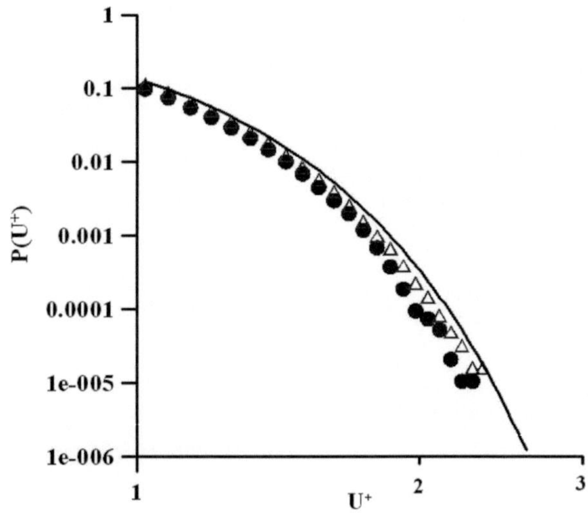

Fig. 9 Distribution functions of maximal velocities for incident wave (*triangles*) and shoreline displacement (*circles*) with the incident wave spectrum width $l = 0.7$; solid line correspondences to the Rayleigh distribution

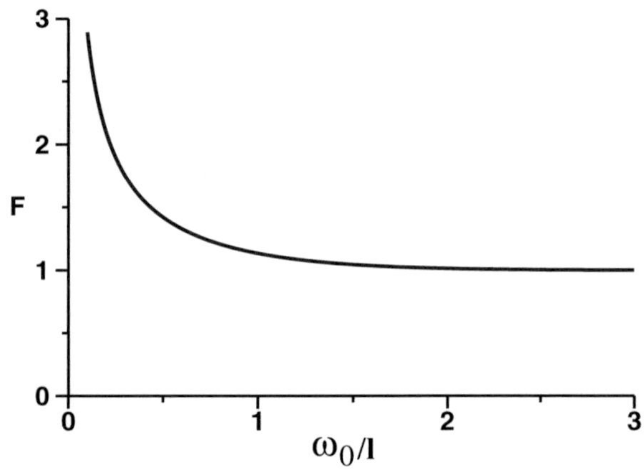

Fig. 10 Function of an influence of the incident wave spectrum width on a runup height of the wave

$$F(z) = \sqrt{1 + \frac{\exp(-z^2/2)}{\sqrt{\pi/2}z\mathrm{erfc}(-z/\sqrt{2})}}. \tag{26}$$

The function $F(z)$ is shown in Fig. 10. It tends to one ($F = 1$) for the narrowband process ($l \ll \omega_0$) and the significant runup height of the wave can be described by the formula for the runup of a sine wave (Didenkulova et al. 2007b). Significant runup height grows with the increasing of the spectrum width, especially when $l > \omega_0$. Thus, Gaussian approximation in a problem of the wave runup on a beach works not only for the case of $l \ll \omega_0$, but also for $l < \omega_0$, when the distribution function differs from Gaussian.

Previous analysis used the wave field presenting as the superposition of the independent spectral components. Such approach is very popular to describe random water waves. Meanwhile, the wave field in shallow water contains many coherent wave components, and an idea to present it as a random assembly of the solitary waves is very popular, see, for instance (Brocchini and Gentile 2001). The runup of solitary wave on a plane beach is well studied (Synolakis 1987) and the runup amplitude can be expressed through soliton amplitude

$$\frac{R}{h} = 2.8312\frac{1}{\sqrt{\alpha}}\left(\frac{A}{h}\right)^{5/4}. \tag{27}$$

In fact, this formula can be derived from (10) taking into account the relation between the soliton amplitude and duration. If the wave field contains random separated solitons, the runup of each individual soliton presents the independent random process and distribution function of runup amplitude can be found analytically if the distribution function of the soliton amplitudes is known. Assuming for simplicity the Rayleigh distribution for soliton amplitude and using (27), exceedance frequency of the runup amplitude is

$$P(R) = \exp\left[-0.378\alpha^{4/5}\frac{(R/h)^{8/5}}{(A/h)^2}\right], \tag{28}$$

and the probability of appearance of big waves on the coast is high. In fact, this formula is valid for independent solitons. More detailed computing of the statistical runup characteristics of the realistic "soliton" wave field is performed in (Brocchini and Gentile 2001).

So, the wave runup on a plane beach leads to increasing of the probability of the large-amplitude waves, and a freak wave phenomenon should be taken into account in the coastal protection.

4 Conclusion

Distribution functions of the maxima wave characteristics at the point of shoreline (displacement and velocity), caused by a wave coming from the open sea, are analyzed in frames of nonlinear shallow water theory. Modeled (Gaussian) spectrum is used for numerical simulations. It is shown that variations of distribution functions for the maximal shoreline displacement and shoreline velocity are weak for $l < \omega_0$. For this case, the significant runup height of the wave can be described by the formula for the runup of a sine wave. For the wideband process, especially for $l > \omega_0$, the significant runup height grows significantly.

Acknowledgments This study was supported by the basic part of the state contract No 2014/133 and RFBR grants (14-02-00983, 14-05-00092, 15-35-20563). Ira Didenkulova acknowledges grant MK-1146.2014.5. Anna Sergeeva thanks Volkswagen Foundation.

References

Brocchini M, Gentile R (2001) Modelling the run-up of significant wave groups. Cont Shelf Res 21:1533–1550

Carrier GF, Greenspan HP (1958) Water waves of finite amplitude on a sloping beach. J Fluid Mech 4:97–109

Carrier GF, Wu TT, Yeh H (2003) Tsunami run-up and draw-down on a plane beach. J Fluid Mech 475:79–99

Chien H, Kao C-C, Chuang LZH (2002) On the characteristics of observed coastal freak waves. Coast Eng J 44(4):301–319

Denissenko P, Didenkulova I, Rodin A, Listak M, Pelinovsky E (2013) Experimental statistics of long wave runup on a plane beach. J Coast Res SI 65:195–200

Didenkulova I (2009) New trends in the analytical theory of long sea wave runup. In: Quak E, Soomere T (eds) Applied wave mathematics: selected topics in solids, fluids, and mathematical methods. Springer, pp 265-296

Didenkulova I (2011) Shapes of freak waves in the coastal zone of the Baltic sea (Tallinn Bay). Boreal Environ Res 16(Suppl. A): 138-148

Didenkulova I, Pelinovsky E (2011) Rogue waves in nonlinear hyperbolic systems (shallow-water framework). Nonlinearity 24:R1–R18

Didenkulova II, Slunyaev AV, Pelinovsky EN, Charif Ch (2006a) Freak waves in 2005. Nat Hazards Earth Syst Sci 6:1007–1015

Didenkulova II, Zahibo N, Kurkin AA, Levin BV, Pelinovsky EN, Soomere T (2006b) Runup of nonlinearly deformed waves on a coast. Dokl Earth Sci 411(8):1241–1243

Didenkulova II, Kurkin AA, Pelinovsky EN (2007a) Run-up of solitary waves on slopes with different profiles. Izv Atmos Ocean Phys 43(3):384–390

Didenkulova I, Pelinovsky E, Soomere T, Zahibo N (2007b) Runup of nonlinear asymmetric waves on a plane beach. In: Kundu A (ed) Tsunami and nonlinear waves, pp 173-188

Dysthe K, Krogstad HE, Muller P (2008) Oceanic rogue waves. Annu Rev Fluid Mech 40:287–310

Kânoğlu U (2004) Nonlinear evolution and runup-rundown of long waves over a sloping beach. J Fluid Mech 513:363–372

Kânoğlu U, Synolakis C (2006) Initial value problem solution of nonlinear shallow water-wave equations. Phys Rev Lett 97:148501

Kharif Ch, Pelinovsky E (2003) Physical mechanisms of the rogue wave phenomenon. Eur J Mech / B-Fluid 22(6):603–634

Kharif Ch, Pelinovsky E, Slunyaev A (2009) Rogue waves in the ocean. Springer, Berlin

Massel SR (1996) Ocean surface waves: their physics and prediction. World Scientific, Singapore

Nikolkina I, Didenkulova I (2011) Rogue waves in 2006–2010. Nat Hazards Earth Syst Sci 11:2913–2924

Nikolkina I, Didenkulova I (2012) Catalogue of rogue waves reported in media in 2006–2010. Nat Hazards 61(3):989–1006

Olagnon M, Athanassoulis GA (eds) (2001) Rogue waves 2000. Ifremer, France

Pedersen G, Gjevik B (1983) Runup of solitary waves. J Fluid Mech 142:283–299

Pelinovsky E, Mazova R (1992) Exact analytical solutions of nonlinear problems of tsunami wave run-up on slopes with different profiles. Nat Hazards 6:227–249

Rosenthal W (2003) Rogue waves: forecast and impact on marine structures. GKSS Research Center, Geesthacht, Germany

Sand SE, Hansen NE, Klinting P, Gudmestad OT, Sterndorff MJ (1990) Freak wave kinematics. In: Torum A, Gudmestad OT (eds) Water wave kinematics. Kluwer, Dordrecht, pp 535–549

Sergeeva AV, Didenkulova II (2005) Runup of irregular waves on a plane beach. Izv Russ Acad Eng Sci 14:98–105

Slunyaev A, Didenkulova I, Pelinovsky E (2011) Rogue waters. Contemp Phys 52(6):571–590

Spielfogel LO (1976) Runup of single waves on a sloping beach. J Fluid Mech 74:685–694

Synolakis CE (1987) The runup of solitary waves. J Fluid Mech 185:523–545

Synolakis CE (1991) Tsunami runup on steep slopes: how good linear theory really is. Nat Hazards 4:221–234

Tadepalli S, Synolakis CE (1994) The Runup of N-waves. Proc R Soc Lond A445:99–112

Tinti S, Tonini R (2005) Analytical evolution of tsunamis induced by near-shore earthquakes on a constant-slope ocean. J Fluid Mech 535:33–64

Torum A, Gudmestad OT (eds) (1990) Water wave kinematics. Kluwer, Dordrecht

Numerical Study for Run-Up of Breaking Waves of Different Polarities on a Sloping Beach

Artem Rodin, Ira Didenkulova and Efim Pelinovsky

Abstract The transformation and run-up of long breaking bell-shaped wave pulses of various polarities is studied numerically in the nonlinear shallow-water theory framework using CLAWPACK software. The considered water basin contains a section of constant depth and a section of a slopping beach. For small-amplitude incident waves regardless their polarity, the results of numerical computations usually coincide with predictions of the nonlinear shallow-water theory for non-breaking waves. Nonlinear effects start to be important when incident wave is located far from the shoreline even for initially small-amplitude waves. With further increase in incident wave amplitude, the wave transforms into the shock wave (bore) before approaching the beach. Run-up characteristics of waves of different polarities are compared. Nonlinear effects and induced energy dissipation caused by wave breaking during its run-up on a beach are more prominent for negative pulses rather than for positive ones.

1 Introduction

Run-up of non-breaking waves on a plane beach is well studied analytically using the solutions of the nonlinear shallow-water equations. The progress in this direction started from the pioneer work of Carrier and Greenspan (1958) who solved nonlinear shallow-water equations for wave run-up on a plane beach using Legendre transformation. After their work, run-up of incident waves of various shapes on a

A. Rodin (✉) · I. Didenkulova · E. Pelinovsky
Nizhny Novgorod State Technical University n. a. R.E. Alekseev, Nizhny Novgorod, Russia
e-mail: artem@cens.ioc.ee

A. Rodin
Institute of Cybernetics at Tallinn University of Technology, Tallinn, Estonia

I. Didenkulova
Marine Systems Institute, Tallinn University of Technology, Tallinn, Estonia

E. Pelinovsky
Institute of Applied Physics, Nizhny Novgorod, Russia

E. Pelinovsky
Special Research Bureau for Automation of Marine Researches, Yuzhno-Sakhalinsk, Russia

E. Pelinovsky and C. Kharif (eds.), *Extreme Ocean Waves*,
DOI 10.1007/978-3-319-21575-4_9

plane beach has been investigated by many authors (Antuono and Brocchini 2007, 2008, 2010; Carrier et al. 2003; Didenkulova 2009; Didenkulova and Pelinovsky 2008; Didenkulova et al. 2006a, 2007a, b, 2008; Dobrokhotov and Tirozzi 2010; Kânoğlu 2004; Kânoğlu and Synolakis 2006; Madsen and Fuhrman 2008; Madsen and Schaffer 2010; Mazova et al. 1991; Pedersen and Gjevik 1983; Pelinovsky and Mazova 1992; Pritchard and Dickinson 2007; Shermeneva and Shugan 2006; Spielvogel 1975; Synolakis 1991; Synolakis et al. 1988; Synolakis 1987; Tadepalli and Synolakis 1994; Tinti and Tonini 2005). Recently, the same approach has been applied to the irregular wave field (Denissenko et al. 2011, 2013; Didenkulova and Pelinovsky 2011b; Didenkulova et al. 2010, 2011). Irregularity of the beach has been considered in Dutykh et al. (2011), while influence of bottom friction on run-up height has been studied in Bernatskiy and Nosov (2012). Possible resonance effects during wave run-up have been investigated numerically and experimentally in Ezersky et al. (2013a, b), Stefanakis et al. (2011). An attempt to include second dimension has been made by Choi et al. (2008), Didenkulova and Pelinovsky (2009), Didenkulova and Pelinovsky (2011a, b, c), Rybkin et al. (2014), Zahibo et al. (2006), who studied wave run-up in long and narrow bays.

We underline that most of papers cited above consider run-up of waves of positive polarity (crests) having in mind that these waves induce significant flooding of the coast. However, if the leading positive wave contains a precursor in the form of a negative pulse (trough), it leads to substantial increase in the run-up height. This effect was discussed in papers by Tadepalli and Synolakis (1994), Soloviev and Mazova (1994) and was entitled as N-wave effect. The N-wave effect is valid as for waves of weak amplitude (almost linear theory), as for large-amplitude non-breaking waves. However, it has been shown in Didenkulova et al. (2006b), Pelinovsky and Rodin (2011, 2012), Zahibo et al. (2008) for a basin of constant depth that different polarities of the incident wave result in different manifestations of nonlinear effects during wave propagation. The simple explanation for it is the following. The total water depth under the wave trough is always smaller than the one under the crest, and therefore, the nonlinear effects at the wave trough are always stronger than at the wave crest. This effect for the case of non-breaking wave run-up on a beach is discussed in Didenkulova et al. (2014), and in the current study, we extend the contrastive analysis of influence of wave polarity on run-up characteristics performed in Didenkulova et al. (2014) to the case of breaking waves.

The paper is organized as follows. The shallow-water mathematical model is briefly presented in Sect. 2. Run-up characteristics for positive and negative incident waves are described in Sects. 3 and 4, respectively. Contrastive analysis of run-up characteristics for incident wave of different polarities is performed in Sect. 5. Main results are summarized in Sect. 6.

2 Mathematical Model

Nonlinear shallow-water equations with a Manning bottom friction are written in the divergence form:

$$\frac{\partial(Hu)}{\partial t} + \frac{\partial}{\partial x}\left[Hu^2 + \frac{1}{2}gH^2\right] = +g\,H\,dh/dx - \gamma\,H\,u, \qquad \frac{\partial H}{\partial t} + \frac{\partial}{\partial x}[Hu] = 0, \tag{1}$$

$$\gamma = g\,n^2/H^{(4/3)}|u|,$$

where $H(x, t) = h(x) + \eta(x, t)$ is a total water depth, $\eta(x, t)$ is the elevation of water surface above the mean sea level $z = 0$, $u(x, t)$ is depth-averaged horizontal velocity of the water flow, g is the gravity acceleration, $h(x)$ is the unperturbed water depth and n is the Manning coefficient, taken as $n = 0.025\,s/m^{(-1/3)}$ in all our computations which is typical for geophysical problems.

Since nonlinear shallow-water equations are averaged equations over the water depth flow velocity $u(x, t)$, they do not allow description of crest overturning. However, written in the divergence form, Eq. (1) allows taking into account dissipation of wave energy at the front of the shock wave (bore). This assumption is used when the wavelength is large enough compared to the thickness of wave-overturning zone and often applied to the description of wave breaking in the framework of the shallow-water theory (Stoker 1957). All analytical and numerical results cited in Introduction have been obtained within this system of Eq. (1).

The composite geometry of the problem shown in Fig. 1 contains a 250-m-long basin of constant (3.5 m) water depth, which is matched with a plane beach of a slope 1:6. The parameters of the basin are selected in order to match the dimensions of the Large Wave Flume (GWK), Hannover, Germany, where the authors recently carried out a series of experiments on long-wave run-up (Denissenko et al. 2011, 2013; Didenkulova et al. 2013).

The system of Eq. (1) is solved numerically using the CLAWPACK software package (www.clawpack.org/) based on finite volume method (LeVeque 2004). In

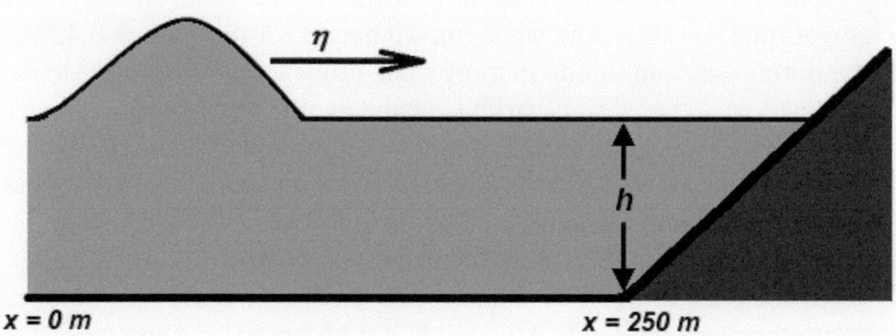

Fig. 1 The composite geometry of the problem

order to describe shock waves and prevent spurious oscillations in the solution which are result of the 2nd order corrections, the simulation algorithm applies numerical viscosity, which should not much affect run-up characteristics. The boundary condition at the left boundary of the computational domain ($x = 0$) corresponds to free wave propagation across the border. On the right boundary, the boundary condition $H(x, t) = 0$ defines the oscillations of the moving shoreline $x(t)$. In all calculations the spatial step of 0.1 m and time step were adapted automatically to satisfy the Courant's criterion.

Initial conditions correspond to the wave located in a basin of constant depth ($x < 250$ m) propagating onshore:

$$\eta_{in}(x, 0) = A \cosh^{-2}[(x - x_0)/L], \qquad u_{in}(x, 0) = 2\left[\sqrt{g(h + \eta_{in}(x, 0)} - \sqrt{gh}\right]. \tag{2}$$

It is easy to show analytically that the rigorous solution of the system (1) for initial conditions (2) represents Riemann wave (Pelinovsky and Rodin 2011, 2012):

$$H(x, t) = H_0\left[x - V(H)t\right], \qquad V = 3\sqrt{gH} - 2\sqrt{gh}, \tag{3}$$

where $H_0(x) = h + \eta_{in}(x, 0)$ is an initial wave shape.

In all our computations, characteristic half-wavelength is $L = 11$ m that corresponds to a half-wave period of 2 s. Incident pulse is located in the point $x_0 = 50$ m. Wave amplitude A is varied from 0.05 to 3.5 m for wave crests and from −0.05 to −3.49 m for troughs in order to keep the water layer continuous.

3 Run-Up of Breaking Waves of Positive Polarity on a Sloping Beach

In the first set of calculations, initial wave amplitude was rather small, so that the wave climbed the beach and reflected from it without breaking; see Fig. 2 for wave amplitude of 0.1 m. However, even when nonlinearity is weak, the nonlinear effects are still present and lead to wave steepening while it approaches the slope, which is clearly seen at time $t = 30$ s. The wave run-up height for this case is 0.43 m, which exceeds the initial wave amplitude in more than 4 times. The travel time to the coast is 40 s. Reflected wave has a sign-variable shape as it is predicted by the analytical theory. Amplitude of reflected wave is less than amplitude of incident wave (0.08 m at the time moment $t = 70$ s, Fig. 2) due to wave transformation and spreading in space. The reflected wave is also affected by the nonlinear effects and contains steep front at the maximum crest. It is important that reflected wave has a weak positive tail after the trough due to weak resonance between the coast and the point matching the slope with the constant depth. This tail is not observed in asymptotic analytical considerations by Didenkulova et al. (2006a, 2007b), where the mentioned resonance is neglected.

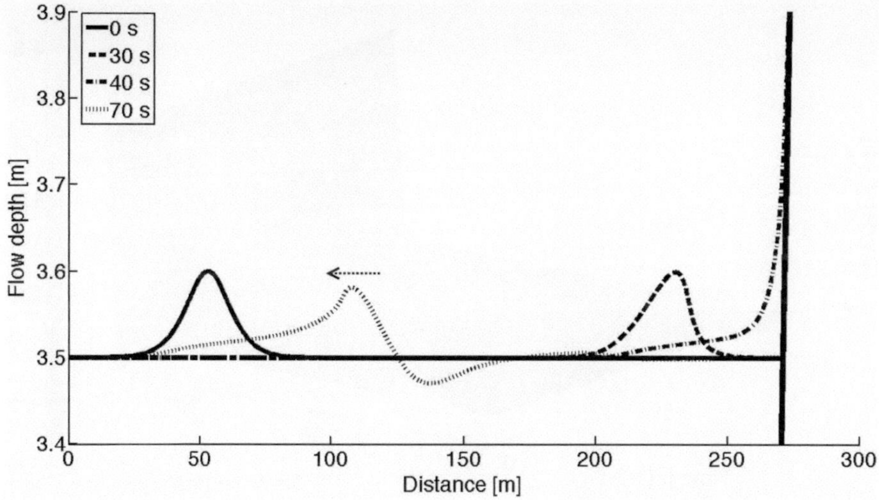

Fig. 2 Run-up and reflection of a weakly nonlinear positive pulse ($A = 0.1\,\mathrm{m}$) from a sloping beach

With the growth of wave amplitude, nonlinear effects become more visible. Pulse with amplitude $A = 0.5\,\mathrm{m}$ breaks before approaching the beach and transforms into the shock wave (Figs. 3 and 4). Such shock wave propagates faster than linear wave; hence, maximal run-up height is achieved at slightly earlier time $t = 37.5\,\mathrm{s}$. Its value is 1.44 m, and run-up ratio (R/A, where R is run-up height) is 2.88, which is less than in the previous non-breaking case. Decrease in the run-up ratio is related to the wave

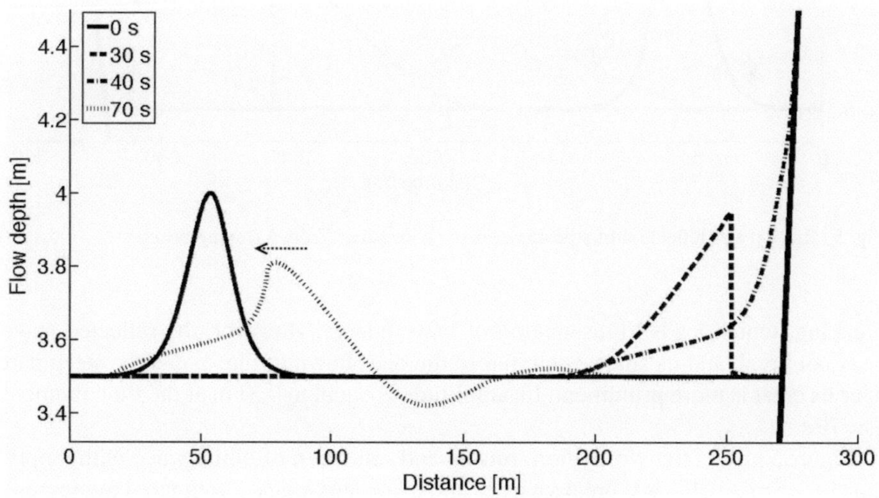

Fig. 3 Run-up and reflection of a positive pulse ($A = 0.5\,\mathrm{m}$) from a sloping beach

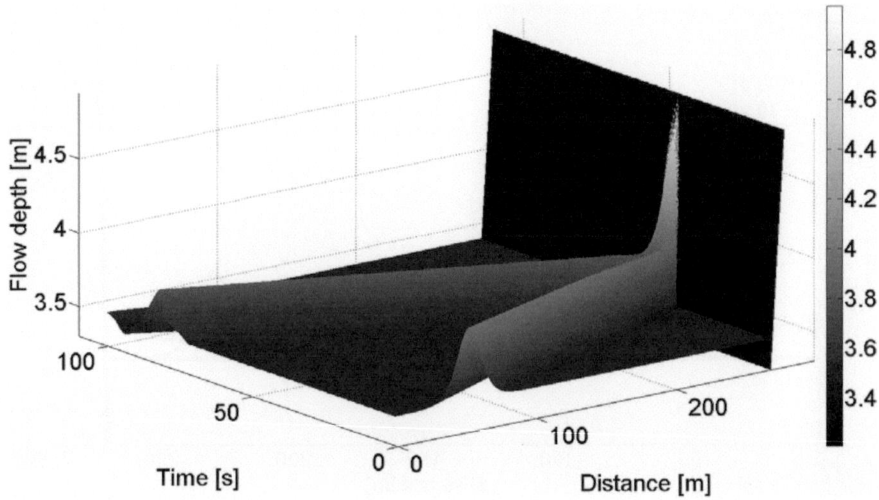

Fig. 4 Run-up and reflection of a positive pulse ($A = 0.5$ m) from a sloping beach: x-t diagram

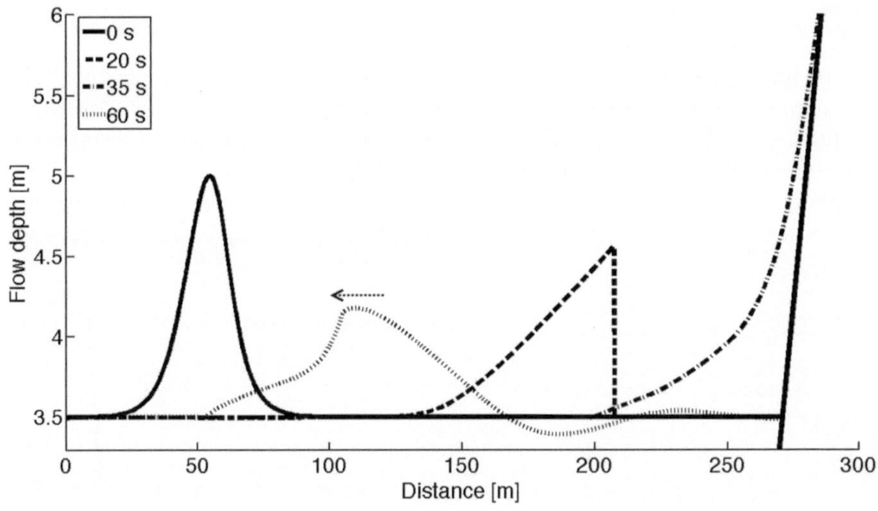

Fig. 5 Run-up and reflection of a positive pulse ($A = 1.5$ m) from a sloping beach

breaking, which leads to dissipation of wave energy. However, the reflected wave does not break and its shape is similar to the one shown in Fig. 2, just the steep part near its crest is more prominent. Its amplitude is equal to 0.31 m at the time moment $t = 70$ s.

Figure 5 shows transformation, run-up and reflection of initial wave with amplitude of 1.5 m. Such wave breaks earlier and dissipates quicker compared to previous cases. It also propagates faster than in the previous case, so that its maximal run-up

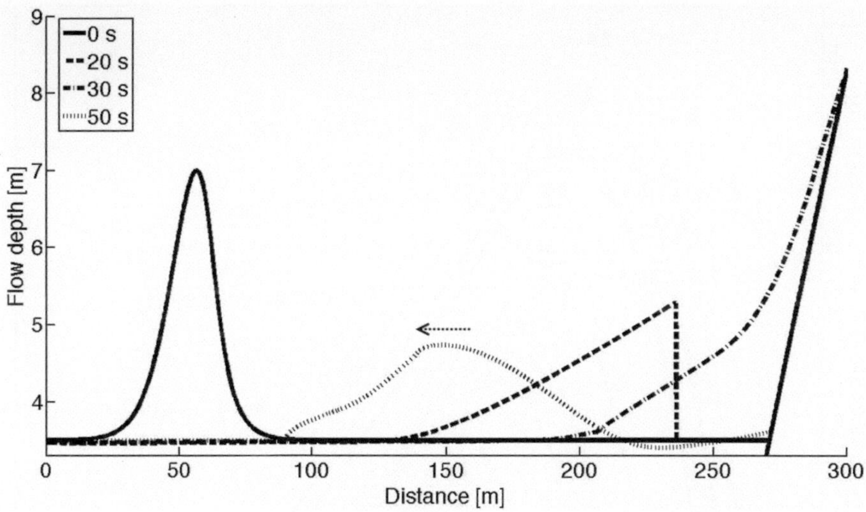

Fig. 6 Run-up and reflection of a positive pulse ($A = 3.5$ m) from a sloping beach

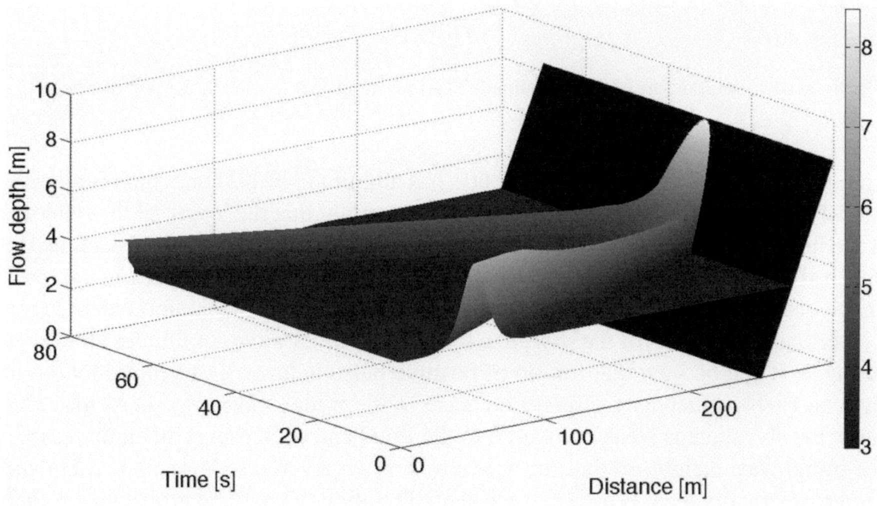

Fig. 7 Run-up and reflection of a positive pulse ($A = 3.5$ m) from a sloping beach: x-t diagram

height of 2.83 m is achieved at $t = 34.5$ s. Dissipation reduces the run-up ratio to 1.88. Reflected wave decays rapidly. Its amplitude is 0.67 m at $t = 70$ s, which is almost 45 % of the initial wave amplitude.

When the initial wave amplitude is extremely high and equal to the water depth (3.5 m), the wave breaks almost instantaneously, and its shape becomes triangular and its front moves quicker than in the linear case (Figs. 6 and 7). The wave reaches its maximal run-up height of 4.84 m in 30.5 s, and the run-up ratio decreases significantly

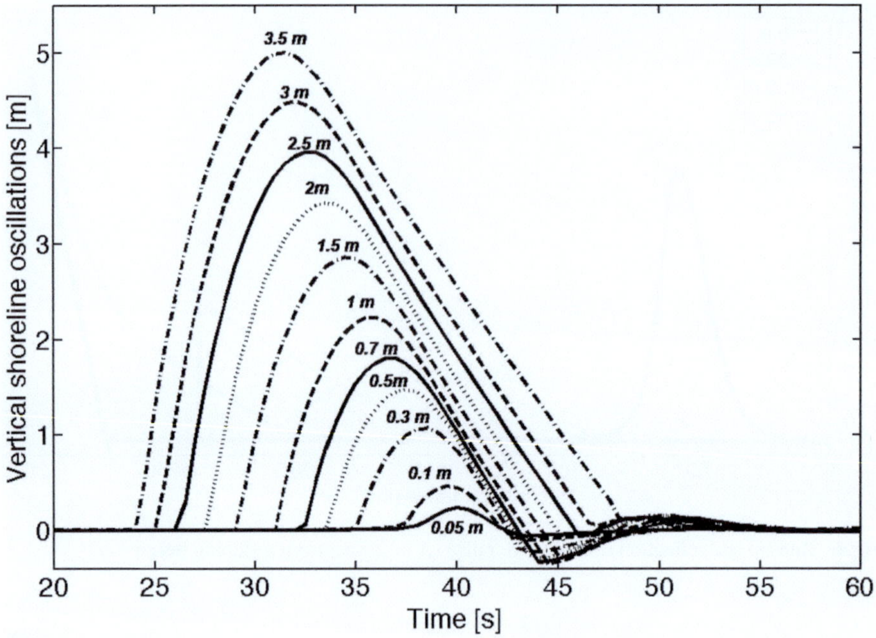

Fig. 8 Vertical oscillations of the shoreline for various amplitudes of the initial positive pulse

up to 1.38. The reflected wave at $t = 60$ s has amplitude of 1.24 m, which is approximately 35 % of initial wave amplitude. It is important that the length of the reflected wave increases with an increase in its amplitude, as the shock wave approaching the beach also has a longer wavelength than the incident wave. It also results in a smooth shape of a reflected wave in contrast to the large-amplitude breaking incident wave.

Vertical oscillations of the shoreline for various initial wave amplitudes are shown in Fig. 8. It can be seen that run-up of positive pulse induces significant flooding of the coast which increases with an increase in the initial wave height. At the same time, the ebb stage is weak compared to the flood and decreases with an increase in the initial wave height, so that for waves of very large amplitude, such as 2.5 m and larger, practically there is no ebb stage at all. Duration of the flood is also longer than for the ebb, and it increases with an increase in the initial wave amplitude.

Detailed analysis of run-up characteristics for both positive and negative pulses is performed below in Sect. 5. Here, we just show the comparison between results of numerical simulations of weakly nonlinear non-breaking wave run-up and predictions of the analytical theory (see Fig. 9), where the analytical solution is obtained following procedure described in Didenkulova et al. (2006a), Didenkulova (2009); Didenkulova et al. (2007b). Figure 9 shows a good agreement between numerical and analytical results.

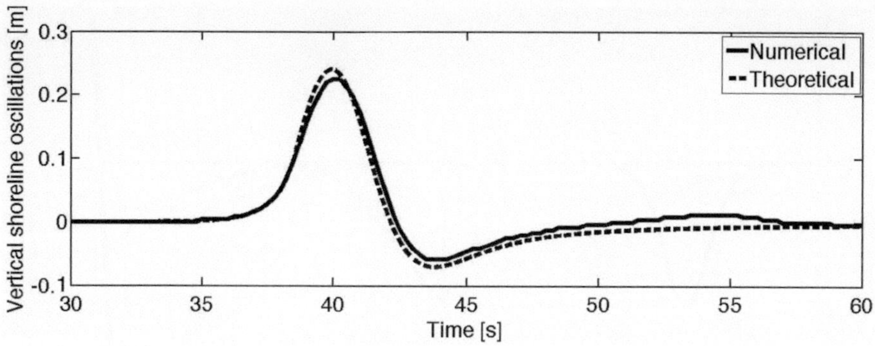

Fig. 9 Comparison of numerical simulations with predictions of the analytical theory for a 0.05 m weakly nonlinear wave of positive polarity

4 Run-Up of breaking Waves of Negative Polarity on a Sloping Beach

Now let us consider run-up of a wave of negative polarity (trough). Figure 10 demonstrates how the wave trough of small-amplitude (0.1 m) climbs the same sloping beach, described above. In general, this process is sign-inverted with respect to the run-up of positive pulse. Maximum run-up height of 0.17 m is approximately twice less than run-down amplitude (0.33 m), but it is higher than initial wave amplitude. The run-down ratio is equal to 3.3 which is less than the corresponding run-up ratio for a positive pulse of the same amplitude. The reflected wave is also inverted with

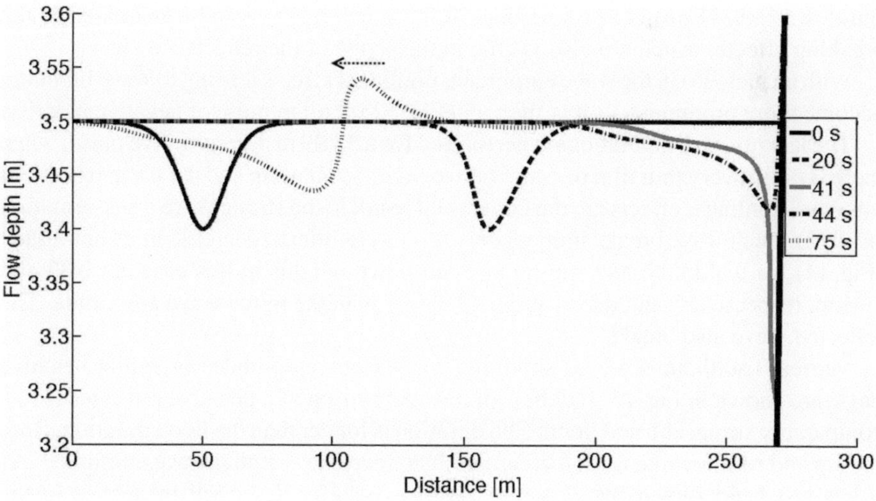

Fig. 10 Run-up and reflection of a negative pulse ($A = 0.1$ m) from a sloping beach

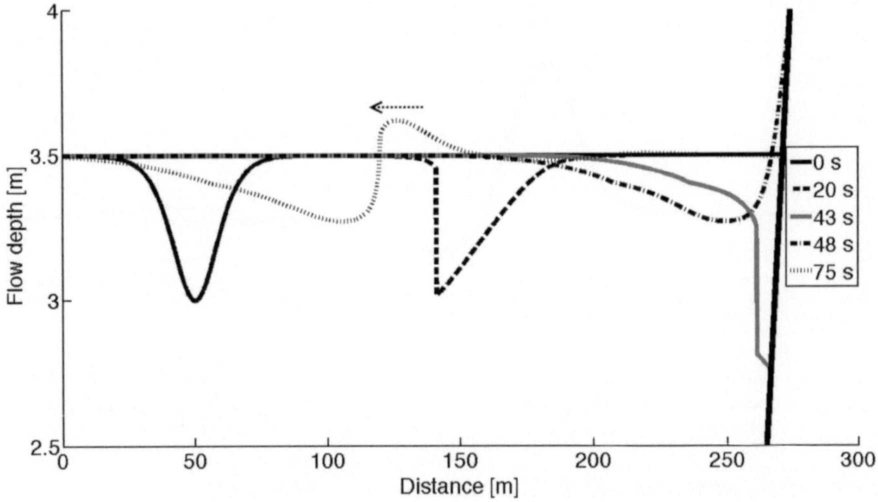

Fig. 11 Run-up and reflection of a negative pulse ($A = 0.5\,\text{m}$) from a sloping beach

respect to the positive amplitude case: it starts from the trough followed by the wave crest. However, due to its shape with large amplitude difference at the wave front, its steepness increases significantly as a result of nonlinear effects.

Figure 11 illustrates run-up of a negative pulse with amplitude of 0.5 m. Its maximum run-up and run-down heights are 0.53 and 0.83 m, respectively. It is important to mention that the time to the maximum run-up/run-down height has increased compared to the previous small-amplitude case. This is explained by the nonlinear speed of the trough, which is less than the linear speed of long-wave propagation. The run-down ratio (1.66) is also less than in the previous case, and it is related to the breaking effects, which are also visible in the shape of the reflected wave.

With an increase in the wave amplitude, nonlinear effects leading to wave breaking become more prominent, as it is shown in Fig. 12 for a 1 m pulse of negative polarity.

The last run of computations is performed for a 3.499 m deep negative pulse, when there is only a very thin film of water between the sea bottom and the wave trough. In this case, nonlinear effects are the largest and result in the strong wave transformation (Fig. 13). Such wave breaks immediately with a significant decrease in its amplitude (Fig. 14), so that maximum run-up and run-down heights in this case are 0.83 and 1.49 m, respectively, and this is about 40 % less than the initial wave amplitude. The reflected wave also breaks.

Vertical oscillations of the shoreline for various amplitudes of initial negative pulse are shown in Fig. 15. It demonstrates that run-up of a negative pulse leads to a comparably strong ebb and flood. Ebb duration is longer than the flood duration. Both run-up and run-down heights increase with an increase in initial wave amplitude. As it has just been shown for pulses of positive polarity described in Sect. 3, in the weak amplitude case, numerical results for non-breaking wave run-up are close to the predictions of the analytical theory (see Fig. 16).

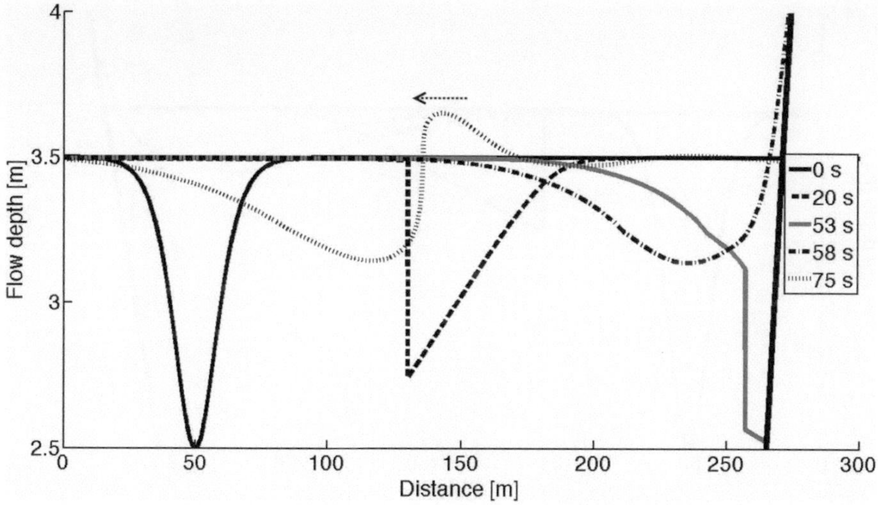

Fig. 12 Run-up and reflection of a negative pulse ($A = 1$ m) from a sloping beach

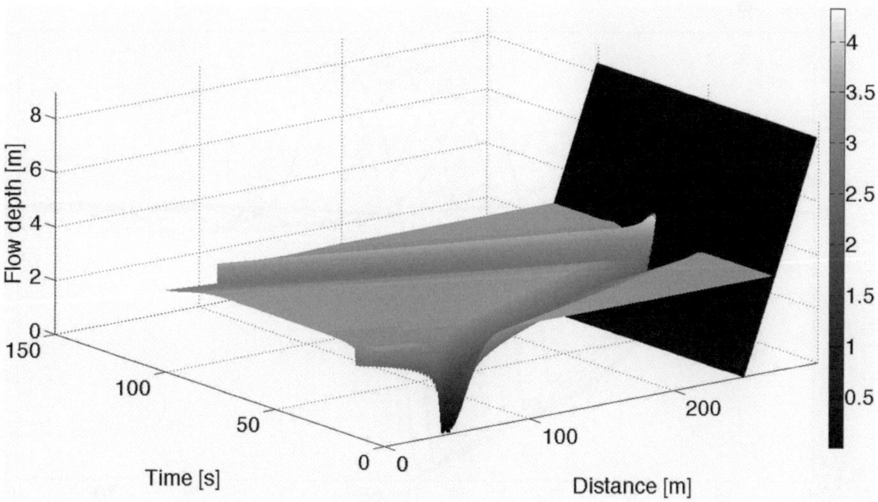

Fig. 13 Run-up and reflection of a negative pulse ($A = 3.499$ m) from a sloping beach: x-t diagram

5 Contrastive Analysis of Run-up Characteristics for Breaking Waves of Different Polarity

As it has been pointed out above, run-up and run-down heights increase with an increase in the initial wave amplitude. Figure 17 shows these values versus initial wave amplitude. It is clearly seen that all these functions are strongly nonlinear

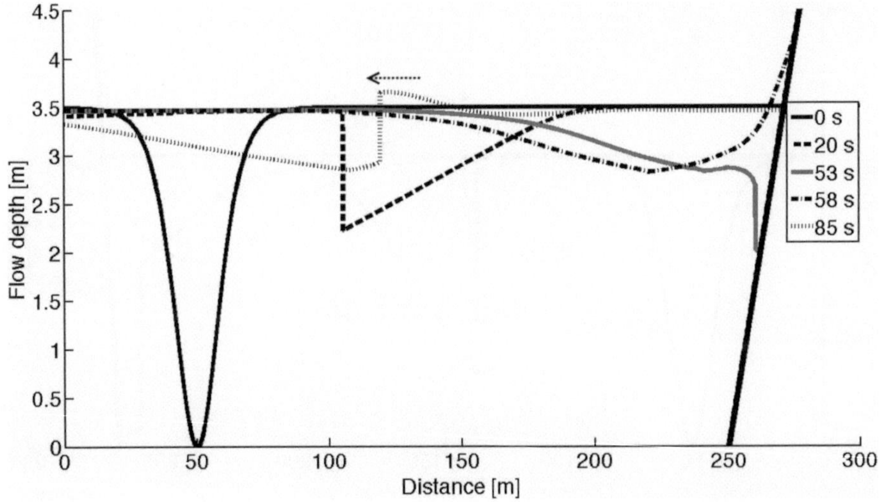

Fig. 14 Run-up and reflection of a negative pulse ($A = 3.499\,\mathrm{m}$) from a sloping beach

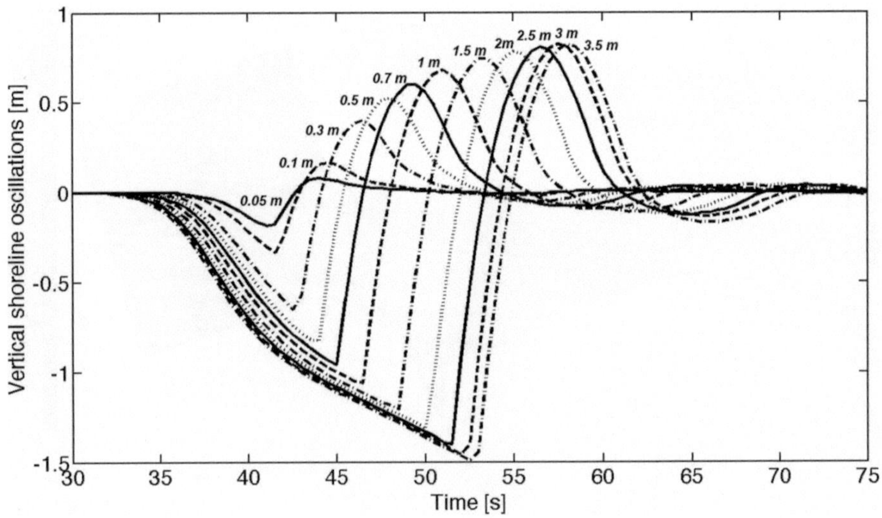

Fig. 15 Vertical oscillations of the shoreline for various amplitudes of the initial negative pulse

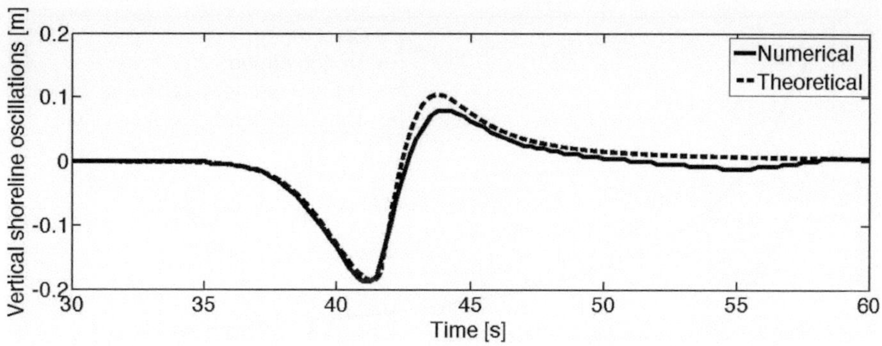

Fig. 16 Comparison of numerical simulations with predictions of the analytical theory for a 0.05 m weakly nonlinear wave of negative polarity

in contrast to the predictions of the analytical theory, which is related to energy dissipation due to wave breaking. For positive pulses, the run-up height (blue solid line) increases with an increase in the wave amplitude, while the run-down height (green dash line) behaves non-monotonically and tends to zero for large-amplitude waves. For the run-up of wave trough, as expected, the run-down height exceeds the run-up height, but these values are comparable. Similar effect was found also for non-breaking waves (Didenkulova et al. 2014). It is also seen that for large-amplitude negative waves, run-down and run-up heights tend to be constant.

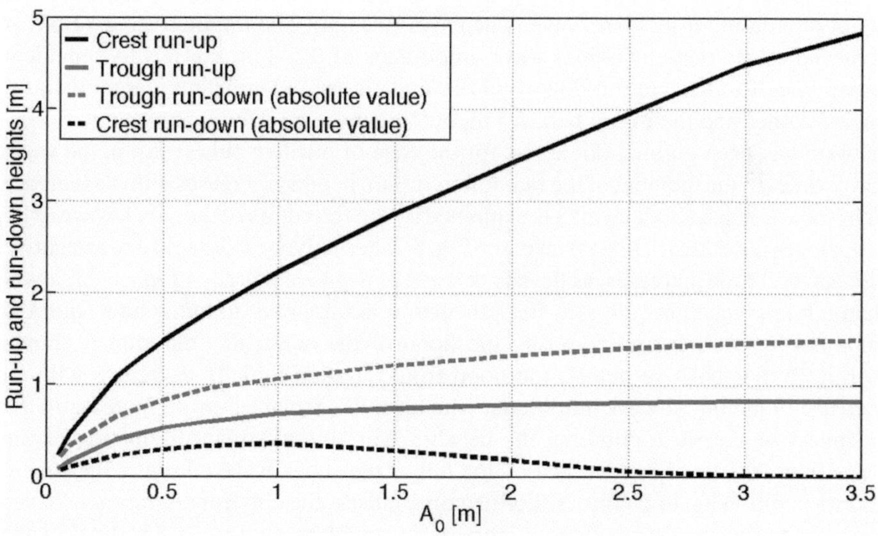

Fig. 17 Run-up and run-down heights versus initial wave amplitude

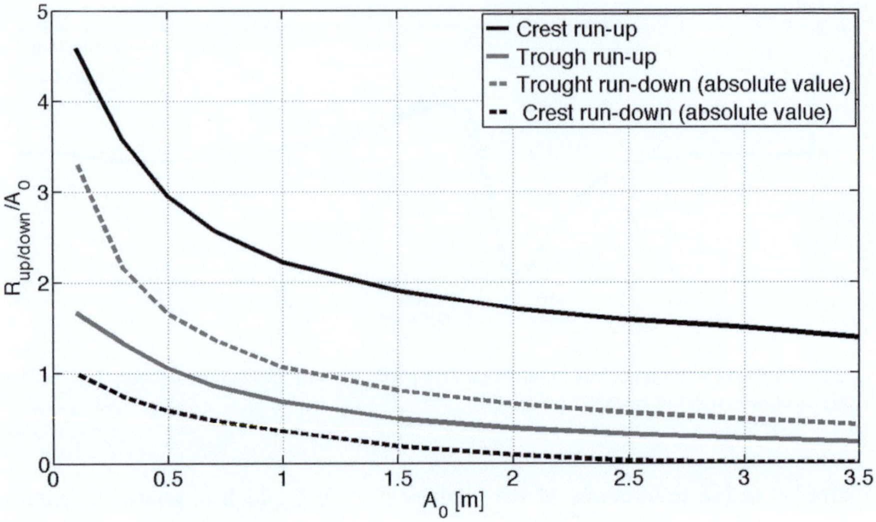

Fig. 18 Run-up and run-down ratios versus initial wave amplitude

Figure 18 displays run-up amplification or run-up ratio (the ratio of run-up or run-down heights to initial wave amplitude) versus initial amplitude. As it is expected, all ratios decrease monotonically with an increase in initial wave amplitude due to the wave breaking effects.

The ratio of run-down and run-up heights is demonstrated in Fig. 19. This function is not monotonic if incident wave is negative (trough). The minimal value (1.55) is achieved in the range of initial wave amplitudes of 0.5–1 m. For positive incident waves, the difference in run-down and run-up heights is significantly larger than for negative ones and their ratio behaves monotonically.

As it has been pointed out above for the case of positive pulse run-up, the wave travel time (to the moment of the maximum run-up height) decreases with an increase in initial wave amplitude, which is explained by the formation of the shock wave, who propagates faster than a linear wave; see Fig. 8. Alternatively, for the case of a negative pulse, travel time increases, as the shock wave of negative polarity propagates slower than a linear wave; see Fig. 15. For practice, it is important to know how soon the maximal inundation occurs during the flood of the coast; this duration is shown in Fig. 20. For positive pulses, the flood front duration initially decreases with an increase in the initial wave amplitude, which can be explained by an increase in the steepness of the wave climbing the beach. Then, due to strong dissipation of the wave front, the flood is governed by the tail of the shock wave climbing the beach, and its duration again increases. For negative pulses, the ebb front duration behaves monotonically and grows with an increase in initial amplitude, as the shock wave arrives later than a smooth negative part of approaching wave.

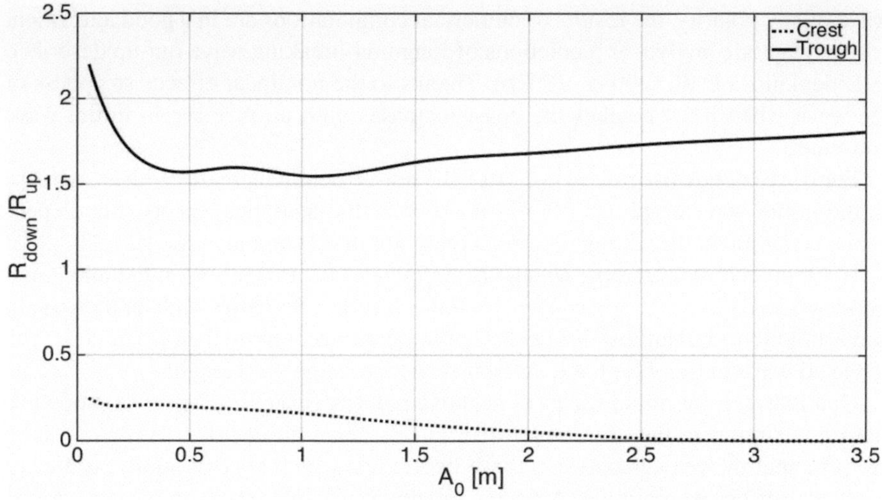

Fig. 19 Ratio of run-down and run-up heights R_{down}/R_{up} versus initial wave amplitude

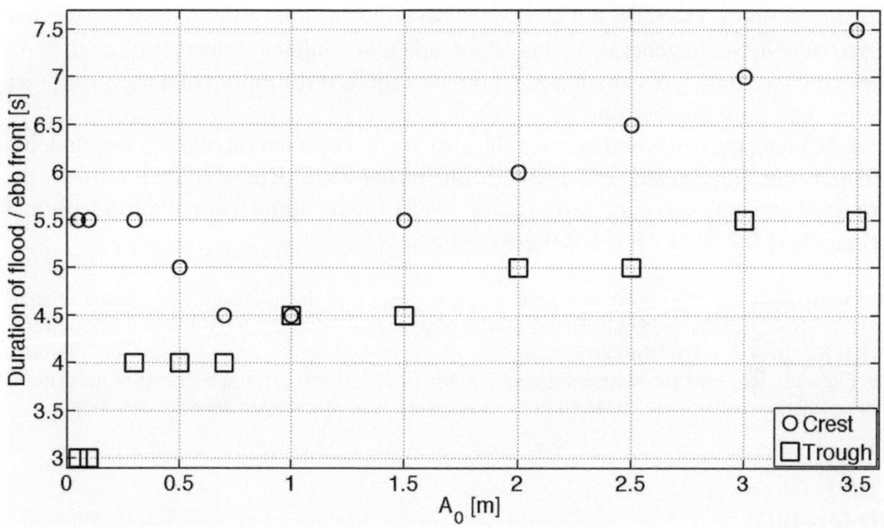

Fig. 20 Duration of flood / ebb front versus initial wave amplitude

6 Conclusion

Run-up of long breaking waves of different polarities is studied in the framework of the nonlinear shallow-water theory. The geometry of the basin consists of the section of constant depth matched with a sloping beach. Initial pulses of solitary shape are located far from the shoreline. In the case of weakly nonlinear incident wave of any polarity, the results of numerical computations are in a good agreement with asymptotic analytical predictions of long non-breaking wave run-up described in Didenkulova et al. (2006a, 2007b). Thanks to the nonlinear effects, steepness of the wave when it approaches the coast increases with an increase in initial wave amplitude.

For waves of positive polarity (crests), the run-up height increases with an increase in the initial wave amplitude, but in contrast to the analytical theory, this dependence is not linear. Breaking effects decrease amplitude and increase wavelength of waves approaching the slope, which also influences the travel time to the maximum flood and leads to its decrease. The ebb stage duration decreases with an increase in wave amplitude and almost disappears for large waves of more than 2.5 m high. The reflected wave in this case has a smooth shape and large wavelength.

Qualitatively, for initial waves of negative polarity (troughs), the process of wave run-up is similar to the one for positive pulses, but sign-inverted. However, there are principal differences between these two cases, which become more prominent for waves of large amplitude. After the breaking, negative wave propagates slower than the linear one, and the shock wave is formed at the back slope of the wave. It influences the maximal values of run-up and run-down heights and travel time to the maximum flood. The reflected wave can also break if initial amplitude of the wave is high enough. So, in general, we may conclude that nonlinear effects leading to wave breaking and energy dissipation are more pronounced for run-up of a negative pulse rather than for a positive one.

Of course, the discussed here results also depend on a length of the constant depth basin, beach slope, shape and length of the initial wave. However, qualitatively, the obtained dependence of the wave run-up height on the initial wave amplitude should remain also for other values of these parameters.

Acknowledgments The results presented in this paper are obtained with the support of State Contract No 2014/133 and grants RFBR (14-02-00983, 14-05-00092, 15-35-20563, 15-55-45053), MK-1146.2014.5 and SF0140007s11. Authors also acknowledge the support from CENS through the European Regional Development Fund (ERDF). Some aspects of the appearance of the extreme run-up characteristics are considered in the framework of a Volkswagen grant.

References

Antuono M, Brocchini M (2007) The boundary value problem for the nonlinear shallow water equation. Stud Appl Math 119:71–91

Antuono M, Brocchini M (2008) Maximum run-up, breaking conditions and dynamical forces in the swash zone: a boundary value approach. Coast Eng 55:732–740

Antuono M, Brocchini M (2010) Solving the nonlinear shallow-water equations in physical space. J Fluid Mech 643:207–232

Bernatskiy A, Nosov M (2012) The role of bottom friction in models of nonbreaking Tsunami wave runup on the shore. Izv Atmos Ocean Phys 48:427–431

Carrier GF, Greenspan HP (1958) Water waves of finite amplitude on a sloping beach. J Fluid Mech 4:97–109

Carrier GF, Wu TT, Yeh H (2003) Tsunami run-up and draw-down on a plane beach. J Fluid Mech 475:79–99

Choi BH, Pelinovsky E, Kim DC, Didenkulova I (2008) Two- and three-dimensional computation of solitary wave runup on non-plane beach. Nonlinear Process Geophys 15:489–502

Denissenko P, Didenkulova I, Pelinovsky E, Pearson J (2011) Influence of the nonlinearity on statistical characteristics of long wave runup. Nonlinear Process Geophys 18:967–975

Denissenko P, Didenkulova I, Rodin A, Listak M, Pelinovsky E (2013) Experimental statistics of long wave runup on a plane beach. J Coast Res SI65:195–200

Didenkulova I (2009) New trends in the analytical theory of long sea wave runup. In: Quak E, Soomere T (eds) Applied wave mathematics: selected topics in solids, fluids, and mathematical methods. Springer, Berlin, pp 265–296

Didenkulova I, Pelinovsky EN (2008) Run-up of long waves on a beach: the influence of the incident wave form. Oceanology 48:1–6

Didenkulova I, Pelinovsky E (2009) Non-dispersive traveling waves in inclined shallow water channels. Phys Lett A 373(42):3883–3887

Didenkulova I, Pelinovsky E (2011a) Nonlinear wave evolution and runup in an inclined channel of a parabolic cross-section. Phys Fluids 23(8):086602

Didenkulova I, Pelinovsky E (2011b) Rogue waves in nonlinear hyperbolic systems (shallow-water framework). Nonlinearity 24:R1–R18

Didenkulova I, Pelinovsky E (2011c) Runup of tsunami waves in U—shaped bays. Pure Appl Geophys 168:1239–1249

Didenkulova I, Zahibo N, Kurkin AA, Levin BV, Pelinovsky EN, Soomere T (2006a) Runup of nonlinearly deformed waves on a coast. Dokl Earth Sci 411:1241–1243

Didenkulova I, Zahibo N, Kurkin A, Pelinovsky EN (2006b) Steepness and spectrum of a nonlinearly deformed wave on shallow waters. Izv Atmos Ocean Phys 42:773–776

Didenkulova I, Kurkin A, Pelinovsky E (2007a) Run-up of solitary waves on slopes with different profiles. Izv Atmos Ocean Phys 43:384–390

Didenkulova I, Pelinovsky E, Soomere T, and Zahibo N (2007b). In: Kundu A (ed) Runup of nonlinear asymmetric waves on a plane beach. Tsunami and nonlinear waves, Springer, Heidelberg, pp 175–190

Didenkulova I, Pelinovsky E, Soomere T (2008) Run-up characteristics of tsunami waves of "unknown" shapes. Pure Appl Geophys 165:2249–2264

Didenkulova II, Sergeeva AV, Pelinovsky EN, Gurbatov SN (2010) Statistical estimates of characteristics of long-wave run-up on a beach. Izv Atmos Ocean Phys 46:530–532

Didenkulova I, Pelinovsky E, Sergeeva A (2011) Statistical characteristics of long waves nearshore. Coast Eng 58:94–202

Didenkulova I, Denissenko P, Rodin A, Pelinovsky E (2013) Effects of wave asymmetry on its runup on a beach. J Coast Res SI 65:207–212

Didenkulova II, Pelinovsky EN, Didenkulov OI (2014) Run-up of long solitary waves of different polarities on a plane beach. Izv Atmos Ocean Phys 50(5):532–538

Dobrokhotov SY, Tirozzi B (2010) Localized solutions of one-dimensional non-linear shallow-water equations with velocity c = (x)1/2. Russ Math Surv 65:177–179

Dutykh D, Labart C, Mitsotakis D (2011) Long wave runup on random beaches. Phys Rev Lett 107:184504

Ezersky A, Abcha N, Pelinovsky E (2013a) Physical simulation of resonant wave run-up on a beach. Nonlinear Process Geophys 20:35–40

Ezersky A, Tiguercha D, Pelinovsky E (2013b) Resonance phenomena at the long wave run-up on the coast. Nat Hazards Earth Syst Sci 13:2745–2752

Kânoğlu U (2004) Nonlinear evolution and runup-drawdown of long waves over a sloping beach. J Fluid Mech 513:363–372

Kânoğlu U, Synolakis C (2006) Initial value problem solution of nonlinear shallow water-wave equations. Phys Rev Lett 97:148501

LeVeque RJ (2004) Finite volume methods for hyperbolic problems. Cambridge University Press, Cambridge
Madsen PA, Fuhrman DR (2008) Run-up of tsunamis and periodic long waves in terms of surf-similarity. Coast Eng 55:209–223
Madsen PA, Schaffer HA (2010) Analytical solutions for tsunami run-up on a plane beach: single waves, N-waves and transient waves. J Fluid Mech 645:27–57
Mazova RK, Osipenko NN, Pelinovsky EN (1991) Solitary wave climbing a beach without breaking. Rozpr Hydrotech 54:71–80
Pedersen G, Gjevik B (1983) Runup of solitary waves. J Fluid Mech 142:283–299
Pelinovsky E, Mazova R (1992) Exact analytical solutions of nonlinear problems of tsunami wave run-up on slopes with different profiles. Nat Hazards 6:227–249
Pelinovsky EN, Rodin AA (2011) Nonlinear deformation of a large-amplitude wave on shallow water. Dokl Phys 56:305–308
PelinovskyE N, Rodin AA (2012) Transformation of a strongly nonlinear wave in a shallow-water basin. Izv Atmos Ocean Phys 48:383–390
Pritchard D, Dickinson L (2007) The near-shore behaviour of shallow-water waves with localized initial conditions. J Fluid Mech 591:413–436
Rybkin A, Pelinovsky E, Didenkulova I (2014) Nonlinear wave run-up in bays of arbitrary cross-section: generalization of the Carrier-Greenspan approach. J Fluid Mech 748:416–432
Shermeneva M, Shugan I (2006) The computations of the wave runup on gentle beach. Lett J Tech Phys 32:33–38
Soloviev SL, Mazova RK (1994) On the influence of sign of leading tsunami wave on run-up height on the coast. Sci Tsunami Hazards 12:2531
Spielvogel LO (1975) Runup of single waves on a sloping beach. J Fluid Mech 74:685–694
Stefanakis TS, Dias F, Dutykh D (2011) Local run-up amplification by resonant wave interactions. Phys Rev Lett 107:124502
Stoker JJ (1957) Water waves: mathematical theory with applications. Interscience Publishers, New York
Synolakis CE (1991) Tsunami runup on steep slopes: how good linear theory really is? Nat Hazards 4:221–234
Synolakis CE, Deb MK, Skjelbreia JE (1988) The anomalous behavior of the run-up of cnoidal waves. Phys Fluids 31:3–5
Synolakis CE (1987) The runup of solitary waves. J Fluid Mech 185:523–545
Tadepalli S, Synolakis CE (1994) The runup of N-waves. Proc R Soc Lond A445:99–112
Tinti S, Tonini R (2005) Analytical evolution of Tsunamis induced by near-shore earthquakes on a constant-slope ocean. J Fluid Mech 535:33–64
Zahibo N, Pelinovsky E, Golinko V, Osipenko N (2006) Tsunami wave runup on coasts of narrow bays. Int J Fluid Mech Res 33:106–118
Zahibo N, Didenkulova I, Kurkin A, Pelinovsky E (2008) Steepness and spectrum of nonlinear deformed shallow water wave. Ocean Eng 35:47–52

Tsunami Waves Generated by Cliff Collapse: Comparison Between Experiments and Triphasic Simulations

S. Viroulet, A. Sauret, O. Kimmoun and C. Kharif

Abstract Although many tsunamis arise from underwater earthquakes, some are induced by submarine and subaerial landslides. For example, the collapse of an unstable cliff into the sea can generate a tsunami wave near the coast and exhibit extreme run-up. As a result, those tsunamis develop significant hazards to the population. The threat caused by such tsunamis strongly depends on the topography and location of the unstable cliff. Predicting the height of the tsunami wave generated by a subaerial or submarine collapse requires experimental investigations, analytical modeling, and numerical simulations. In this chapter, we discuss recent experimental results on granular collapse in water and the influence of the physical parameters on the amplitude of the tsunami waves. We also present triphasic numerical simulations based on a finite-volume method where different rheologies are considered for the slide. We also emphasize the difficulty of such simulations and compare the numerical results to experimental measurements.

1 Introduction

A tsunami is an ocean wave that can propagate over long distances and can lead to considerable damage along the coast. Among all natural disasters, tsunamis can

S. Viroulet (✉)
School of Mathematics and Manchester Centre for Nonlinear Dynamics, University of Manchester, Manchester, UK
e-mail: sylvain.viroulet@manchester.ac.uk

A. Sauret
Surface du Verre Et Interfaces, UMR 125 CNRS/Saint-Gobain, Aubervilliers, France
e-mail: alban.sauret@saint-gobain.com

O. Kimmoun · C. Kharif
Aix-Marseille Université, CNRS, Ecole Centrale Marseille, IRPHE UMR 7342, Marseille, France
e-mail: olivier.kimmoun@centrale-marseille.fr

C. Kharif
IRPHE, Ecole Centrale Marseille, 49 rue F. Joliot-Curie, 13384 Marseille cedex 13, France
e-mail: kharif@irphe.univ-mrs.fr

E. Pelinovsky and C. Kharif (eds.), *Extreme Ocean Waves*,
DOI 10.1007/978-3-319-21575-4_10

be one of the deadliest. Indeed, if the wave amplitude reaches 10 m in height when arriving near the coast, the subsequent flooding can devastate land over several kilometers.

The most common mechanism responsible for tsunami generation is a submarine earthquake (Ioualalen et al. 2007), but other geological events can also cause tsunamis. For instance, the impact of asteroids (Gisler et al. 2004; Kharif and Pelinovsky 2005), volcanic eruptions (Francis 1985; Choi et al. 2003), and landslides result in tsunamis that lead to important local hazards. Because the wavelengths associated with tsunamis generated by a landslide are significantly lower than those from tsunamis induced by a submarine earthquake, the dispersive effects are more important during the propagation of the former (Glimsdal et al. 2013). For this reason, it is common to call the waves generated by a landslide “impulse waves” rather than tsunamis. However, despite the larger dissipation, the amplitude of those impulse waves depends directly on the volume and configuration of the landslide, and can become much larger than tsunami wave generated by submarine earthquake (Fritz 2001).

Over the past twenty years, various studies have been performed to characterize the mechanisms involved in the generation of tsunamis by landslides. In some particular cases, the propagation of the generated waves can be described by the Saint-Venant equations (Popinet 2011) and in a more general framework, the Boussinesq equations are able to capture the propagation of an initial wave (Shi et al. 2011, 2012). Although the propagation is now well described, the initial generation of the waves remains poorly understood because of its complexity. Many experimental studies have characterized waves generated by a landslide. The simplest approach used a solid block to model the landslide (Scott 1845). The mathematical description was obtained fifty years later by Korteweg and De Vries (1895). To obtain a more realistic situation, a number of studies have later considered a solid triangular block, with a width equal to that of the channel, that slides on a slope and impacts into water (Heinrich 1992; Watts 1997, 1998, 2000). These studies quantified the amount of energy transferred from the solid block to the waves and derived scaling laws to estimate the amplitude of the generated waves. Walder et al. (2003) studied the tsunami generation by a solid block sliding in air on a varying slope allowing for a smooth transition between the slope and the bottom of the channel and avoid the impact between the solid block and the bottom of the channel.

Large-scale experiments were carried out by Liu et al. (2005) to study the run-up and rundown of waves generated by a 3D slide. This study showed that the run-up is significantly larger for subaerial slides and that it is controlled by the initial size and position of the slide. In parallel, Panizzo et al. (2005) have performed similar 3D experiments to predict the amplitude of the generated waves for the same parameters as Walder et al. (2003). Experiments at the 1:500 scale were performed by Sælevik et al. (2009) to estimate the amplitude of waves generated by the collapse of a cliff in the Akneset fjord in Norway. This study examined the influence of the length of the slide on the generated wave train. They used several solid blocks connected to each other, and observed that the total volume of the slide is the predominant parameter that controls the amplitude of the first wave, whereas the total length influences the dispersive tail. They showed that the shorter the solid assembly, the larger the wave train following the first wave. Other configurations have also been studied to characterize the generation of tsunamis by a landslide. They consist in approximating the slide by a Gaussian or elliptical shape instead of a triangular or trapezoidal block (Grilli and Watts 2005; Enet et al. 2003, 2005, 2007). These experiments constitute validating cases for many numerical simulations of the generation of tsunamis by submarine landslides. Risio et al. (2009) considered the same geometry as Enet et al. (2007), but in order to study tsunami generation by a subaerial landslide. Contrary to previous studies, they focused on the water rise along the coast perpendicularly to the slide. They observed that the maximum run-up does not take place near the slide but at 2–3 times from its width.

Whereas the generation of tsunamis by a solid has been extensively studied since the 1970s, the generation of tsunamis by a granular slide is more recent and offers a more realistic approach. During the collapse of the slide, modeling of the interactions between the granular material and the generated wave is of crucial importance to estimate the initial amplitude of the tsunami. Various experiments using granular materials were carried out by Fritz (2002), Fritz (2001); Fritz et al. (2003a, b, 2004, 2009). These bi-dimensional experiments were performed in a $11\text{m} \times 0.5\text{m} \times 1\text{m}$ channel and the granular medium was initially accelerated using a pneumatic piston. Using linear regression, the authors determined the main characteristics of the generated waves based on several dimensionless parameters such as the Froude number $\mathrm{Fr} = \mathrm{V}/\sqrt{\mathrm{g\,H}}$ (where V is the velocity of the slide, H is the water depth, and g is the gravitational acceleration). These experimental studies were followed by the experiments of Zweifel et al. (2006) that studied the effects of the density of the granular material. Heller et al. (2008) considered the scale effects of tsunamis generated by landslides. They found that the interfacial tension and viscous effects decrease the amplitude of the generated wave but only have a small influence on the speed of their propagation. More recently, Mohammed (2010) and Mohammed and Fritz (2012) performed large-scale experiments on tsunami generation by a 3D sliding granular material. The granular mass used in this study reaches a value of 1.4 tons and the Froude number at the impact varied between 1 and 4. They were able to obtain the evolution of the amplitude taking into account the direction of propagation of the wave compared to the direction of collapse.

Several analytical studies considered the influence of different parameters in tsunami generation by landslides. Murty (1979) calculated the amplitude of the wave generated by a submarine landslide assuming that the potential energy of the slide is entirely transferred to the solitary wave. This assumption of total transfer of energy was also used by Monaghan and Kos (2000) to predict the elevation of the free surface in the historical experiment of John Scott Russell. Later, Didenkulova et al. (2010) studied the generation and propagation of tsunami induced by a deformable landslide on a varying topography. They found that amplitude of the generated wave does not vary monotonically with the distance. Wang et al. (2011) suggested a formulation for the generation of tsunamis by strong submarine landslide, taking into account the effects of the wave, the viscous friction with the fluid and the solid friction with the bottom of the sliding solid. More recently, scaling laws predicting the evolution of a wave have been obtained by Viroulet et al. (2013) for a solid slide and by Heller and Hager (2014) for a deformable slide.

Depth-averaged models have been widely used in numerical simulations of the tsunami propagation (Tinti and Bortolucci 2000). However, because these models neglect the vertical acceleration, they are not well suited to capture the generation of the initial wave by a landslide. Assier-Rzadkieaicz et al. (1997) used experimental results obtained with granular material to develop a simulation of the wave generation by a submarine landslide. Their code solved the Navier-Stokes equations by using a Bingham rheology to model the flow of granular material. Using the same rheology, Watts et al. (2003) simulated wave generated by submarine landslide and showed that the non-linearity and dispersion of the generated waves need to be taken into account in numerical models. These results have led to a numerical code, *Geowave*, which is the combination between two open-source codes: TOPICS, *Tsunami Open and Progressive Initial Condition Systems* Watts et al. (2003), that gives the initial conditions, and FUNWAVE, that solves the nonlinear Boussinesq equations for the propagation (Shi et al. 2011). Another numerical method particularly used to study tsunamis relies on the potential method, known as the *Boundary Elements Method* (BEM). Grilli and Watts (1999) developed a 2D code (2D-FNPF) to study the generation by a strong submarine landslide with a semi-elliptical shape. A 3D extension of this code was also performed by Grilli et al. (2002). However, depending on the initial depth of the submarine landslide, and in the case of a subaerial landslide, a strong vorticity is generated during the collapse and the possible fragmentation of the free surface can not be simulated using this method.

To take into account the interactions between the slide and the free surface, multi-phase codes have recently been developed. Kelfoun et al. (2010) modeled the tsunami generation on the island of "La Réunion" by solving the Saint-Venant equations for both, the fluid and the slide with a Mohr-Coulomb friction law for the rheology of the slide. Other numerical codes have been developed by solving the Navier-Stokes equations for the three phases air/water/slide. Mader and Gitting (2002) used the SAGE code that relies on an adaptive mesh to solve the Navier-Stokes equations with a Godunov scheme. Quecedo et al. (2004) modeled the generation by a deformable landslide using a level-set method and a viscoplastic rheology.

More recently, Abadie et al. (2010) used a finite volume method to solve the Navier-Stokes equation and a Volume of Fluid method (VoF) to track the interfaces. Several types of rheology can be implemented in this code (THETIS). The authors applied this method to study the possible collapse of the Cumbre Vieja, a volcano located in the Canary Islands (Abadie et al. 2012). They coupled the multi-phasic tsunami generation, simulated with the code THETIS, with the propagation mechanism, simulated with the code FUNWAVE. This numerical study highlights the importance of the 3D effects during both the generation and the propagation phases. Their study showed that 2D numerical simulations largely overestimate the amplitude of the wave.

These different studies illustrate that a good modeling of the couplings between the slide and the wave is crucial to predict the hazards induced by landslide generated tsunamis with the most realistic model involving a granular medium to reproduce the

Fig. 1 A view of the cliff at Cap Canaille located just above the mediterranean sea in Cassis, France (picture: S. Viroulet)

deformable slide. However, the interactions of the landslide with the slope have not been considered, as most of the past studies considered situations at Froude number $Fr > 1$, where the friction with the slope can be neglected. In particular, during the collapse of a cliff initially located just above the water surface, the velocity of the landslide is small compared to the speed of propagation of the generated wave and the interactions need to be characterized. This situation is observed for instance in Cassis (France) where the possible collapse of a cliff of more than 350 m high would occur directly above the water as illustrated in Fig. 1 (Averbukh et al. 2013).

2 Experimental Investigations

2.1 Experimental Setup

We have performed systematic experiments to study the interactions between dry granular collapse and water that generates impulse waves. An aquarium tank of dimension 220 cm in length, 40 cm in height and 20 cm in width was built so that it is long enough to allow the propagation of the wave while also allowing the observation of the interactions between granular collapse and the generated waves.

To characterize the influence of the granular medium on both the generated wave and the granular collapse, we have used four different media of varying size, polydispersity, and sphericity. We considered three sizes of spherical glass beads (1.5, 4, and 10 mm) and density $\rho_b = 2500$ kg.m^{-3}. We also used aquarium sand with an average diameter of 4 mm and density of $\rho_s = 2300$ kg.m^{-3} that models a non-spherical and polydisperse medium. The reservoir of granular material was located on one side of the aquarium and was limited by a door that opens vertically. The velocity of the door was fast enough, around 2 m.s^{-1}, to assume that the shape of the granular heap was not modified before the collapse. The granular material flows subsequently down the slope, where the same granular material has been glued to the surface to ensure a no-slip boundary condition (Pouliquen 1999; Cassar et al. 2005).

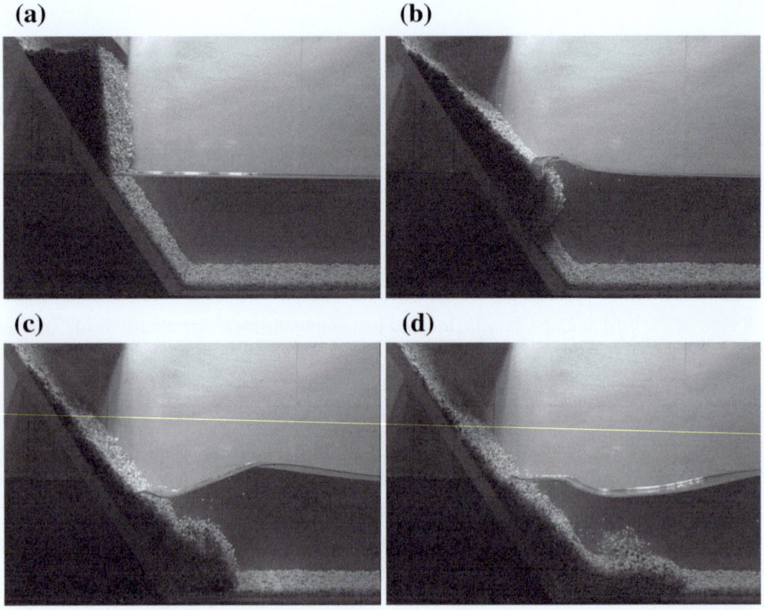

Fig. 2 Granular collapse of $m = 2$ kg of aquarium sand from air into water. The time step between two pictures is 0.2 s

The evolution of the granular collapse and the generation of the wave were recorded using a high speed camera (Vision Research Phantom V641) at a rate of 100 frames per second. In addition, we measured the evolution of the amplitude of the waves generated during the propagation using 4 resistive gauges placed at 0.45, 0.75, 1.05, and 1.35 m away from the vertical door (Viroulet et al. 2013a, b). An example of the results obtained with the high speed camera is shown in Fig. 2 for a collapse of aquarium sand.

2.2 Evolution of Amplitude During the Propagation of the Wave

We first focus on the propagation of the leading wave, which is followed by an oscillatory wave train. Because the wave generation is a complex phenomenon, prior to any systematic study, we ensured that the experiments were fully reproducible. Figure 3(a–d) report an example of the evolution of the wave amplitude at the four resistive gauges, for $m = 2$ kg of 4 mm glass beads on a slope of 45°. Whereas the free surface elevation remains relatively small ($\eta_{\max} < 2$ cm), the difference in elevation between the experiments is smaller than 1 mm. Therefore, the slight difference in compaction of the initial granular media or the variation of the velocity

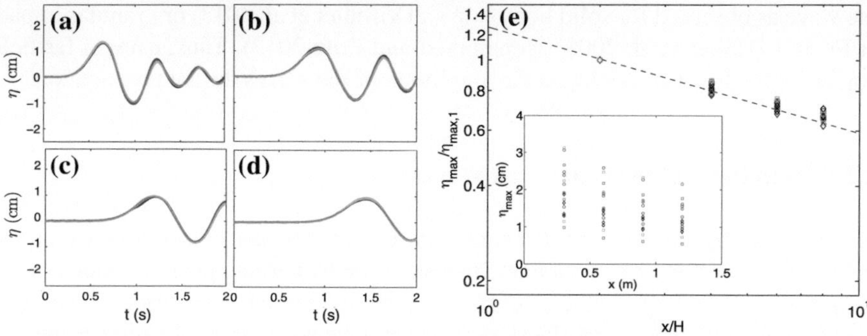

Fig. 3 Time evolution of the free surface elevation measured at the four resistive gauges placed at **a** 0.45 m, **b** 0.75 m, **c** 1.05 m, and **d** 1.35 m from the lifting gate. The granular collapse starts at time $t = 0$. The three curves (*red*, *green*, *blue*) correspond to replicate of the experiment performed with $m = 2$ kg of 4 mm glass beads sliding on a slope of 45°. **e** Evolution of the rescaled amplitude of the leading wave as a function of the rescaled distance of propagation for various masses (1, 2 or 3 kg), various slopes (35°, 40°, 45°, 50°, 55°), and four granular materials (glass beads of diameter 1.5, 4, and 10 mm and aquarium sands). Inset: Measured dimensional amplitude of the leading wave as a function of the distance of propagation

of the lifting gate do not significantly modify the experimental measurement. This result confirmed that our experimental setup can reliably be used to determine the amplitude of the wave generated by a granular collapse into water.

For each experiment, we rescale the amplitude of the wave at the probes by the amplitude η_{max_1} at the first probe (see Fig. 3e). To explain the dependence of the tsunami amplitude on the distance from the source x, we consider the weakly dispersive linear shallow water approximation. In this approximation, the free surface elevation η is described by the linearized Korteweg-de-Vries equation (Whitham 2011; Pelinovsky et al. 2000):

$$\frac{\partial \eta}{\partial t} + c\,\frac{\partial \eta}{\partial x} + \frac{c\,H^2}{6}\,\frac{\partial^3 \eta}{\partial x^3} = 0. \tag{1}$$

A solution to this equation is

$$\eta(x,t) = Q\left(\frac{2}{ct\,H^2}\right)^{1/3} \mathrm{Ai}\left[\left(\frac{2}{ct\,H^2}\right)^{1/3}(x - ct)\right], \tag{2}$$

where Q is a constant corresponding to the cross section of the volume of displaced water, H is the water depth, $c = \sqrt{gH}$ and Ai is the Airy function. Thus, during the propagation, the amplitude of the leading wave decays as $t^{-1/3}$, which leads to a dependence as $x^{-1/3}$. This scaling law is consistent with the experimental data regardless of the initial mass or slope (see Fig. 3e). Note also that the viscosity and the interfacial tension are expected to have negligible effects on the propagation of the wave as observed for solid body impact (Viroulet et al. 2013) or granular impact at Fr $>$ 1 (Heller et al. 2008; Mohammed and Fritz 2012). Thus, using a far-field approximation, we can estimate the amplitude of the waves during propagation.

2.2.1 Amplitude of the Generated Waves

An important parameter to predict tsunamis generated by landslides is the volume of the slide. We performed a systematic investigation by varying the initial mass of the slide from 0.5 to 3 kg for the four granular media previously described. The other experimental parameters, i.e. the slope angle and the water depth are kept constant at 45° and 15 cm, respectively. We observe that the amplitude of the first wave increases with the mass of the slide (see Fig. 4a). Fitting the experimental measurements leads to a power law $\eta_{max_1} \propto m^{0.84}$ for the glass beads and $\eta_{max_1} \propto m^{0.95}$ for the aquarium sand. The experiments performed with the 4 mm and 10 mm glass beads lead to similar values. The exponent of the power law for the 1.5 mm glass beads is the same but the prefactor is slightly lower. Surface tension effects are likely a cause of this discrepancy. Indeed, the size of the beads should be compared with the capillary length in the system, $L_c = \sqrt{\sigma/\rho\, g} \simeq 2.7$ mm (where σ and ρ are the interfacial tension and the density of the water, respectively, g is the gravitational constant). Therefore, unlike in the propagation phase of the wave, the effects of the interfacial tension can be significant during the generation. We can reasonably assume that these effects are less important with the aquarium sand or large glass beads (10 mm) than for experiments with glass beads of 1.5 mm.

Predicting the influence of the mass of the slide on the amplitude of the first wave remains challenging. Indeed, an analysis based on the initial potential energy of the slide transferred to the first wave is not satisfying. When the mass of the slide is large (typically $>$ 2 kg in our setup), the leading wave starts to propagate while some

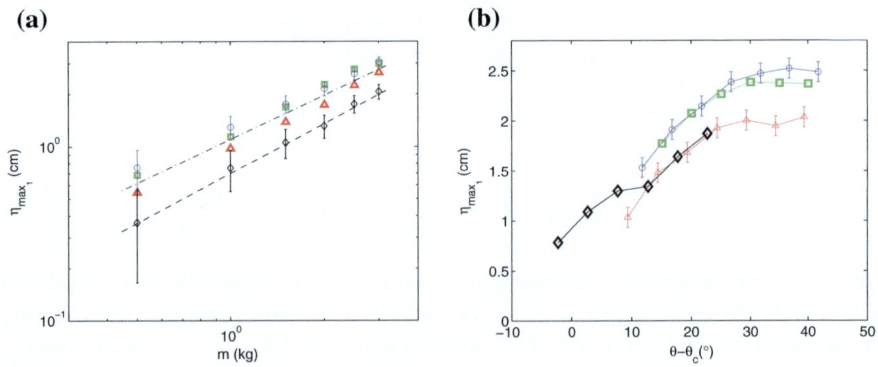

Fig. 4 **a** Evolution of the maximum amplitude at the first probe as a function of the mass of granular material. **b** Evolution of the amplitude of the leading wave as a function of $\theta - \theta_c$. (△: 1.5 mm glass beads, ○: 4 mm glass beads, □: 10 mm glass beads and ◇: aquarium sand)

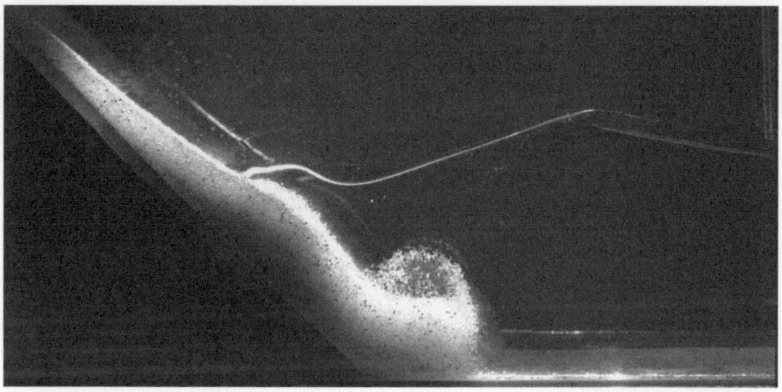

Fig. 5 Image of a granular collapse into water of 3 kg of 1.5 mm glass beads on a slope of 45° taken at $t = 0.45$ s

granular media is still out of the water. Therefore, all the mass of the material does not contribute to the generation of the first wave. This situation is illustrated in Fig. 5 where the first wave has left the region where the generation takes place while some granular material is still flowing down the slope.

The present observation can be related to the work of Sælevik et al. (2009) on the influence of the length of the slide which was solid in their case. Indeed, they observed that the length influences the dispersive wave train. In our situation, for a larger mass of granular material, the collapsing time is longer and an excess of granular material comes into play while generating the wave train following the first wave. The same explanation shows that the evolution of the amplitude of the first wave with the slide mass is almost independent of the material used. Indeed, the generation of the first wave is mainly caused by the impact into water and is roughly the same regardless of the granular media used.

An important parameter in geophysical situations is the angle of the slope. However, comparing the results for different granular materials is not straightforward as they flow at different angles. Therefore, we measured the angle of repose and avalanche of the four granular media (Börzsönyi et al. 2008). We find that the critical angles of avalanche θ_c differ by 17° between the glass beads and the aquarium sand. The evolution of maximum elevation of the free surface as a function of $\theta - \theta_c$ is illustrated in 4b (for more details see Viroulet et al. (2014)). We show that the amplitude of the first wave only depends on $\theta - \theta_c$. Again, the interfacial tension may explain the difference of the amplitude for the 1.5 mm glass beads. In this figure, we also observe a saturation of the amplitude of the first wave for $\theta - \theta_c \geq 30°$ for all granular media (note that for the aquarium sand these large values can not be achieved with our experimental setup).

We calculate the energy of the wave train associated to the first three waves (the energy contained in the following wave train can be reasonably neglected). The evolution of this energy E as a function of $\theta - \theta_c$ is shown in Fig. 6a. The experimental measurements collapse on a main curve and a transition is observed at $\theta - \theta_c \simeq 10°$. Note that these low values can not be achieved with our device for glass beads. Again,

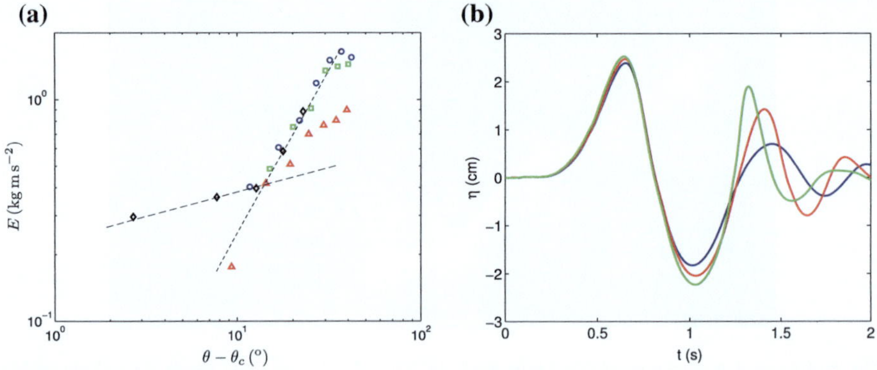

Fig. 6 **a** Evolution of the energy E in the wave train as a function of $\theta - \theta_c$ (△ : 1.5 mm glass beads, ○ : 4 mm glass beads, □ : 10 mm glass beads and ◊: aquarium sand). **b** Elevation of the free surface at the first probe for slopes of (-) 50°, (-) 55° and (-) 60°, a water depth equal to 0.15 m and 2 kg of 4 mm glass beads

the evolution of the energy of the wave train generated with 1.5 mm beads is slightly different from other granular materials, which confirms that the interfacial tension effects are important when the typical size of the granular material is smaller than the capillary length.

For all granular media, we observe that the total energy of the wave train increases with the slope and remains the same for $\theta - \theta_c > 30°$. This result seems contradictory with previous results on the amplitude of the first wave but can be explained using the measurement of the probes. Indeed, for large slope angles, the amplitude of the second wave becomes more important. In Fig. 6b, we report the evolution of the free surface at the first probe for three different slopes and 4 mm glass beads. Beyond a certain angle, the amplitude of the first wave does not increase anymore while the amplitude of the second wave increases. The energy and the amplitude of the first wave reach a saturation. However, the rest of the energy is found in the second wave which becomes as dangerous as the leading wave. Figure 6a shows that the last point obtained with the 4 mm glass beads becomes lower in energy for large slope angles. This is due to the large curvature of the second wave, which in this configuration breaks before reaching the probe, thereby decreasing the total energy of the wave train. When the slope increases and the energy of the first wave does not increase, the rest of the energy is transferred to the wave train and more particularly in the second wave.

3 Numerical Simulations

Our experiments have shown that taking into account the properties of granular material are crucial to predict the amplitude of the generated waves. The physical mechanisms involved in the collapse of an initially dry granular medium in water are particularly difficult to model. For these reasons, the numerical modeling of a granular collapse from air into water remains challenging. The modeling of a submarine collapse by solving the dynamics of each grain together with the fluid dynamics is very recent (Topin et al. 2012), and the next digital step is to add an air/water interface to the granular collapse. Therefore, the modeling of the impact of a dry granular material in water and the generated wave train still requires enormous efforts. Here, we use the software Gerris (Popinet 2003, 2009) to run numerical simulations of a situation similar to the experiments. The granular material is modeled as a fluid with non-Newtonian properties. The aim of these simulations is to reproduce the evolution of the first wave using the Navier–Stokes equations for triphasic simulations.

3.1 Parameters of the Simulations

The numerical domain is divided into three subdomains corresponding to 1 m × 1 m squares connected to each other. The slope is modeled by a solid wedge with a no-slip boundary condition. The water depth is equal to 15 cm for all the simulations. The density and viscosity of the air and water are taken for a temperature of 20 °C and a pressure 1013 hPa. To simulate different rheologies for the granular media, we have performed numerical simulations for two fluids: a Newtonian fluid ($\rho = 1500\,\mathrm{kg\,m^3}$, $\mu = 5\,\mathrm{kg.m^{-1}\cdot s^{-1}}$) and a Bingham fluid ($\rho = 1500\,\mathrm{kg.m^3}$, $\mu = 5\,\mathrm{kg.m^{-1}\cdot s^{-1}}$ and $\tau_0 = 100\,\mathrm{kg.m^{-1}\cdot s^{-2}}$).

The mesh refinement occurs at the interface between water and other fluids (air and granular material) and a constant mesh is maintained in the water and granular phases. The smallest mesh size is about 1.9 mm for the simulations presented in this article. The boundary conditions are a no-slip at the bottom, a symmetry condition at the left and right ends of the domain, and an open boundary on the top of the domain. All simulations are performed on a single processor with a computing time ranging from 6 to 12 h.

To characterize the evolution of the waves during their propagation, we set four probes at the same distances as in the experiments (0.45, 0.75, 1.05, and 1.35 m) from the point where the granular material enters the water.

3.2 Comparison of Numerical and Experimental Results

We consider the particular situation of 2 kg of glass beads of 1.5 mm in diameter sliding on a slope of 45° to quantitatively compare experiments and numerical simulations. Figure 7 shows the experimental results and the numerical simulations for two different rheologies of the slide.

We observe that the numerical simulations are in good agreement with the experiments at early times. The amplitude of the first wave during the impact is slightly larger when using a slide made of a fluid rather than a Bingham fluid. When the first wave leaves the generation region and starts to propagate, the apparition of a vortex at the edge of the slide is visible in the simulation using a Newtonian fluid. We can also observe the counter flow which slows down the collapse (see Fig. 7e). These observations are less visible in the simulation using a Bingham fluid, where the flow is slower and seems to become solid after the impact. Finally, the second wave starts to appear in a similar manner than observed in the experiments (Fig. 7g, h, i), although the slides have moved significantly less.

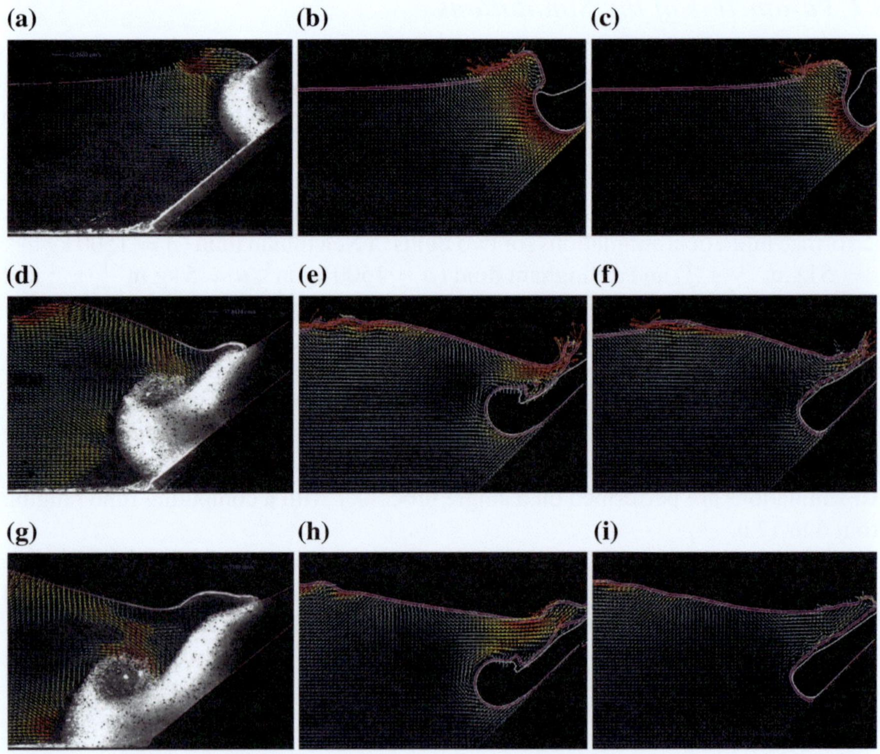

Fig. 7 Comparison between the experiments (**a**, **d**, **g**) and the numerical results obtained with Gerris using a newtonian fluid (**b**, **e**, **h**) and a Bingham rheology (**c**, **f**, **i**)

As mentioned above, numerical simulations of the granular material interacting with the fluid are really complex. For instance, the collapse of a fluid starting from air into water generates a huge vorticity and bubbles which are particularly difficult to simulate and very expensive in CPU time. Therefore, the main purpose of the present simulations is to reproduce the evolution of the amplitude of the first wave. We accordingly set the parameters of the different fluids used to simulate the slide. Apart from the density, which corresponds to the equivalent density in the experiments ($\rho_{eq} = 0.6\,\rho_g \simeq 1500$ kg.m^{-3}), the viscosity and the yield stress of the slide are not based on physical properties but rather fitted to the experiments.

3.3 Evolution of the Amplitude During Propagation

We compare the evolution of the wave amplitude between our experiments and our numerical simulations using the probes. We first need to define a rheology for the granular material and study its influence on the generated wave. We compared the wave amplitudes obtained in the experiments and in the numerical simulations for a mass $m = 2$ kg for the four types of granular materials on a slope of 45°. The simulation with a Newtonian rheology produces results in fair agreement with the experiments performed with the 4 mm glass beads, while the approximation of a Bingham fluid gives results consistent with the 1.5 mm glass beads experiments. Three configurations were considered by varying only the angle of the slope (35°, 45° and 60°). Figure 8 shows the results at the different probes for these configurations.

The evolution of the amplitude of the first wave for a slope of 45° is well captured by the numerical simulations for both the Newtonian fluid/4 mm glass beads and the Bingham fluid/1.5 mm glass beads. These first results are not surprising, since, as stated previously, the parameters of the fluids used in the numerical simulations were chosen to best match this configuration. If we now consider the results obtained for a slope of 60° we see that, once again, the numerical simulations are in relatively good agreement with the experiments. More specifically, for the Bingham fluid and the 1.5 mm glass beads, the evolution of the free surface obtained experimentally (1.5 mm glass beads) and numerically (Bingham fluid) are in close agreement for the four probes. We observe a small time offset for the second peak at the first probe which decreases during the propagation.

The largest discrepancy is observed between the amplitude of the first trough and the second crest when comparing the Newtonian fluid and the 4 mm glass beads. It shows that approximating the slide by a Newtonian fluid does not successfully reproduce the interactions between the granular material and the generated waves. However, during the propagation, the energy of the first wave is transferred to the dispersive wave train and thus the difference in amplitude at the second peak decreases. For a slope angle of 35°, the numerical simulations overestimate the amplitude of

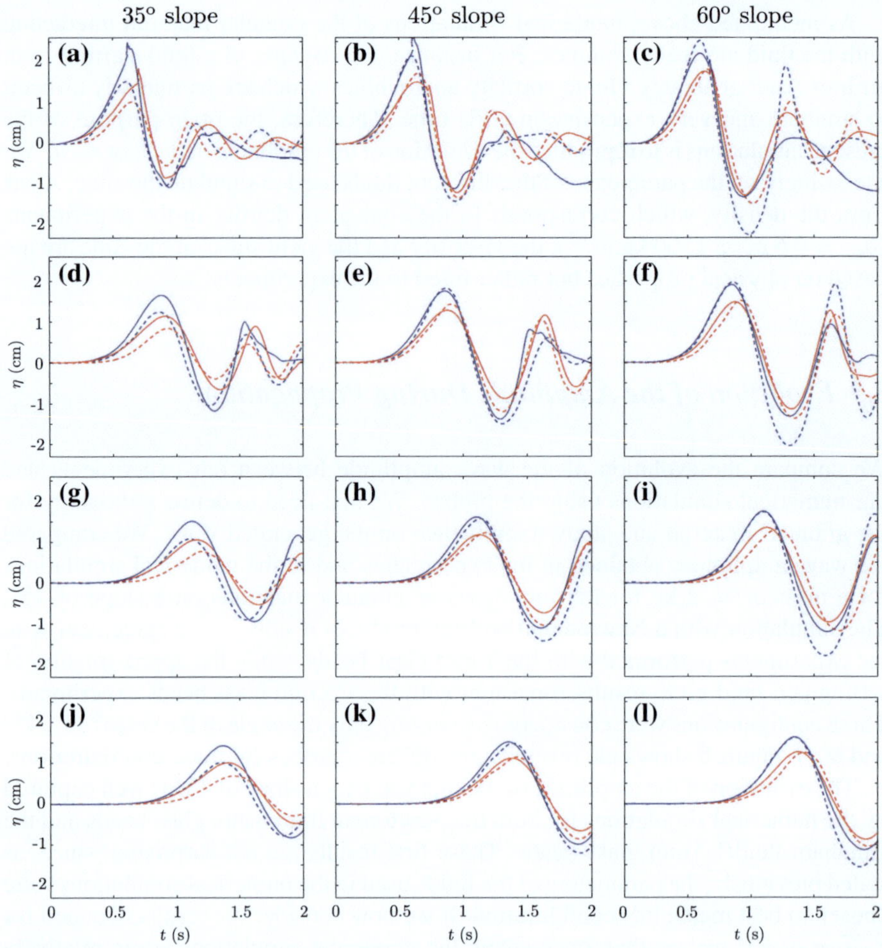

Fig. 8 Comparison of the evolution of the free surface elevation between the experiments and the numerical simulations for a slope of 35°, 45°, and 60° measured at (**a**, **b**, **c**) 0.45 m, (**d**, **e**, **f**) 0.75 m, (**g**, **h**, **i**) 1.05 m and (**j**, **k**, **l**) 1.35 m from the lifting gate. The *blue* and *red dashed lines* are experiments with 4 and 1.5 mm glass beads, respectively. The *blue* and *red* continuous line are the numerical results obtained with Gerris simulating the collapse of a Newtonian and Bingham fluid into water, respectively

the first wave and the largest discrepancies are observed in this configuration. These results are in agreement with the fact that the velocity of the slide is the lowest in this configuration. Therefore, the interactions between the granular materials and the generated waves are the most important.

4 Conclusion

From our systematic experimental study, we were able to highlight the importance of the internal properties of the granular material during its collapse into water at small Froude number. We also emphasized the importance of the interplay between the granular media and the fluid during the generation of the waves. In particular, in the configuration studied experimentally, the friction on the slope becomes dominant. Therefore, the amplitude of the generated waves does not only depend on the slope angle but on the difference between the slope angle and the avalanching angle of the granular material used. This angular dependence has also been observed for the runout distance of the granular media, which is always smaller for a granular collapse from air into water rather than for a collapse in a single fluid (air or water). We also reported that the energy transferred by the slide to the generated waves does not exceed 10 % for all initial configurations. Using numerical simulations, we then show that the approximation of a viscous fluid for the granular medium gives satisfactory results for the prediction of the amplitude of the first wave in some particular configurations, but does not satisfactory reproduce the wave train generated and the evolution of the slide. For this reason, an important experimental and numerical effort is required to understand the physical mechanisms that take place during a subaerial collapse. For instance, a numerical study using the coupling of a discrete method to simulate the granular flow (DEM) and a continuous method for the air/water phases (CFD) would lead to important informations on the energy transfer to the waves during a granular collapse into water.

Acknowledgments The authors thank Olivier Pouliquen, Pascale Aussillous, and Pierre-Yves Lagrée for many enlightening discussions. We thank Emilie Dressaire and Parmesh Gajjar for the careful reading of the manuscript and helpful comments

References

Abadie S, Morichon D, Grilli S, Glockner S (2010) Numerical simulation of waves generated by landslides using a multiple-fluid Navier-Stokes model. Coast Eng 57(9):779–794

Abadie SM, Harris JC, Grilli ST, Fabre R (2012) Numerical modeling of tsunami waves generated by the flank collapse of the cumbre vieja volcano (la palma, canary islands): tsunami source and near field effects. J Geophys Res Ocean 117(C5)

Assier-Rzadkiewicz S, Mariotti C, Heinrich P (1997) Numerical simulation of submarine landslides and their hydraulic effects. J Waterway Port Coast Ocean Eng 123(4):149–157

Averbukh E, Dussouillez P, Kharif C, Khvostova O, Kurkin A, Rochette P, Soomere T (2013) Exposure of a coastal city to a landslide tsunami: a case study of Cassis. France. Estonian J Eng 19(2):124–142

Börzsönyi T, Halsey TC, Ecke RE (2008) Avalanche dynamics on a rough inclined plane. Phys Rev E 78(1):011306

Cassar C, Nicolas M, Pouliquen O (2005) Submarine granular flows down inclined plane. Phys Fluids 17(10):103301

Choi BH, Pelinovsky E, Kim KO, Lee JS (2003) Simulation of the trans-oceanic tsunami propagation due to the 1883 Krakatau volcanic eruption. Nat Hazards Earth Syst Sci 3(5):321–332
Di Risio M, Bellotti G, Panizzo A, De Girolamo P (2009) Three-dimensional experiments on landslide generated waves at a sloping coast. Coast Eng 56:659–671
Didenkulova I, Nikolkina I, Pelinovsky E, Zahibo N (2010) Tsunami waves generated by submarine landslides of variable volume: analytical solution for a bassin of variable depth. Nat Hazards Earth Syst Sci 10(11):2407–2419
Enet F, Grilli S (2005) Tsunami landslide generation: modelling and experiments. In: Proceedings of 5th international on ocean wave measurement and analysis
Enet F, Grilli S (2007) Experimental study of tsunami generation by three-dimensional rigid underwater landslides. J Waterway Port Coast Ocean Eng 133:442–454
Enet F, Grilli ST, Watts P (2003) Laboratory experiments for tsunamis generated by underwater landslides: comparison with numerical modeling. In: Proceedings of the 13th offshore and polar engineering conference
Francis PW (1985) The origin of the 1883 Krakatau tsunami. J Volcanol Geotherm Res 25(3):349–363
Fritz HM (2002) Initial phase of landslide generated impulse waves. Ph.D. thesis, Swiss Federal Institute of Technology Zürich
Fritz HM (2001) Lituya Bay case: rockslide impact and wave run-up. Sci Tsunami Hazards 19(1):3–22
Fritz HM, Hager WH, Minor HE (2003a) Landslide generated impulse waves; part 1: instantaneous flow fields. Exp Fluids 35(6):505–519
Fritz HM, Hager WH, Minor HE (2003b) Landslide generated impulse waves; part 2: hydrodynamic impact craters. Exp Fluids 35(6):520–532
Fritz HM, Hager WH, Minor HE (2004) Near field characteristics of landslide generated impulse waves. J Waterway Port Coast Ocean Eng 130(6):287–302
Fritz HM, Mohammed F, Yoo J (2009) Lituya Bay landslide impact generated mega-tsunami 50th anniversary. Pure Appl Geophys 166(1–2):153–175
Gisler GR, Weaver RP, Mader CL, Gittings ML (2004) Two and Three dimensional simulation of asteroid ocean impacts. Sci Tsunami Hazards 21(2):119
Glimsdal S, Pedersen GK, Harbitz CB, Lovholt F (2013) Dispersion of tsunamis: does it really matter? Nat Hazards Earth Syst Sci 13:1507–1526
Grilli ST, Vogelmann S, Watts P (2002) Development of a 3D numerical wave tank for modeling tsunami generation by underwater landslides. Eng Anal Bound Elem 26(4):301–313
Grilli ST, Watts P (1999) Modeling of waves generated by a moving submerged body. Applications to underwater landslides. Eng Anal Boundy Elem 23(8):645–656
Grilli ST, Watts P (2005) Tsunami generation by submarine mass failure part I: modeling, experimental validation, and sensitive analysis. J Waterway Port Coast Ocean Eng 131(6):283–297
Heinrich P (1992) Nonlinear water waves generated by submarine and aerial landslides. J Waterway Port Coast Ocean Eng 118(3):249–266
Heller V, Hager WH (2014) A universal parameter to predict subaerial landslide tsunamis? J Mar Sci Eng 2:400–412
Heller V, Hager WH, Minor HE (2008) Scale effects in subaerial landslide generated impulse waves. Exp Fluids 44(5):691–703
Ioualalen M, Asavanant J, Kaewbanjak N, Grilli ST, Kirby JT, Watts P (2007) Modeling the 26 December 2004 Indian ocean tsunami: case study of impact in Thailand. J Geophys Res Oceans 112(C7)
Kelfoun K, Giachetti T, Labazuy P (2010) Landslide-generated tsunamis at Réunion Island. J Geophys Res Earth Surf 115:F4
Kharif C, Pelinovsky E (2005) Asteroid impact tsunami. Comptes Rendus Physique 6(3):361–366
Korteweg DJ, De Vries G (1895) On the change of form of long waves advancing in a rectangular canal, and on new type of long stationary waves. Phil Mag 39(240):422–443

Liu PF, Wu TR, Raichlen F, Synolakis CE, Borrero JC (2005) Runup and rundown generated by three-dimensional sliding masses. J Fluid Mech 536:107–144
Mader CL, Gitting ML (2002) Modeling the 1958 Lituya Bay mega-tsunami. II. Sci Tsunami Hazards 20(5):241–250
Mohammed F (2010) Physical modeling of tsunamis generated by three-dimensional deformable granular landslides. Ph.D. thesis, Georgia Institute of Technology, December 2010
Mohammed F, Fritz HM (2012) Physical modeling of tsunamis generated by three-dimensional deformable granular landslides. J Geophys Res Oceans 117:C11015
Monaghan JJ, Kos A (2000) Scott Russell's wave generator. Phys Fluids 12(3):622–630
Murty TS (1979) Submarine slide generated water waves in Kitimat Inlet, British Columbia. J Geophys Res Oceans 84(C12):7777–7779
Panizzo A, De Girolamo P, Petaccia A (2005) Forecasting impulse waves generated by subaerial landslides. J Geophys Res Oceans 110:C12025
Pelinovsky E, Talipova T, Kharif C (2000) Nonlinear-dispersive mechanism of the freak wave formation in shallow water. Phys D : Nonlinear Phenom 147(1–2):83–94
Popinet S (2003) Gerris: a tree-based adaptive solver for the incompressible Euler equations in complex geometries. J Comput Phys 190(2):572–600
Popinet S (2009) An accurate adaptive solver for surface-tension-driven interfacial flows. J Comput Phys 228(16):5838–5866
Popinet S (2011) Quadtree-adaptive tsunami modelling. Ocean Dyn 61(9):1261–1285
Pouliquen O (1999) Scaling laws in granular flows down rough inclined planes. Phys Fluids 11(3):542–548
Quecedo M, Pastor M, Herreros MI (2004) Numerical modeling of impulse wave generated by fast landslides. Int J Numer Methods Eng 59(12):1633–1656
Sælevik G, Jensen A, Pedersen G (2009) Experimental investigation of impact generated tsunami; related to a potential rock slide. Western Norway. Coast Eng 56(9):897–906
Scott Russel J (1844) Report on Waves. In: Report of the 14th meeting of the British Assocaition for the Advancement of Science, vol 311, pp. 390
Shi F, Kirby JT, Harris JC, Geiman JD, Grilli ST (2012) A High-order adaptive time-stepping TVD solver for Boussinesq modeling of breaking waves and coastal inundation. Ocean Model 43:36–51
Shi F, Kirby JT, Tehranirad B, Harris JC, Grilli ST (2011) FUNWAVE-TVD, Fully nonlinear Boussinesq wave model with TVD solver, documentation and user's manual (Version 2.1). Technical report, Center for Applied Coastal Research, University of Delaware
Tinti S, Bortolucci E (2000) Energy of water waves induced by submarine landslides. Pure Appl Geophys 157(3):281–318
Topin V, Monerie Y, Perales F, Radjai F (2012) Collapse dynamics and runout of dense granular materials in a fluid. Phys Rev Lett 109(18):188001
Viroulet S, Cébron D, Kimmoun O, Kharif C (2013) Shallow water waves generated by subaerial solid landslides. Geophys J Int 193(2):747–762
Viroulet S, Sauret A, Kimmoun O (2014) Tsunami generated by a granular collapse down a rough inclined plane. Europhys Lett 105(3):34004
Viroulet S, Sauret A, Kimmoun O, Kharif C (2013a) Effondrement granulaire dans l'eau: application à la génération de tsunami. In: 21eme Congrès Francais de Méecanique, Bordeaux, France
Viroulet S, Sauret A, Kimmoun O, Kharif C (2013b) Granular collapse into water: toward tsunami landslides. J Visualization 16(3):189–191
Walder JS, Watts P, Sorensen OE, Janssen K (2003) Tsunamis generated by subaerial mass flows. J Geophys Res Solid Earth 108(B5)
Wang Y, Liu PLF, Mei CC (2011) Solid landslide generated waves. J Fluid Mech 675:529–539
Watts P (1997) Water waves generated by underwater landslides. Ph.D. thesis, California Institute of Technology
Watts P (1998) Wavemaker curves for tsunamis generated by underwater landslides. J Waterway Port Coast Ocean Eng 124(3):127–137

Watts P (2000) Tsunami features of solid block underwater landslides. J Waterway Port Coast Ocean Eng 126(3):144–152

Watts P, Grilli ST, Kirby JT, Fryer GJ, Tappin DR (2003) Landslide tsunami case studies using a Boussinesq model and a fully nonlinear tsunami generation model. Nat Hazards Earth Syst Sci 3(5):391–402

Whitham GB (2011) Linear and nonlinear waves, vol 42. John Wiley & Sons

Zweifel A, Hager WH, Minor HE (2006) Plane impulse waves in reservoirs. J Waterway Port Coast Ocean Eng 132(5):358–368

An Analytical Model of Large Amplitude Internal Solitary Waves

Nikolay I. Makarenko and Janna L. Maltseva

Abstract The problem on steady internal waves in a weakly stratified two-layered fluid is studied analytically. We consider the model with homogeneous fluid in the lower layer and with exponentially stratified fluid in the upper layer. The long-wave approximation is constructed by means of implementing a scaling procedure with a small Boussinesq parameter. Extreme configurations of solitary waves such as broad table-top waves are discussed.

Keywords Stratified fluid · Solitary waves

1 Introduction

It is well known at present (Helfrich and Melville 2006; Pelinovsky et al. 2007) that ocean internal waves can exhibit huge magnitude comparable with total fluid depth. One of the most interesting extreme forms of nonlinear internal waves is intimately related with the broadening effect. Broadening occurs when the phase speed of solitary wave is close to the propagation speed of smooth internal bore. Turner and Vanden-Broeck (1988) found this effect numerically by calculation interfacial table-top solitary waves in a two-fluid system.

Two-layer approximation is a standard model of sharp pycnocline in a stratified fluid. In this description, the density is constant in each layer but has a discontinuity at interface. At the same time, in several cases one should take into account slight continuous stratification of fluid layer in spite of the density jump at interface. In this paper, we consider a theoretical model of strongly nonlinear interfacial waves in a two-layer fluid which has a constant density in lower layer, and the density depends exponentially on the height in upper layer. The newly proposed model equation generalizes the models suggested by Ovsyannikov et al. (1985) and Miyata (1985) for

N.I. Makarenko (✉) · J.L. Maltseva
Lavrentyev Institute of Hydrodynamics, Novosibirsk 630090, Russia
e-mail: makarenko@hydro.nsc.ru

J.L. Maltseva
e-mail: maltseva@hydro.nsc.ru

E. Pelinovsky and C. Kharif (eds.), *Extreme Ocean Waves*,
DOI 10.1007/978-3-319-21575-4_11

a system with constant densities in both layers, as well as the "2.5-layer" model considered by Voronovich (2003). We demonstrate that perturbed model is well consistent with the known results about table-top waves in the perfect two-fluid system. We have also found the range of parameters where continuous stratification in upper layer essentially affects the shape of solitary waves. In this view, the resulting model takes into account the influence of weak stratification outside of the pycnocline.

The method of derivation involves asymptotic analysis of the Dubreil–Jacotin—Long equation which is equivalent to the fully nonlinear Euler equations. Long-wave scaling procedure uses small Boussinesq parameter which characterizes small slope of the density profile in upper layer. Our method combines approaches suggested originally by Ovsyannikov et al. (1985) and Miyata (1985) for a two-layer fluid with expansion procedure developed by Long (1965) and Benney and Ko (1978) in the case of continuous stratification. Previously, this method was used in theoretical study of table-top solitary waves and internal bores in a weakly stratified fluid without homogeneous layers (Makarenko 1999; Maltseva 2003). Lamb and Wan (1998) investigated numerically flat-crested waves in a continuously stratified fluid, and Grue et al. (2000) observed breaking and broadening of such solitary waves in laboratory experiments.

2 Basic Equations

We consider a 2D motion of inviscid inhomogeneous two-layered fluid which is weakly stratified under gravity. The Euler equations describing steady flows are

$$\begin{aligned} &\rho(uu_x + vu_y) + p_x = 0, \\ &\rho(uv_x + vv_y) + p_y = -\rho g, \\ &u_x + v_y = 0, \quad u\rho_x + v\rho_y = 0, \end{aligned} \tag{1}$$

where ρ is the fluid density, u and v are the velocity components, p is the pressure and g is the gravity acceleration. The flow is confined between the flat bottom $y = -h_1$ and the rigid lid $y = h_2$ (Fig. 1), and the interface $y = \eta(x)$ separates the layers (the value $\eta = 0$ gives equilibrium level of this interface).

We introduce the stream function ψ as usually by $u = \psi_y, \ v = -\psi_x$, so the mass conservation implies the dependence $\rho = \rho(\psi)$, and the pressure p can be found from the Bernoulli equation

$$\frac{1}{2}|\nabla\psi|^2 + \frac{1}{\rho(\psi)}p + gy = b(\psi).$$

It is supposed that the fluid velocity (u, v) tends to the upstream velocity $(c_j, 0)$ as $x \to -\infty$ where c_j is the wave speed with respect to jth layer (j = 1, 2). In this case, boundary conditions at the bottom, at the interface, and at the lid take the forms, respectively

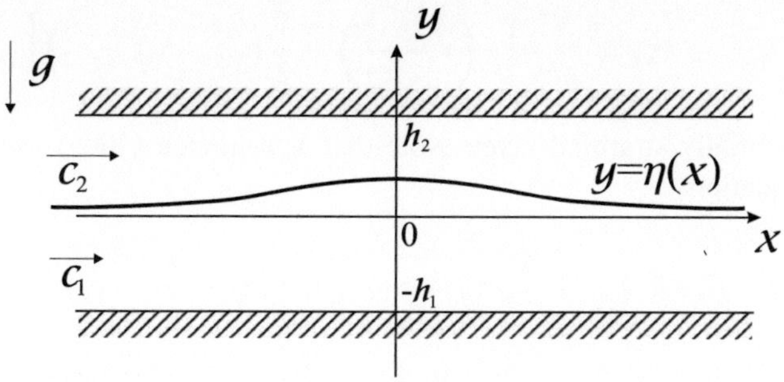

Fig. 1 Scheme of motion

$$\psi = -c_1 h_1 \ (y = -h_1), \qquad \psi = 0 \ (y = \eta), \qquad \psi = c_2 h_2 \ (y = h_2). \qquad (2)$$

It is well known (Yih 1980) that the system (1) can be reduced to the Dubreil–Jacotin—Long (DJL) equation for a stream function

$$\rho(\psi)\Delta\psi + \rho'(\psi)\left(gy + \frac{|\nabla\psi|^2}{2}\right) = H'(\psi).$$

Here the function $H(\psi) = \rho(\psi)b(\psi)$ involves the Bernoulli function $b(\psi)$ and the density $\rho(\psi)$ which are specified by the condition $\rho(\psi) = \rho_\infty(\psi/c_j)$ in jth layer. In this paper, we consider the upstream density profile

$$\rho_\infty(y) = \begin{cases} \rho_1 & (-h_1 < y < 0), \\ \rho_2 \exp(-N^2 y/g) & (0 < y < h_2), \end{cases} \qquad (3)$$

where N = const is the Brunt—Väisälä frequency, and the constants ρ_1 and ρ_2 are such that $\rho_2 < \rho_1$. The special case $N = 0$ gives a familiar two-fluid system with a piece-wise constant fluid density ρ and the Bernoulli function b. In general case $N \neq 0$, the function $b(\psi)$ is constant in lower layer $-h_1 < y < \eta(x)$ only, $b(\psi) = c_1^2/2$. At the same time, we have in upper layer $\eta(x) < y < h_2$

$$b(\psi) = \frac{1}{2}c_2^2 + \frac{g\psi}{c_2} + \frac{g^2}{N^2}\left(1 - e^{\frac{N^2\psi}{gc_2}}\right).$$

As a consequence, we obtain the Laplace equation for a stream function in the homogeneous layer

$$\psi_{xx} + \psi_{yy} = 0, \qquad (4)$$

and ψ should satisfy the nonlinear equation

$$\psi_{xx} + \psi_{yy} = \frac{N^2}{gc_2}\left\{g\left(y - \frac{\psi}{c_2}\right) + \frac{1}{2}\left(\psi_x^2 + \psi_y^2 - c_2^2\right)\right\} \tag{5}$$

in the exponentially stratified layer. Note that Voronovich (2003) used the linear Helmholtz equation

$$\psi_{xx} + \psi_{yy} + \frac{N^2}{c^2}(\psi - cy) = 0 \qquad (c = c_1 = c_2) \tag{6}$$

which is a simplified version of (5). By this approach, the nonlinearity appears due to interfacial boundary condition

$$[\rho(\psi)(|\nabla\psi|^2 + 2gy - 2b(\psi)] = 0, \qquad y = \eta(x), \tag{7}$$

where the square brackets mean the discontinuity jump. The condition (7) provides continuity of the pressure p everywhere in the flow domain. As Long (1965) noted, nonlinear inertial terms of Eq. (5) are significant in the case when perturbation method uses several parameters in addition to small quantity $\Delta\rho/\rho$. Therefore, we take into account all the nonlinearities appearing from exact Euler equations. In addition, we reformulate the condition (7) by taking into account conservation of total momentum due to the flow force integral

$$\int_{-h_1}^{h_2} (p + \rho\psi_y^2)\, dy = \text{const.}$$

Excluding the pressure p by the Bernoulli equation, we obtain the relation

$$\rho_1 \int_{-h_1}^{\eta(x)} \left(c_1^2 + \psi_{1y}^2 - \psi_{1x}^2 - 2gy\right) dy + \tag{8}$$

$$+\rho_2 \int_{\eta(x)}^{h_2} e^{-\frac{N^2\psi_2}{gc_2}} \left\{c_2^2 + \psi_{2y}^2 - \psi_{2x}^2 - \frac{2g^2}{N^2}\left(e^{\frac{N^2\psi_2}{gc_2}} - 1\right) + 2g\left(\frac{\psi_2}{c_2} - y\right)\right\} dy = C,$$

where the constant C depends on parameters of upstream flow as follows:

$$C = \rho_1 g h_1^2 + 2\rho_1 c_1^2 h_1 + 2\rho_2 c_2^2 h_2 + 2\rho_2 g\left(\frac{c_2^2}{N^2} + \frac{g^2}{N^4}\right)\left(1 - \frac{N^2 h_2}{g} - e^{-\frac{N^2 h_2}{g}}\right).$$

It is easy to check by direct calculation that the integral relation (8) is equivalent to the boundary condition (7) which is rather simple. However, the Eq. (8) provides more effectively the derivation of model describing solitary waves of finite amplitude.

3 The Model Equation

In order to formulate approximate model in a dimensionless form, we introduce now some certain scales and parameters. The density profile (3) is clarified by the Boussinesq parameters σ and μ defined by the formulae

$$\sigma = \frac{N^2 h_2}{g}, \qquad \mu = \frac{\rho_1 - \rho_2}{\rho_2}. \tag{9}$$

Here, the constant σ characterizes the slope of density profile in a continuously stratified layer, and μ is the dimensionless density jump at interface. As usual, both these parameters are small in the case of slight stratification. However, we expect that the interfacial mode dominates modes of internal waves in stratified layer, when $\mu \gg \sigma$. Therefore, we fix the constant μ and use the parameter σ as the perturbation parameter.

Further, we define the densimetric Froude number F_j as the scaled phase speed c_j in jth layer,

$$F_j^2 = \frac{\rho_j c_j^2}{g(\rho_1 - \rho_2)h_j} \quad (j = 1, 2). \tag{10}$$

The pair $F = (F_1, F_2)$ is a distinctive mark for a two-fluid system in the presence of a velocity jump at a sharp interface. We demand the difference between c_1 and c_2 to be moderate in order to avoid the Kelvin–Helmholtz instability, at least in accordance with long-wave limit criteria (Ovsyannikov et al. 1985):

$$|c_1 - c_2| < \sqrt{\frac{g(\rho_1 - \rho_2)(\rho_1 h_2 + \rho_2 h_1)}{\rho_1 \rho_2}}.$$

In addition to the Froude numbers F_j, we use also the parameter λ given by the formula

$$\lambda^2 = \left(\frac{N h_2}{c_2}\right)^2 = \frac{\sigma g h_2}{c_2^2} = \frac{\sigma}{\mu F_2^2}.$$

The constant λ characterizes the inverse densimetric Froude number defined for a continuously stratified fluid in upper layer. Finally, the parameter $r = h_1/h_2$ is the layer thickness ratio for the fluid at rest.

The derivation procedure involves the slow independent variable $\xi = \sqrt{\sigma}\, x/h_2$, the dimensionless variables $(\bar{y}, \bar{\eta}) = (y, \eta)/h_2$ and the stream function $\psi = c_j h_j \bar{\psi}$ $(j = 1, 2)$, scaled separately in jth layer. We are seeking for the function $\bar{\psi}$ expanded in powers on σ as

$$\bar{\psi} = \psi^{(0)} + \sigma\, \psi^{(1)} + O(\sigma^2)$$

where the coefficients $\psi^{(k)}$ are determined by the Eqs. (4) and (5) under boundary conditions (2) transformed to appropriate dimensionless form. Thus, in lower layer $-r < y < \eta$ we obtain the coefficients

$$\psi^{(0)} = \frac{y-\eta}{r+\eta}, \qquad \psi^{(1)} = -\frac{1}{6}\left(\frac{1}{r+\eta}\right)_{\xi\xi}\left\{(y+r)^3 - (r+\eta)^2(y+r)\right\}$$

(bar is omitted in notations of dimensionless variables here and later). Respectively, in upper layer $\eta < y < 1$ the coefficients of perturbation series are

$$\begin{aligned}
\psi^{(0)} &= y - \eta\,\frac{\sin\lambda(1-y)}{\sin\lambda(1-\eta)},\\
\psi^{(1)} &= \frac{\sin\lambda(1-y)}{2\lambda}\left(\frac{\eta}{\sin\lambda(1-\eta)}\right)_{\xi\xi}\left\{(1-\eta)\cot\lambda(1-\eta) - (1-y)\cot\lambda(1-y)\right\} +\\
&\quad + \frac{1}{6}\eta^2\left\{\frac{\sin\lambda(\eta-y)+\sin\lambda(1-\eta)-\sin\lambda(1-y)}{\sin^3\lambda(1-\eta)} + \frac{\sin^2\lambda(1-y)}{\sin^2\lambda(1-\eta)} - \frac{\sin\lambda(1-y)}{\sin\lambda(1-\eta)}\right\} +\\
&\quad + \frac{\eta(\eta-y)}{2}\,\frac{\sin\lambda(1-y)}{\sin\lambda(1-\eta)}.
\end{aligned}$$

Finally, we use the dimensionless version of integral relation (8) in the form

$$\mu r^3 F_1^2 \int_{-r}^{\eta} (\psi_y^2 - \sigma\psi_\xi^2)\,dy + \int_{\eta}^{1} e^{-\sigma\psi}\left\{\mu F_2^2(1+\psi_y^2-\sigma\psi_\xi^2) - 2\sigma^{-1}\left(e^{\sigma\psi}-1\right) + 2(\psi - y)\right\} dy =$$

$$= (1+\mu)\eta^2 - \mu r F_1^2(\eta - r) + 2\mu F_2^2 + 2\left(\lambda^{-2}+\sigma^{-2}\right)\left(1-\sigma-e^{-\sigma}\right).$$

Substituting the power expansion for ψ and truncating the terms $O(\sigma^2)$ we obtain the nonlinear ordinary differential equation for the wave shape $\eta(x)$ which depends on dimensionless variables x with scale unit h_2:

$$\left(\frac{d\eta}{dx}\right)^2 = \frac{\eta^2\left(A_0 + A_1\eta + A_2\eta^2 + A_3\eta^3\right)}{B_0 + B_1\eta + B_2\eta^2 + B_3\eta^3 + B_4\eta^4}. \tag{11}$$

Here, the coefficients A_j and B_j are trigonometric polynomials depending on $s_n = \sin n\lambda(1-\eta)$ and $c_n = \cos n\lambda(1-\eta)$ with integer or half-integer n,

$$\begin{aligned}
A_0 &= 18r\lambda s_1^2\left\{\left[2(F_1^2-1) - \sigma F_2^2\right]s_1^2 + \lambda F_2^2 s_2\right\},\\
A_1 &= 2\lambda F_2^2\left\{s_1^2\left[\lambda(9-2\sigma r)s_2 - s_1^2(6r\lambda^2+9\sigma)\right]\right.\\
&\quad \left. - 2s_{1/2}^2\left[r\sigma\lambda s_1 + 3r\sigma\lambda^2(1+2c_1)\right]\right\} - 36\lambda s_1^4,\\
A_2 &= 4\lambda^2 F_2^2 s_{1/2}^2\left\{3\sigma\lambda(r-1)(1+2c_1) - 4(3\lambda s_1^2 + \sigma s_2)c_{1/2}^2 - \sigma s_1\right\},
\end{aligned}$$

$$
\begin{aligned}
A_3 &= 12\lambda^3\sigma F_2^2 s_{1/2}^2(1+2c_1),\\
B_0 &= 12\lambda r^3 F_1^2 s_1^4 + 9r F_2^2(2\lambda - s_2)s_1^2,\\
B_1 &= 9F_2^2\left\{r\lambda(2\lambda - s_2)s_2 - \left[2(r-1)\lambda + s_2\right]s_1^2\right\},\\
B_2 &= 9\lambda F_2^2\left\{\lambda\left[r(s_1^2-3)+2\right]s_2 + 4s_1^4 + 2r\lambda^2 - 2(r\lambda^2+3)s_1^2\right\},\\
B_3 &= -9\lambda^2 F_2^2\left\{(c_1^2+2)s_2 + 2\lambda(r-1)c_1^2\right\},\\
B_4 &= -18\lambda^3 F_2^2 c_1^2.
\end{aligned}
$$

Note that λ in the above formulas is $\lambda = O(\sqrt{\sigma})$ for a fixed $F_2 \neq 0$. This is the case when continuous stratification disappears in upper layer as $\sigma \to 0$. In such a way, the Eq. (11) reduces to the equation

$$
\left(\frac{d\eta}{dx}\right)^2 = \frac{3\eta^2\left[\eta^2 + \left(F_2^2 - rF_1^2 - 1 + r\right)\eta + r\left(F_1^2 + F_2^2 - 1\right)\right]}{r^3F_1^2(1-\eta) + F_2^2(r+\eta)}. \tag{12}
$$

as $\lambda \to 0$. This simplified version of model equation was earlier obtained by Ovsyannikov et al. (1985) and, in the case $c_1 = c_2$, by Miyata (1985) (see also Choi and Camassa (1999)) for a two-fluid system with constant density in the both layers. Recently, this model was thoroughly compared and validated with experimental observations and numerical calculations of solitary waves in the framework of fully nonlinear Euler equations (Camassa et al. 2006).

For a fixed $\lambda \neq 0$ and small $\sigma > 0$ the Froude number F_2 has the order $F_2 = O(\sqrt{\sigma})$, and the Eq. (11) takes the limiting form

$$
r^3F_1^2\eta_x^2 = 3\eta^2\left(r(F_1^2 - 1) - \eta\right) \tag{13}
$$

as $\sigma \to 0$. This is the Boussinesq—Rayleigh equation known as the model of surface solitary waves in homogeneous fluid layer with dimensionless depth r. In the case under consideration, the Froude number $F_1 = c_1/\sqrt{g_1 h_1}$ is defined by the reduced gravity acceleration $g_1 = (\rho_1 - \rho_2)g/\rho_1$. This curious analogy also agrees with the Eq. (12) which results to Eq. (13) as $F_2 \to 0$.

4 Solitary Waves

The parametric range of solitary waves is obtained as the domain in (F_1, F_2)-plane being supercritical with respect to the spectrum of small amplitude sinusoidal waves determined by linearized Eqs. (2), (4), (5) and (7). The dispersion relation of linear waves is

$$
\Delta(k; F) = 0
$$

where k is a dimensionless wave-number and the function Δ is defined by the formula

$$\Delta = F_1^2\, rk\sqrt{\sigma}\, \coth rk\sqrt{\sigma} + F_2^2 \left(\sqrt{\lambda^2 - \frac{\sigma^2}{4} - k^2\sigma}\, \cot \sqrt{\lambda^2 - \frac{\sigma^2}{4} - k^2\sigma} - \frac{\sigma}{2} \right) - 1.$$

This dispersion relation determines real wave-numbers k if and only if the Froude point $F = (F_1, F_2)$ belongs to the shaded area on the Fig. 2.

The boundary of this spectral domain is determined by the equation

$$F_1^2 + F_2^2 \left(\sqrt{\frac{\sigma}{\mu F_2^2} - \frac{\sigma^2}{4}}\, \cot \sqrt{\frac{\sigma}{\mu F_2^2} - \frac{\sigma^2}{4}} - \frac{\sigma}{2} \right) = 1. \tag{14}$$

Note that a perfect two-fluid system with constant densities ρ_1 and ρ_2 in the layers has one-modal spectrum bounded by the unit circle $F_1^2 + F_2^2 = 1$. The curve (14) transforms to this circle in a weak-stratification limit $\sigma \to 0$ by fixing μ. However, it is clear that this transformation from a non-bounded spectral domain to compact spectrum is not uniform in F_1 as $\sigma \to 0$.

Equation (11) has solitary-wave solutions when the point (F_1, F_2) is located between the spectrum and the curve B_λ which gives the broadening limit of internal waves. The line $\sqrt{r}F_1 + F_2 = \sqrt{1+r}$ (the dashed line on Fig. 2b) is the bore diagram for Eq. (12), it gives a limit form of the curve B_λ as $\lambda \to 0$. Small amplitude solitary waves bifurcate from upstream flow at the spectrum boundary (14), these waves are described by weakly nonlinear version of the Eq. (11)

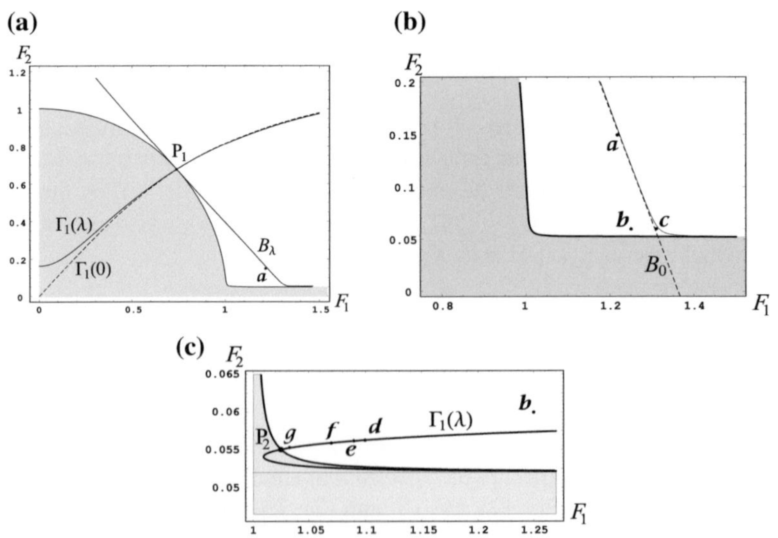

Fig. 2 The spectrum and supercritical domain $\sigma = 0.00008,\ \mu = 0.003,\ r = 1.2$

$$\eta_x^2 = \eta^2 \left[\gamma_0 + \gamma_1 \eta + O(\eta^2)\right]. \tag{15}$$

Here the lowest-order coefficient γ_0 is positive in supercritical domain $\Delta(0; F) > 0$ since

$$\gamma_0 = \frac{3\Delta(0; F)}{r^2 F_1^2 + F_2^2} + O(\sigma^2),$$

and the coefficient $\gamma_1(F, \lambda)$ depends on parameters F_j and λ in a complicated way.

The Fig. 2b presents an enlarged fragment of Fig. 2a near the positive semi-axis OF_1. This is a thin spectral layer for the perturbed problem, it has the thickness $O(\sqrt{\sigma})$ and therefore disappears as $F_1 > 1$ and $\sigma \to 0$. Similarly, Fig. 2c illustrates scaled fragment of Fig. 3b which is located close to the Froude point $F = (1, 0)$. Branches of the curve $\Gamma_1(\lambda) : \gamma_1(F, \lambda) = 0$ shown on Fig. 2a and c indicate where the balance of nonlinearities η^3 and η^4 is possible in the Eq. (15). It is really true

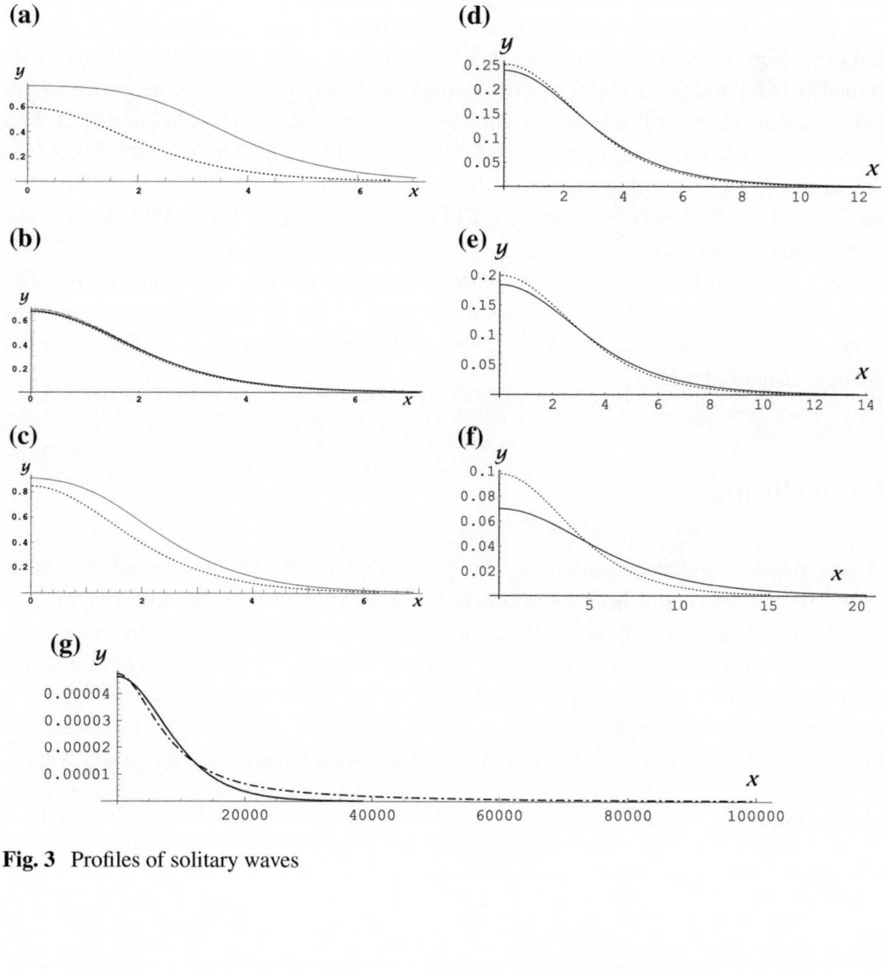

Fig. 3 Profiles of solitary waves

in the vicinity of the point P_1 with the coordinates $F_1 = r/\sqrt{(1+r)r} + O(\lambda)$ and $F_2 = 1/\sqrt{1+r} + O(\lambda)$. This is the same effect as the balance of quadratic and cubic nonlinearities in the KdV–mKdV model (Kakutani and Yamasaki 1978).

In contrast, the neighborhood of the point P_2 with coordinates

$$F_1 = 1 + \frac{1}{2\pi}\sqrt{\frac{\sigma}{r\mu}} + O(\sigma), \qquad F_2 = \frac{1}{\pi}\sqrt{\frac{\sigma}{\mu}} + \frac{\sqrt{r}\sigma}{\pi^2\mu} + O(\sigma^{3/2})$$

demonstrates one more case of exotic behavior of solitary waves. For example, when the point F belongs to the curve $\Gamma_1(\lambda)$ then the fractional scaling of variables $x = \sigma^{1/4}x_0,\ \eta = \sigma^{1/2}\eta_0$ leads from Eq. (11) to the model equation

$$\left(\frac{d\eta_0}{dx_0}\right)^2 = \eta_0^3 \frac{R^2 + 3R(\pi\eta_0 + R)^3 - 3(\pi\eta_0 + R)^4}{r^3(\pi\eta_0 + R)^4 + 3r^2/(2\pi^2\mu^2)} \qquad (R = \sqrt{r/\mu}). \tag{16}$$

This equation describes solitary waves with power decay as $|x_0| \to \infty$ (not exponential decay because of multiplier η_0^3 at right-hand side).

Figure 3a–f presents the profiles of solitary waves calculated for the Eqs. (11), (12), and (13). Here the solid line corresponds to the Eq. (11), the dashed line corresponds to the Eq. (12), and the dotted line corresponds to the Eq. (13). The point (a) gives the broad plateau-shape solitary wave which is precisely described by (11) and (12). The point (b) demonstrates the location of Froude point (F_1, F_2) where all the Eqs. (11), (12) and (13) are in good agreement. The point (c) is out of validity range of Eq. (12) but the solitary wave solution for the Eq. (11) still exists here. The points (d), (e), and (f) demonstrate how the Eq. (13) loses its accuracy by moving along the curve $\Gamma_1(\lambda)$. Solitary wave for Eq. (11) seems to be broadening on the Fig. 3d–f. Indeed, it is the lost of exponential decay near the point P_2. The Fig. 3g illustrates good agreement of Eq. (11) (solid line) and Eq. (16) (dashed line) for the point (g); here Eq. (13) fails to approximate at all.

5 Conclusion

In this paper, we have considered the problem on permanent internal waves at the interface between a homogeneous layer and exponentially stratified layer in a two-fluid system. An ordinary differential equation describing large amplitude solitary waves has been obtained using the long-wave scaling procedure. This equation extends the model suggested by Ovsyannikov et al. (1985) and Miyata (1985) for a two-layer fluid with two homogeneous layers. Parametric range of solitary waves is characterized, including extreme regimes such as broad plateau-shape solitary waves.

Acknowledgments This work was supported by RFBR (grant No 07-01-00309), INTAS–SB RAS (grant No 06-1000013-9236) and Interdisciplinary Program of SB RAS (Project No 113). N.I.M. acknowledges also the support by RFBR–CNRS (grant No 07-01-92212).

References

Benney DJ, Ko DRS (1978) The propagation of long large amplitude internal waves. Stud Appl Math 59:187–199

Camassa R, Choi W, Michallet H, Rusås PO, Sveen JK (2006) On the realm of validity of strongly nonlinear asymptotic approximations for internal waves. J Fluid Mech 549:1–23

Choi W, Camassa R (1999) Fully nonlinear internal waves in a two-fluid system. J Fluid Mech 396:1–36

Grue J, Jensen A, Rusås PO, Sveen JK (2000) Breaking and broadening of internal solitary waves. J Fluid Mech 413:181–217

Helfrich KR, Melville WK (2006) Long nonlinear internal waves. Annu Rev Fluid Mech 38:395–425

Kakutani T, Yamasaki N (1978) Solitary waves on a two-layer fluid. J Phys Soc Jpn 45:674–679

Lamb K, Wan B (1998) Conjugate flows and flat solitary waves for a continuously stratified fluid. Phys Fluids 10:2061–2079

Long RR (1965) On the Boussinesq approximation and its role in the theory of internal waves. Tellus 17(1):46–52

Makarenko NI (1999) Conjugate flows and smooth bores in a weakly stratified fluid. J Appl Mech Technol Phys 40(2):249–257

Maltseva JL (2003) Limiting forms of internal solitary waves. JOMAE Trans ASME 125(1):76–79

Miyata M (1985) An internal solitary wave of large amplitude. La Mer 23(2):43–48

Ovsyannikov LV, Makarenko NI, Nalimov VI, et al. (1985) Nonlinear problems of the theory of surface and internal waves. Nauka, Novosibirsk [in Russian]

Pelinovsky EN, Polukhina O, Slyunaev A, Talipova T (2007) Internal solitary waves. In: Grimshaw R (ed) Solitary waves in fluids. WIT Press, Southhampton, pp 85–110

Turner REL, Vanden-Broeck J-M (1988) Broadening of interfacial solitary waves. Phys Fluids 31:2486–2490

Voronovich AG (2003) Strong solitary internal waves in a 2.5- layer model. J Fluid Mech 474:85–94

Yih CS (1980) Stratified flows. Academic Press, New York

Symbolic Computation for Nonlinear Wave Resonances

E. Tobisch (Kartashova), C. Raab, Ch. Feurer, G. Mayrhofer and W. Schreiner

Abstract Extreme ocean waves are characterized by the energy concentration in a few chosen waves/modes. Frequency modulation due to the nonlinear resonances is one of the possible processes yielding the appearance of independent wave clusters which keep their energy. Energetic behavior of these clusters is defined by (1) integer solutions of the resonance conditions, and (2) coupling coefficients of the dynamical system on the wave amplitudes. General computation algorithms are presented which can be used for arbitrary three-wave resonant system. Implementation in Mathematica is given for planetary ocean waves. Short discussion concludes the paper.

1 Introduction

Resonance is a common thread which runs through almost every branch of physics; without resonance we would not have radio, television, music, etc. Resonance causes an object to oscillate; sometimes the oscillation is easy to see (vibration in a guitar string), but sometimes this is impossible without measuring instruments (electrons in an electrical circuit). A well-known example with Tacoma Narrows Bridge (at the time it opened for traffic in 1940, it was the third longest suspension bridge in the world) shows how disastrous resonances can be: on the morning of 7 November, 1940, the four-month-old Tacoma Narrows Bridge began to oscillate dangerously up and down, tore itself apart and collapsed. Although designed for winds of 120 mph, a wind of only 42 mph caused it to collapse. The experts did agree that somehow

E. Tobisch (Kartashova) (✉)
Institute for Analysis, Johannes Kepler University Linz, Altenbergerstr. 69, 4040 Linz, Austria
e-mail: Lena.Kartashova@risc.uni-linz.ac.at

C. Raab
Radon Institute for Computational and Applied Mathematics (RICAM), Johannes Kepler University Linz, Altenbergerstr. 69, 4040 Linz, Austria

Ch. Feurer · G. Mayrhofer · W. Schreiner
Research Institute for Symbolic Computation (RISC), Johannes Kepler University Linz, Altenbergerstr. 69, 4040 Linz, Austria

E. Pelinovsky and C. Kharif (eds.), *Extreme Ocean Waves*,
DOI 10.1007/978-3-319-21575-4_12

the wind caused the bridge to resonate, and nowadays, wind tunnel testing of bridge designs is mandatory.

Another famous example is the experiments of Tesla who studied in 1898 experimentally vibrations of an iron column and noticed that at certain frequencies specific pieces of equipment in the room would start to jiggle. Playing with the frequency he was able to move the jiggle to another part of the room. Completely fascinated with these findings, he forgot that the column ran downward into the foundation of the building, and the vibrations were being transmitted all over Manhattan. The experiments had started sort of a small earthquake in his neighborhood with smashed windows, swayed buildings, and panicky people in the streets. For Tesla, the first hint of trouble came when the walls and floor began to heave (Cheney 1989). He stopped the experiment as soon as he saw police rushing through the door.

The difference between resonances in a human made system and in some natural phenomena is very simple. We can change the form of a bridge and stop the experiment by switching off electricity but we can not change the direction of the wind, the form of the Earth atmosphere, or the sizes of an ocean. What we can try to do is *to predict* drastic behavior of a real physical system by computing its resonances. While linear resonances in different physical systems are comparatively well studied, to compute characteristics of nonlinear resonances and to predict their properties is quite a nontrivial problem, even in the one-dimensional case. Thus, the notorious Fermi-Pasta-Ulam numerical experiments with a nonlinear *1D*-string (carried out more then 50 years ago) are still not fully understood (Berman and Israilev 2005). On the other hand, nonlinear wave resonances in continuous *2D*-media like ocean, space, atmosphere, plasma, etc. are well studied in the frame of wave turbulence theory (Zakharov et al. 1992) and provide a sound basis for qualitative and sometimes also quantitative analysis of corresponding physical systems. The notion of nonlinear wave interactions is crucial in the wave turbulence theory (Zakharov et al. 2004). Excluding resonances allows to describe a nonlinear wave system statistically, by wave kinetic equations and power-law energy spectra of turbulence (Zakharov and Filonenko 1967), and to observe this behavior in numerical experiments (Pushkarev and Zakharov 2000). Direct computations with Euler equations (modified for gravity water waves, Zakharov et al. 2005) show that the existence of resonances in a wave system yields some additional effects which are not covered by the statistical description. The role of resonances in the evolution of water wave turbulent systems has been studied profoundly by a great number of researchers. One of the most important conclusions (for gravity water waves) made recently in Tanaka (2007) is the following: *The four-wave resonant interactions control the evolution of the spectrum at every instant of time, whereas non-resonant interactions do not make any significant contribution even in a short-term evolution.*

The behavior of a resonant wave system can be briefly described (Kartashova 1998) as follows: (1) not all waves take part in resonant interactions, (2) resonantly interacting waves form a few independent small wave clusters, such that there is no energy flow between these clusters, (3) including some small but non-zero resonance width into consideration *does not* destroy the clusters. A model of laminated wave turbulence (Kartashova 2006a) allows to describe statistical and resonant regimes simultaneously while methods to compute resonances numerically are presented in Kartashova (2006b) (idea) and in Kartashova and Kartashov (2006, 2007a, b) (implementation). Our main purpose here is to study the possibilities of a symbolic implementation of these general algorithms using the computer algebra system Mathematica.

The implemented software can be executed with local installations of Mathematica and the corresponding method libraries; however, we have also developed a Web interface that allows to run the methods from any computer in the Internet via a conventional Web browser. The implementation strategy is simple and based on generally available technologies; it can serve as a blueprint for other mathematical software with similar features.

We take as our principal example the barotropic vorticity equation in a rectangular domain with zero boundary conditions which describes oceanic planetary waves, and show how (a) to compute interaction coefficients of corresponding dynamical systems, (b) to solve resonant conditions, (c) to construct the topological structure of the solution set, and (d) to use the software via a Web interface over the Internet. A short discussion concludes the paper.

2 Mathematical Background

Wave turbulence takes place in physical systems with nonlinear dispersive waves that are described by evolutionary dispersive NPDEs. The role of the evolutionary dispersive NPDEs in the theoretical physics is so important that the notion of dispersion is used for *a physical* classification of PDEs into dispersive and non-dispersive. The well-known mathematical classification of PDEs into elliptic, parabolic, and hyperbolic equations is based on the form of equations and can be applied to the second-order PDEs on an arbitrary number of variables. On the other hand, the physical classification is based on *the form of solutions* and can be applied to PDEs of arbitrary order and arbitrary number of variables. In order to construct the physical classification of PDEs, two preliminary steps are to be made (1) to divide all variables into two groups—time- and space-like variables (t and x correspondingly); and (2) to check that the *linear part* of the PDE under consideration has a wave-like solution in the form of Fourier harmonic

$$\psi(x,t) = A \exp i[kx - \omega t],$$

with amplitude A, wavenumber k, and wave frequency ω. The direct substitution of this solution into the linear PDE shows then that ω is an explicit function on k, for instance,

$$\psi_t + \psi_x + \psi_{xxx} = 0 \quad \Rightarrow \omega(k) = k - 5k^3.$$

If ω as a function on k is real-valued and such that $\mathrm{d}^2\omega/\mathrm{d}k^2 \neq 0$, it is called *a dispersion function* and the corresponding PDE is called evolutionary dispersive PDE. If the dimension of the space variable x is more that 1, i.e. $\vec{x} = (x_1, \ldots, x_p)$, $\vec{k}$ is called the wave-vector and the dispersion function $\omega = \omega(\vec{k})$ depends on the coordinates of the wave-vector. This classification *is not complementary* to a standard mathematical one. For instance, though hyperbolic PDEs normally do not have dispersive wave solutions, the hyperbolic equation $\psi_{tt} - \alpha^2\psi_{xx} - \beta^2\psi = 0$ has them.

In the huge amount of application areas of NPDEs (classical and quantum physics, chemistry, medicine, sociology, etc.) a nonlinear term of the corresponding NPDE can be regarded as small. This is symbolically written as

$$L(\psi) = -\varepsilon N(\psi) \tag{1}$$

where L and N are linear and nonlinear parts of the equation correspondingly and ε is a small parameter defined explicitly by the physical problem setting. It can be shown that in this case the solution ψ of (1) can be constructed as a combination of the Fourier harmonics with amplitudes A depending on the time variable and possessing two properties formulated here for the case of quadratic nonlinearity:

- **P1**. The amplitudes of the Fourier harmonics satisfy the following system of nonlinear ordinary differential equations (ODEs) written for simplicity in the real form

$$\begin{aligned} \dot{A}_1 &= \alpha_1 A_2 A_3 \\ \dot{A}_2 &= \alpha_2 A_1 A_3 \\ \dot{A}_3 &= \alpha_3 A_1 A_2 \end{aligned} \tag{2}$$

  with coefficients α_j being functions on wavenumbers;
- **P2**. The dispersion function and wavenumbers satisfy the *resonance conditions*

$$\begin{cases} \omega(\vec{k}_1) \pm \omega(\vec{k}_2) \pm \omega(\vec{k}_3) = 0, \\ \vec{k}_1 \pm \vec{k}_2 \pm \vec{k}_3 = 0. \end{cases} \tag{3}$$

The transition from (1) to (2) can be performed by some standard methods (for instance, multi-scale method Nayfeh 1981) which also yields the explicit form of resonance conditions.

Keeping in the mind our main problem—to find a solution of (1)—one has to take care of the initial and boundary conditions. This is done in the following way: the case of periodic or zero boundary conditions yields *integer wavenumbers*, otherwise they are real. Correspondingly, one has to find all integer (or real) solutions of (3), substitute corresponding wavenumbers into the coefficients α_j and then look for the solutions of (2) with given initial conditions.

One can see immediately a big problem which appears as soon as one has to solve a NPDE with periodical or zero boundary conditions. Indeed, dispersion functions take different forms, for instance,

$$\omega^2 = k^3, \;\; \omega^2 = k^3 + \alpha k, \;\; \omega^2 = k, \;\; \omega = \alpha/k, \;\; \omega = m/n(n+1) \;\cdots, \text{etc.}$$

with $\vec{k} = (m, n)$, $k = \sqrt{m^2 + n^2}$, and α being a constant. This means that (3) corresponds to a system of Diophantine equations of many variables, normally 6–9, with cumulative degrees 10–16. Those have to be solved usually for the integers of the order $\sim 10^3$, which means that computations have to be performed with integers of order 10^{48} and more. Original algorithms to solve these systems of equations have been developed based on some profound results of number theory (Kartashova 2006b) and implemented numerically Kartashova and Kartashov (2006, 2007a, b).

Further on, an evolutionary dispersive NPDE with periodic or zero boundary conditions is called *three-term mesoscopic system* if it has a solution of the form

$$\tilde{\psi} = \sum_{j=1}^{\infty} A_j \exp i[\vec{k}_j \vec{x}_j - \omega t]$$

and there exists at least one triple $\{A_{j_1}, A_{j_2}, A_{j_3}\} \in \{A_j\}$ such that **P1** and **P2** keep true with some nonzero coefficients $\alpha_j, \; \alpha_j \neq 0 \; \forall j = 1, 2, 3$.

3 Equations for Wave Amplitudes

3.1 Method Description

The barotropic vorticity equation describing ocean planetary waves has the form (Kartashova and Reznik 1992)

$$\frac{\partial \Delta \psi}{\partial t} + \beta \frac{\partial \psi}{\partial x} = -\varepsilon J(\psi, \Delta \psi) \qquad (4)$$

with boundary conditions

$$\psi = 0 \;\; \text{for} \;\; x = 0, L_x; \;\; y = 0, L_y.$$

Here β is a constant called Rossby number, ε is a small parameter and the Jacobean has the standard form

$$J(a, b) = \frac{\partial a}{\partial x}\frac{\partial b}{\partial y} - \frac{\partial a}{\partial y}\frac{\partial b}{\partial x}. \qquad (5)$$

First we give a basic introduction on how a PDE can be turned into a system of ODEs by a multi-scale method. Using operator notation, our problem (4) is viewed as a perturbed version of the linear PDE $L(\psi) = 0$. We pick a solution of this equation, say ψ_0, which is a superposition of several waves φ_j, i.e. $\psi_0 = \sum_{j=1}^{s} A_j\varphi_j$, each being a solution itself. To construct a solution of the original problem, we make the amplitudes time-dependent. As the size of the nonlinearity in (1) is just of order ε, the amplitudes will vary only on time-scales $1/\varepsilon$ times slower than the waves. Hence we define an additional time-variable $t_1 := t\varepsilon$ called "slow time" to handle this time scale. So we look for approximate solutions of (1) that have the following form

$$\psi_0(t, t_1, \vec{x}) = \sum_{j=1}^{s} A_j(t_1)\varphi_j(\vec{x}, t),$$

which for $\varepsilon = 0$ is an exact solution. The exact solution of the equation is written as power series in ε around ψ_0, i.e., $\psi = \sum_{k=0}^{\infty} \psi_k \varepsilon^k$. In our computation, it is truncated up to maximal order m which in our case is $m = 1$, i.e.

$$\psi(t, t_1, \vec{x}) = \psi_0(t, t_1, \vec{x}) + \psi_1(t, t_1, \vec{x})\varepsilon.$$

Plugging $\psi(t, t_1, \vec{x})$ one has to keep in mind that, since $t_1 = \varepsilon t$, we now have $\frac{\mathrm{d}}{\mathrm{d}t} = \frac{\partial}{\partial t} + \varepsilon\frac{\partial}{\partial t_1}$ due to the chain rule. Equations are formed by comparing the coefficients of ε^k. For $k = 0$, this gives back the linear equation, but we keep the equation for $k = 1$. In particular, for (4) we arrive at

$$\begin{aligned} \frac{\partial \Delta\psi_0}{\partial t} + \beta\frac{\partial \psi_0}{\partial x} &= 0, \\ \frac{\partial \Delta\psi_0}{\partial t_1} + \frac{\partial \Delta\psi_1}{\partial t} + \beta\frac{\partial \psi_1}{\partial x} &= -J(\psi_0, \Delta\psi_0). \end{aligned}$$

In order to (2), we have to get rid of all other variables. This is done by integrating against the φ_j's, i.e. $\langle ., \varphi_j\rangle_{L^2(\Omega)}$, and averaging over (fast) time, i.e. $\lim_{T\to\infty} \frac{1}{T}\int_0^T \mathrm{d}t$.

3.2 The Implementation

This method was implemented in Mathematica with order $m = 1$ in mind only. So it will not be immediately applicable to higher orders without some (minor) adjustments. The ODEs are constructed done by the function

```
ODESystem[L(ψ), N(ψ), ψ,
   {x₁,..,xₙ}, t, domain, jacobian, m, s, A, linwav,
   {λ₁,..,λₚ}, paramvalues].
```

Basically, this function takes the problem together with the solution of the linear equation as input and computes the list of ODEs for the amplitudes as output. Its arguments are in more detail:

- `L(ψ), N(ψ)`: Linear and nonlinear part of equation (1), each applied to a symbolic function parameter. Derivatives have to be specified with `Dt` instead of `D` and the nonlinear part has to be a polynomial in the derivatives of the function.
- ψ: symbol used for function in `L(ψ), N(ψ)`
- `{x`$_1$`,...,x`$_n$` }, t`: list of symbols used for space-variables, and symbol for time-variable.
- `domain`: The domain on which the equation is considered has to be specified in the form `{{x`$_1$`,minx`$_1$`,maxx`$_1$`},..., {x`$_n$`,minx`$_n$`,maxx`$_n$`}}`, where the bounds on `x`$_i$ may depend on `x`$_1$`,...,x`$_{i-1}$ only.
- `jacobian`: For integration the (determinant of the) Jacobian must also to be passed to the function. This is needed in case the physical domain does not coincide with the domain of the variables above; it can be set to `1` otherwise.
- `m, s`: maximal power of ε and number of waves considered
- `A`: symbol used for amplitudes
- `linwav`: General wave of the linear equation is assumed to have separated variables, i.e. $\varphi(\vec{x}, t) = B_1(x_1) \dots B_n(x_n) \exp(i\theta(x_1, \dots, x_n, t))$, and has to be given in the form `{B`$_1$`(x`$_1$`),..., B`$_n$`(x`$_n$`), θ(x`$_1$`,...,x`$_n$`,t)}`.
- `{ λ`$_1$`,...,λ`$_p$`}`: list of symbols of parameters the functions in `linwav` depend on
- `paramvalues`: For each of the `s` waves explicit values of the parameters $\{\lambda_1,\dots,\lambda_p\}$ have to be passed as a list of `s` vectors of parameter values.

```
ODESystem[linearpart_,nonlinearpart_,fun_Symbol,vars_List,
    t_Symbol,domain_List,jacobian_,ord_Integer,num_Integer,
    A_Symbol,linwav_List,params_List,paramvalues_List] :=
  Module[{B,theta,eq,k},
    eq = PerturbationEqns[linearpart,nonlinearpart,
                          fun,vars,t,ord];
    eq = PlugInGenericWaveTuple[eq,fun,vars,t,A,B,theta,num]
            /. fun[1]->(0&);
    eq = Table[Resonance2[eq,linwav,vars,t,params,A,B,theta,
                 num,paramvalues,k],
             {k,num}];
    Map[Integrate[Simplify[#,And@@(Function[B,B[[2]]<B[[1]]<
                    B[[3]]]/@domain)]*jacobian,
                 Sequence@@domain]&,
        eq,{2}]
    ]
```

Internally this function is divided into three subroutines briefly described below.

3.2.1 Perturbation Equations, General Form

The first of the subroutines is

PerturbationEqns[L(ψ), N(ψ), ψ, {x$_1$, ...,x$_n$}, t, m].

As mentioned before, we approximate the solution of our problem by a polynomial of degree m in ε. This subroutine works for arbitrary m. In the first step, we construct equations by coefficient comparison. Additional time-variables will be created automatically and labeled t[1], ...,t[m]. The output is a list of $m+1$ equations corresponding to the powers $\varepsilon^0, \dots, \varepsilon^m$. The implementation is quite straightforward. First set $\psi = \sum_{k=0}^{m} \psi_k(t, t_1, \dots, t_m, x_1, \dots, x_n)\varepsilon^k$ in (1), where $t_k = \varepsilon^k t$, i.e. $\frac{\mathrm{d}}{\mathrm{d}t} = \frac{\partial}{\partial t} + \sum_{k=1}^{m} \varepsilon^k \frac{\partial}{\partial t_k}$. Then extract the coefficients of $\varepsilon^0, \dots, \varepsilon^m$ on both sides and assemble the equations. Finally replace $\varepsilon^k t$ by t_k again.

```
PerturbationEqns[linearpart_,nonlinearpart_,fun_Symbol,
    vars_List,time_Symbol,ord_Integer] :=
  Module[{i,j,e,eq},
    eq = ((linearpart == -e*nonlinearpart)
          /. {fun->Sum[e^i*fun[i][time,Sequence@@Table[e^j*
                              time,{j,ord}],Sequence@@
                              DeleteCases[vars,time]],
                      {i,0,ord}]});
    eq = (eq /. ((Dt[#, __]->0)& /@ Join[vars,{time,e}]));
    eq = (Equal@@#)& /@
          Transpose[Take[CoefficientList[#,e],1+ord]& /@
                      (List@@eq)];
    eq /. Table[e^j*time->time[j],{j,ord}]
    ]
```

3.2.2 Perturbation Equations, Given Linear Mode

In step two, we set $\psi_0(t, t_1, \vec{x}) = \sum_{j=1}^{s} A_j(t_1)\varphi_j(\vec{x}, t)$ as described above. This is done by the function

PlugInGenericWaveTuple[eq, ψ, {x$_1$, ...,x$_n$}, t, A, B, θ , s] where the first argument is the output of the previous step. The symbols B and θ have to be passed for labeling the shape and phase functions, respectively. The output consists of two parts. The first part of the list formulates the assumption $L(\varphi_j) = 0$ explicitly for each of the waves. This is not used in subsequent computations, but is provided as a way to check the assumption. The second part of the list is the equation corresponding to the coefficients of ε from the previous step, with ψ_0 as above. As the task of this step is so short the implementation does not need further explanation.

```
PlugInGenericWaveTuple[eq_List,fun_Symbol,vars_List,
  t_Symbol,A_Symbol,B_Symbol,theta_Symbol,num_Integer] :=
  Module[{i,j,waves,n=Length[DeleteCases[vars,t]]},
    waves = Table[A[j][Slot[2]]*
                  Product[B[i][j][Slot[i+2]],{i,n}]*
                  Exp[I*theta[j][Sequence@@Table[Slot[i+2],
                                    {i,n}],Slot[1]]],
                {j,num}];
   {Table[eq[[1]] /. fun[0]->Function[Evaluate[waves[[j]]]],
          {j,num}],
    Expand /@
```

```
        (eq[[2]] /. fun[0]->Function[Evaluate[Total[waves]]])
    }]
```

3.2.3 Time and Scale Averaging

Step three is the most elaborate. Under the assumption that interchange of averaging over time and inner product is justified, an integrand

$$h = \lim_{T\to\infty} \frac{1}{T} \int_0^T \psi_0 \overline{\varphi_k}\, \mathrm{d}t$$

is computed that when integrated over the domain yields

$$\int_\Omega h = \lim_{T\to\infty} \frac{1}{T} \int_0^T \langle \psi_0, \varphi_k \rangle_{L^2(\Omega)}\, \mathrm{d}t.$$

Resonance conditions posed on the phase functions are explicitly used by

```
Resonance[eq, linwav, {x1,..,xn}, t,
     {λ1,..,λp}, A, B, θ, s, cond, k]
```

which receives the output from the previous step in `eq`. Here `cond` specifies the resonance condition in terms of the θ_j, which have to be entered as `θ[j][x1,..,xn,t]` respectively. The last argument is the index of the wave φ_k in the integral above. Alternatively `Resonance2` uses explicit parameter settings `paramvalues` for the waves instead of `cond`. This has been necessary because the general `Resonance` does not give useable results (see Sect. 3.3 for more details). The main work in this step is to find out which terms do not contribute to the result. We exploit the fact that oscillating terms vanish when averaged over time by simply omitting those summands of $\langle \psi_0, \varphi_k \rangle_{L^2(\Omega)}$ that have a factor $\exp(i\theta)$ with some time-dependent phase θ. The code for `Resonance` is not shown here, but is quite similar to `Resonance2`.

```
Resonance2[eq_List,linwav_List,vars_List,t_Symbol,params_List,
    A_Symbol,B_Symbol,theta_Symbol,num_Integer,
    paramvalues_List,testwave_Integer] :=
  Module[{e,i,j,n=Length[DeleteCases[vars,t]]},
    e = Expand[(List@@Last[eq])*
               Exp[-I*theta[testwave][Sequence@@
                                            DeleteCases[vars,t],
                                        t]]];
    e = e /.
         Table[
           theta[j] ->
             (Evaluate[(linwav[[n+1]] /.
                         (Rule@@#& /@
                           Transpose[{params,paramvalues[[j]]}]
                         )
                       ) /. Append[Table[
```

```
                                          DeleteCases[vars,t][[i]]
                                            -> Slot[i],
                                          {i,n}],
                                        t -> Slot[n+1]]
                    ]&
            ),
          {j,num}];
    e = MapAt[
          (Function[theta,If[FreeQ[theta,t],theta,0]
               ]
             [Simplify[#]]
          )&,
          e,
          Position[e,Exp[_]]];
    e = Equal@@
          (e*Conjugate[A[testwave]][t[1]]*
            Product[Conjugate[B[i]
                                 [testwave]
                                 [DeleteCases[vars,t][[i]]]
                               ],
                    {i,n}]
          ) /.
            Flatten[
              Table[B[i][j] ->
                      Function[
                        Evaluate[DeleteCases[vars,t][[i]]],
                        Evaluate[linwav[[i]] /.
                                   (Rule@@#& /@
                                     Transpose[
                                       {params,paramvalues[[j]]
                                       }]
                                   )]],
                    {i,n},{j,num}]]
    ]
```

The integration of h is done by Mathematica and can be quite time-consuming. So `ODESystem` simplifies the integrand first to make integration faster. Still the expressions involved can be quite complicated. This is the most time-consuming part during construction of the ODEs.

3.3 Obstacles

Mathematica sometimes does not seem to take care of special cases and consequently has problems with evaluating expressions depending on symbolic parameters. We give two simple examples to illustrate this issue:

- Orthogonality of sine-functions.
  Indeed, it holds that

$$\forall m, n \in \mathbb{N} : \int_0^{2\pi} \sin(mx)\sin(nx)dx = \pi\delta_{m,n}.$$

Computing this in Mathematica by

```
Integrate[Sin[m*x]Sin[n*x], {x,0,2π},
    Assumptions → m∈Integers && n∈Integers]
```

yields 0 independently of m, n instead.

- Computation of a limit.
  Mathematica evaluates an expression

$$\forall n \in \mathbb{Z} : \lim_{x\to n} \frac{\sin(x\pi)}{x} = \pi\delta_{n,0}$$

and similar expressions in two different ways getting two different answers. On the one hand
`Limit[Sin[(m-n)π]/(m-n), m→n,`
`Assumptions → m∈Integers && n∈Integers]`
gives 0. On the other hand, however, when the condition $m, n \in \mathbb{Z}$ is not used for computing, the result Mathematica yields the correct answer π, as with
`Limit[Sin[(m-n)π]/(m-n), m→n]`.

Unfortunately, these issues prevented us from obtaining a nice formula for the coefficients in symbolic form by `Resonance`. So we just compute results for explicit parameter settings using `Resonance2`.

3.4 Results

3.4.1 Atmospheric Planetary Waves

For the validation of our program, we consider the barotropic vorticity equation on the sphere first. Here numerical values of the coefficients α_i are available (Table 1, Kartashova and L'vov 2007). The equation looks quite similar

$$\frac{\partial \Delta\psi}{\partial t} + 2\frac{\partial\psi}{\partial\lambda} = -\varepsilon J(\psi, \Delta\psi).$$

However in spherical coordinates ($\phi \in [-\frac{\pi}{2}, \frac{\pi}{2}]$, $\lambda \in [0, 2\pi]$) the differential operators are different:

$$\Delta = \frac{\partial^2}{\partial\phi^2} + \frac{1}{\cos(\phi)^2}\frac{\partial^2}{\partial\lambda^2} - \tan(\phi)\frac{\partial}{\partial\phi},$$

$$J(a, b) = \frac{1}{\cos(\phi)}\left(\frac{\partial a}{\partial\lambda}\frac{\partial b}{\partial\phi} - \frac{\partial a}{\partial\phi}\frac{\partial b}{\partial\lambda}\right).$$

The linear modes have in this case the following form (Pedlosky 1987)

$$P_n^m(\sin(\phi))\exp\left(i(m\lambda + \frac{2m}{n(n+1)}t)\right), \tag{6}$$

where $P_n^m(\mu)$ are the associated Legendre polynomials of degree n and order $m \le n$, so again they depend on the two parameters m and n. Also resonance conditions on the parameters look different in this case.

Now we compute the coefficient α_3 in (2). In Kartashova and L'vov (2007), we find the following equation for the amplitude A_3

$$n_3(n_3+1)\frac{\partial A_3}{\partial t_1}(t_1) = 2iZ(n_2(n_2+1) - n_1(n_1+1))A_1(t_1)A_2(t_1),$$

so $\alpha_3 = 2iZ\frac{n_2(n_2+1)-n_1(n_1+1)}{n_3(n_3+1)}$. Parameter settings and corresponding numerical values for Z were taken from the table below (see Kartashova and L'vov 2007). For this equation and $s = 3$, results produced by our program have the form $c_1\overline{A_3}\dot{A}_3 = c_2A_1A_2\overline{A_3}$, so $\alpha_3 = c_2/c_1$.

Testing all resonant triads from the Table 1 from Kartashova and L'vov (2007), we see that the coefficients differ merely by a constant factor of $\pm\sqrt{8}$ which is due to the different scalings of the Legendre polynomials. In our computation, they were normalized s.t. $\int_{-1}^{1} P_n^m(\mu)^2 d\mu = 1$. With three triads, however, results were completely different. Interestingly, these were exactly those triads for which no φ_0 appears in the table.

Furthermore, for the other coefficients in (2), our program computes $\alpha_1 = \alpha_2 = 0$ in all tested parameter settings. This fact can be easily understood in the following way. We checked only resonance conditions but not the conditions for the interaction coefficients to be non-zero which are elaborated enough:

$$m_i \le n_i, \quad n_i \ne n_j \quad \forall i = 1, 2, 3, \quad |n_1 - n_2| < n_3 < n_1 + n_2,$$

and

$$n_1 + n_2 + n_3 \quad \text{is odd.}$$

Randomly taken parameter setting does not satisfy these conditions.

3.4.2 Ocean Planetary Waves

Returning to the original example on the domain $[0, L_x] \times [0, L_y]$, we find explicit formulae for the coefficients in Kartashova and Reznik (1992). According to Sect. 3.3 we can only verify special instances and not general formulae.

Linear modes have now the form (Kartashova and Reznik 1992)

$$\sin\left(\pi\frac{mx}{L_x}\right)\sin\left(\pi\frac{ny}{L_y}\right)\exp\left(i(\frac{\beta}{2\omega}x+\omega t)\right), \tag{7}$$

with $m, n \in \mathbb{N}$ and $\omega = \frac{\beta}{2\pi\sqrt{(\frac{m}{L_x})^2+(\frac{n}{L_y})^2}}$.

Parameter settings solving the resonance conditions were computed as in Sect. 4. Unfortunately results do not match and we have no explanation for that. In particular the condition $\frac{\alpha_1}{\omega_1^2}+\frac{\alpha_2}{\omega_2^2}+\frac{\alpha_3}{\omega_3^2}=0$ stated in Kartashova and Reznik (1992) does not hold for the results of our program since we got $\alpha_1=\alpha_2=0$ in all tested parameter settings, just as in the spherical case.

For example, if we try the triad `{{2,4},{4,2},{1,2}}` where $L_x = L_y = 1$ our program computes $\alpha_3 = \frac{32\sqrt{5}}{11}\pi\left(\sin(3\sqrt{5}\pi) - i(1+\cos(3\sqrt{5}\pi))\right)$, whereas the general formula yields $\alpha_3 = \frac{19+7\sqrt{5}}{11}\pi\sin(3\sqrt{5}\pi)$. However, if we use a triad with $q=1$, e.g. `{{24,18},{9,12},{8,6}}`, both agree on $\alpha_1=\alpha_2=\alpha_3=0$.

4 Resonance Conditions

The main equation to solve is

$$\frac{1}{\sqrt{\left(\frac{m_1}{L_x}\right)^2+\left(\frac{n_1}{L_y}\right)^2}}+\frac{1}{\sqrt{\left(\frac{m_2}{L_x}\right)^2+\left(\frac{n_2}{L_y}\right)^2}}=\frac{1}{\sqrt{\left(\frac{m_3}{L_x}\right)^2+\left(\frac{n_3}{L_y}\right)^2}}$$

for all possible $m_i, n_i \in \mathbb{Z}$ with the scales L_x and L_y (also $\in \mathbb{Z}$) and then to check the condition $n_1 \pm n_2 = n_3$. In the following argumentation, it will be seen that L_x and L_y can be assumed to be free of common factors. Below we refer to L_x and L_y as to *the scale coefficients*.

The first step of the algorithm implemented in Mathematica is to rewrite the equation to $\frac{1}{\sqrt{\tilde{m_1}^2+\tilde{n_1}^2}}+\frac{1}{\sqrt{\tilde{m_2}^2+\tilde{n_2}^2}}=\frac{1}{\sqrt{\tilde{m_3}^2+\tilde{n_3}^2}}$ and transform it in the following way: we factorize the result of each $\tilde{m_i}^2+\tilde{n_i}^2$ and obtain with $\rho_1\cdot\dots\cdot\rho_r$ being the factors of $m_i^2+n_i^2$ and $\alpha_1\cdot\dots\cdot\alpha_r$ their respective powers:

$$m_i^2+n_i^2=\rho_1^{\alpha_1}\cdot\rho_2^{\alpha_2}\cdot\dots\cdot\rho_r^{\alpha_r}.$$

We will now define *a weight* γ_i *of the wave-vector* (m_i, n_i) as the product of the ρ_j's to the quotient of their respective α_j and 2. The weight q_i will be the name of the product of the ρ_j's which have an odd exponent:

$$\sqrt{m_i^2+n_i^2}=\gamma_i\sqrt{q_i}.$$

Our equation then can be re-written as

$$\frac{1}{\gamma_1\sqrt{q_1}} + \frac{1}{\gamma_2\sqrt{q_2}} = \frac{1}{\gamma_3\sqrt{q_3}}$$

and one easily sees that the only way for the equation to possibly hold is $q_1 = q_2 = q_3 = q$ (see Kartashova 2006b for details). Further we call q *an index* of the corresponding wave-vectors. The set of all wave-vectors with the same index is called *a class of index* q and is denoted as Cl_q. Obviously, the solutions of the resonance conditions are to be searched for with separate classes only.

At this point, one can also see that only such scales, L_x and L_y, without common factors are reasonable. If they had a common factor, it would cancel out in the equation.

4.1 Method Description

The following five steps are the main steps of the algorithm:

- *Step 1*: Compute the list of all possible indexes q.
  To compute the list of all indexes q, we use the fact that they have to be square-free and each factor of q has to be different from 3 mod 4 (Lagrange theorem). There exist 57 possible indexes in our computational domains $q \leq 300$:

  $$\{1, 2, 5, 10, 13, 17, 26, 29, 34, 37, 41, 53, 58, 61, 65, 73, 74, 82, 85, 89, 97, 101, 106, 109, 113, 122, 130, 137, 145, 146, 149, 157, 170, 173, 178, 181, 185, 193, 194, 197, 202, 205, 218, 221, 226, 229, 233, 241, 257, 265, 269, 274, 277, 281, 290, 293, 298\}$$

- *Step 2*: Solve the weight equation $\frac{1}{\gamma_1} + \frac{1}{\gamma_2} = \frac{1}{\gamma_3}$.
  For solving the weight equation, we transform it into the equivalent form:

  $$\gamma_3 = \frac{\gamma_1\,\gamma_2}{\gamma_1 + \gamma_2} \tag{8}$$

  The solution triples $\{\gamma_1, \gamma_2, \gamma_3\}$ can now be found by the two for-loops over γ_1 and γ_2 up to a certain maximum parameter and γ_3 is then being founded constructively with formula (8).
- *Step 3*: Compute all possible pairs (m_i, n_i)—if there are any—that satisfy $m_i^2 + n_i^2 = \gamma_i^2\, q$.
  To compute our initial variables m_i, n_i, we use the Mathematica standard function **Sum Of Square Representation [d, x]** which produces a list of all possible representations of an integer x as a sum of d squares, i.e. we can find all possible

pairs (a, b) with $d = 2$ such that they satisfy $a^2 + b^2 = x$. Therefore, checking the condition $m_i^2 + n_i^2 = \gamma_i^2 q$ is easy.

- *Step 4*: Sort out the solutions $\{m1, n1, m2, n2, m3, n3\}$ that do not fulfill the condition $n1 \pm n2 = n3$.
- *Step 5*: Check if by dividing the m_i by L_x and the n_i by L_y there are still exist some solutions.
  Last two steps are trivial.

4.2 The Implementation

Our implementation is quite straightforward and the main program is based on four auxiliary functions shown in the following subsections.

4.2.1 List of Indexes

The function **constructqs[max]** produces the list of all possible indexes q up to the parameter max. The first (obvious) q's $sol = \{1\}$ is given and the function checks the conditions starting with $n = 2$. Every time n satisfies the conditions, it is appended to the list sol. If one condition fails, the next $n = n + 1$ is considered and so on until n reaches the parameter max. Then the list sol is returned:

```
Clear[constructqs];

constructqs[n_, sol_List, max_]; n>max := sol (*6*)
constructqs[n_?SquareFreeQ, sol_List, max_]
:= constructqs[n+1, Append[sol, n], max] (*5*)

constructqs[n_?SquareFreeQ, sol_List, max_];
MemberQ[Mod[PrimeFactorList[n], 4], 3]
:= constructqs[n+1, sol, max] (*4*)

constructqs[n_, sol_List, max_]; !SquareFreeQ[n]
:= constructqs[n+1, sol, max] (*3*)
constructqs[1] := {1} (*2*)

constructqs[max_] := constructqs[3, {1}, max] (*1*)
```

4.2.2 Weight Equation

The function **findγs[γmax]** solves the weight equation in the following way. For a fixed γ_1 and γ_2 running between 1 and γmax, it is checked if γ_3 is an integer. If it is, the triple $\{\gamma_1, \gamma_2, \gamma_3\}$ is added to the list sol which is empty at the initial moment. Once γ_2 reaches γmax, it is set to 1 again and the search starts again with $\gamma_1 = \gamma_1 + 1$. This is done as long as both γ_1 and γ_2 are lower than max. Finally the list sol is returned:

```
findγs[γmax_, γ1_, γ2_, sol_List];
γ1 > γmax := (Clear[γ3],sol) (*6*)

findγs[γmax_, γ1_, γ2_, sol_List]; (γ1 ≤ γmax && γ2>γmax &&
IntegerQ[γ3=(γ1γ2)/(γ1+γ2)])
:= findγs[γmax, γ1+1, 1, Append[sol, {γ1, γ2, γ3}]] (*5*)

findγs[γmax_, γ1_, γ2_, sol_List];
(γ1 ≤ γmax && γ2>γmax &&
!IntegerQ[γ3=(γ1γ2)/(γ1+γ2)])
:= findγs[γmax, γ1 + 1, 1, sol] (*4*)

findγs[γmax_, γ1_, γ2_, sol_List];
(γ1 ≤ γmax && γ2 ≤ γmax && IntegerQ[γ3=(γ1γ2)/(γ1+γ2)])
:= findγs[γmax, γ1, γ2 + 1, Append[sol, {γ1, γ2, γ3}]] (*3*)

findγs[γmax_, γ1_, γ2_, sol_List];
(γ1 ≤ γmax && γ2 ≤ γmax && !IntegerQ[γ3=(γ1γ2)/(γ1+γ2)])
:= findγs[γmax, γ1, γ2 + 1, sol] (*2*)

findγs[γmax_] := findγs[γmax, 1, 1, {}]) (*1*)
```

For **findγs[γmax]** to be executable, the iteration depth of 2^{12} is not sufficient and it was set to ∞.

4.2.3 Linear Condition

The third auxiliary function **makemns** checks whether the linear condition $n_1 \pm n_2 = n_3$ is fulfilled and structures the solution set into a list of pairs $\{\{m_1, n_1\}, \{m_2, n_2\}, \{m_3, n_3\}\}$:

```
Clear[makemns];
makemns[m1_, n1_, m2_, n2_, m3_, n3_] := {} (*3*)
makemns[m1_, n1_, m2_, n2_, m3_, n3_];
(n1 + n2 == n3 || n1 - n2 == n3) :=
  {{m1, n1}, {m2, n2}, {m3, n3}} (*2*)
makemns[mn1_List, mn2_List, mn3_List] :=
  Cases[Flatten[Table[makemns[mn1[[i,1]], mn1[[i,2]],
     mn2[[j,1]], mn2[[j,2]], mn3[[k,1]], mn3[[k,2]]],
     {i, 1, Length[mn1]}, {j, 1, Length[mn2]},
     {k, 1, Length[mn3]}], 2],
     {{x1_,x2_}, {x3_,x4_}, {x5_,x6_}}] (*1*)
```

The function **makemns** is called three times:

In (*1*) from three lists of arbitrarily many pairs {mi, ni}, a three-dimensional array is made combining entries of the three lists with each other. Each entry calls the same program with the parameters of the current combination of {m1, n1, m2, n2, m3, n3}.

In (*2*) and (*3*) it is decided whether the condition $n1 \pm n2 = n3$ is fulfilled. If it is, a solution {{m1, n1},{m2, n2},{m3, n3}} is written in the array. The table is then flattened to the level 2 in order to have a list of solutions. In the end, all empty lists have to be sorted out, done by the function **Cases** which keeps only those cases that have the shape {{x1_,x2_},{x3_,x4_},{x5_,x6_}}.

4.2.4 Scale Coefficients

Finally, the function **respectL[sol, Lx, Ly]** divides each component of the solution by the pair (L_x, L_y) and sorts out the result if any of the six components does not remain an integer:

```
respectL[sol_List, Lx_, Ly_] :=
  Map[solution[#]&,
     Cases[Map[#/{Lx, Ly}&,
     Map[#[[1]]]&, sol], {2}], {{_Integer, _Integer},
    {_Integer, _Integer}, {_Integer,  _Integer}}]]
```

The function **respectL[sol, Lx, Ly]** gets as an input the list of the form {solution[{{m1,n1},{m2,n2},{m3,n3}}],...} and returns the list of the same form.

4.3 Results

All solutions in the computation domain $m, n \leq 300$ have been found in a few minutes. Notice that computations in the domain $m, n \leq 20$ by direct search, without introducing indexes q and classes Cl_q took about 30 min. A direct search in the domain $m, n \leq 30$ has been interrupted after 2 h, since no results were produced.

The number of solutions depends drastically on the scales L_x and L_y, some data are given below (for the domain $m, n \leq 50:$)

$(L_x = 1, L_y = 1)$: 76 solutions;
$(L_x = 3, L_y = 1)$: 23 solutions;
$(L_x = 6, L_y = 16)$: 2 solutions;
$(L_x = 5, L_y = 21)$: 2 solutions;
$(L_x = 11, L_y = 29)$: no solutions (search up to 300, for both $qmax$ and γmax).

Interestingly enough, in all tried possibilities, only an odd q yield solutions.

5 Structure of the Solution Set

5.1 Method Description

The graphical way to present 2D-wave resonances suggested in Kartashova (1998) for three-wave interactions is to regard each 2D-vector $\vec{k} = (m, n)$ as a node (m, n) of integer lattice in the spectral space and connect those nodes which construct one solution (triad, quartet, etc.). Having computed already all the solutions of (3) in Sect. 4, now we are interested in the structure of resonances in spectral space. To each node (m, n) we can prescribe an amplitude $A(m, n, t_1)$ whose time evolution can be computed from the dynamical equations obtained in Sect. 3. Thus, solution set of resonance conditions (3) can be thought of as a collection of triangles, some of them are isolated, some form small groups connected by one or two vertices. Corresponding dynamical systems can be re-constructed from the structure of these groups. For instance, a single isolated triangle corresponding to a solution with wave vectors $(m_1, n_1)(m_2, n_2)(m_3, n_3)$ and wave amplitudes $\{(A1, A2, A3)\}$ corresponds to the following dynamical system:

$$\begin{aligned}
\dot{A_1} &= \alpha_1 A_2 A_3 \\
\dot{A_2} &= \alpha_2 A_1 A_3 \\
\dot{A_3} &= \alpha_3 A_1 A_2
\end{aligned}$$

with α_i being functions of all m_i, n_i (see Sect. 3).

If that two triangles share one common vertex $\{(A1, A2, A3), (A3, A4, A5)\}$, the corresponding dynamical system is

$$\begin{aligned}
\dot{A_1} &= \alpha_1 A_2 A_3 \\
\dot{A_2} &= \alpha_2 A_1 A_3 \\
\dot{A_3} &= \alpha_{3,1} A_1 A_2 + \alpha_{3,2} A_4 A_5 \\
\dot{A_4} &= \alpha_4 A_3 A_5 \\
\dot{A_5} &= \alpha_5 A_3 A_4
\end{aligned}$$

If two triangles have two vertices in common $\{(A1, A2, A3), (A2, A3, A4)\}$, then the dynamical system is quite different:

$$\begin{aligned}
\dot{A_1} &= \alpha_1 A_2 A_3 \\
\dot{A_2} &= \alpha_{2,1} A_1 A_3 + \alpha_{2,2} A_3 A_4 \\
\dot{A_3} &= \alpha_{3,1} A_1 A_2 + \alpha_{3,2} A_2 A_4 \\
\dot{A_4} &= \alpha_4 A_2 A_3 = \frac{\alpha_4}{\alpha_1} \dot{A_1}
\end{aligned}$$

Using the fourth equation, the formulae for $\dot{A_2}$ and $\dot{A_3}$ can be simplified to

$$\dot{A}_4 = \frac{\alpha_4}{\alpha_1}\dot{A}_1 \Rightarrow A_4 = \frac{\alpha_4}{\alpha_1}A_1 + \beta_1$$
$$\dot{A}_2 = A_1 A_3\left(\alpha_{2,1} + \frac{\alpha_{2,2}\alpha_4}{\alpha_1}\right) + \frac{\alpha_4\beta_1}{\alpha_1}$$
$$\dot{A}_3 = A_1 A_2\left(\alpha_{3,1} + \frac{\alpha_{3,2}\alpha_4}{\alpha_1}\right) + \frac{\alpha_4\beta_1}{\alpha_1}.$$

This means that *qualitative dynamics* of the three-term mesoscopic system depends *not on the geometrical structure* of the solution set but on its *topological structure*. Constructing the topological structure of the solution set, we do not consider concrete values of the solution but only the way how triangles are connected. In any finite spectral domain, we can compute all independent wave clusters and write out corresponding dynamical systems thus obtaining complete information about energy transfer through the spectrum. Of course, *quantitative* properties of the dynamical systems depend on the specific values of m_i, n_i (for instance, values of interaction coefficients α_i, magnitudes of periods of the energy exchange among the waves belonging to one cluster, etc.).

5.2 Implementation

To construct the topological structure of a given solution set we need first to find all groups of connected triangles. This is done by the following procedure:

```
FindConnectedGroups[triangles_List] :=
 Block[{groups = {}, tr = triangles, newgroup},
   While[Length[tr] > 0,
     {newgroup, tr} =
       FindConnectedTriangles[{First[tr]}, Rest[tr]];
     groups = Append[groups, newgroup];
   ];
   groups
 ];

FindConnectedTriangles[grp_List,triangles_List]:=
 Module[{points,newGrpMember,tr=triangles},
   points=Flatten[Apply[List,grp,2],1];
   newGrpMember=Cases[tr, _[___,#1,___]]&/@points;
   (tr=DeleteCases[tr, _[___,#1,___]])&/@points;
   newGrpMember=Union[Join@@newGrpMember];
   If[Length[newGrpMember]==0,
     {grp,tr},
     newGrpMember=FindConnectedTriangles[newGrpMember,tr];
     {Join[grp,First[newGrpMember]],
      newGrpMember[[2]]}
   ]
 ];
```

The function `FindConnectedGroups` expects a list of triangles as input, and three different types for data structure can be used. The first type is just a list of three pairs, where each pair contains the coordinates of a node, for example `{{1,2},{3,4},{5,6}}`. An alternative type is like the type before just with another head symbol instead of list, e.g. `Triangle[{1,2},{3,4},{5,6}]`.

The function also works for vertex numbers instead of coordinates, e.g. `Triangle[1, 2, 3]`. In every case, the function returns a partition of the input list where all elements of a list are connected and elements of different lists have no connection to each other.

The function `FindConnectedTriangles` is an auxiliary function which has two parameters. The first list contains allconnected triangles. The second list contains all other triangles which are possibly connected to one of the triangles in the first list. The function `FindConnectedTriangles` returns a pair of lists: the first list contains all triangles which are connected to the selected triangles, the second list contains all remaining.

The input list for `FindConnectedTriangles` is a list of 3-element lists. Before we can use the results produced in Sect. 4 as an input we have to transform the data. This can be easily done by

```
TransformSolution[sol_List]:=
  Flatten[Rest/@sol]/.solution[trs:{___List}]->trs;
```

Some remarks on the implementation

The function `FindConnectedGroups` selects a triangle, which is not yet in a group and calls the function `FindConnectedTriangles`. Since the returned first list always contains at least one triangle, the length of the list `tr` decreases in every loop call, hence the `FindConnectedGroups` terminates. The question left is how to find all triangles connected with a certain triangle. This has been done in the following way. First we search for all triangles which share at least one node with this triangle. Then we restart the search with all triangles found. For efficiency reasons, it is better to perform the search with all triangles we found in one step together. If in one step no further triangles are found then we are ready and return the list of connected triangles and the remaining list. In each step, we remove all triangles we found from the list of triangles which are not declared as connected. This increases the speed because the search is faster if there are less elements to compare. More importantly, this prevents us to search in loops and finds some triangles more than once. In general, search in a loop can be the reason for a termination problem but due to shrinking the list of triangles to search for in every step the termination can be guaranteed.

5.3 Results

In Fig. 1 the geometrical structure of the solution set is shown, for the case $m_i, n_i \leq 50$ and $L_x = L_y = 1$.

Below we show all the topological elements of this solution set.

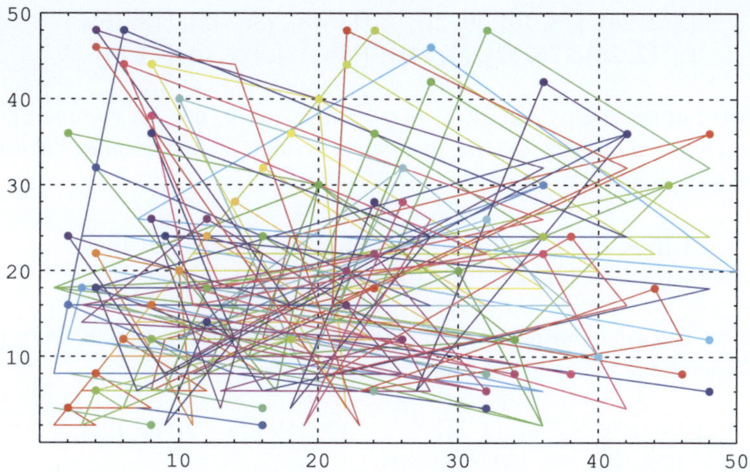

Fig. 1 The geometrical structure of the result in domain $D = 50$

1. Twenty-one groups contain only one triangle (obviously, they have isomorphic dynamical systems):

{{3, 18}, {36, 6}, {2, 12}} {{4, 46}, {14, 44}, {23, 2}}
{{6, 44}, {36, 26}, {13, 18}} {{6, 48}, {42, 24}, {3, 24}}
{{8, 26}, {16, 22}, {13, 4}} {{9, 24}, {48, 18}, {16, 6}}
{{14, 28}, {28, 14}, {7, 14}} {{18, 36}, {36, 18}, {9, 18}}
{{22, 16}, {26, 8}, {11, 8}} {{22, 20}, {28, 10}, {11, 10}}
{{22, 44}, {44, 22}, {11, 22}} {{22, 48}, {42, 32}, {21, 16}}
{{24, 18}, {9, 12}, {8, 6}} {{26, 28}, {28, 26}, {19, 2}}
{{28, 42}, {42, 28}, {21, 14}} {{28, 46}, {50, 20}, {7, 26}}
{{36, 22}, {42, 4}, {11, 18}} {{36, 30}, {15, 18}, {10, 12}}
{{38, 24}, {42, 16}, {21, 8}} {{44, 18}, {46, 12}, {23, 6}}
{{48, 36}, {18, 24}, {16, 12}}

2. Further nine groups contain also one triangle, but in each triangle two points coincide (again, they have isomorphic dynamical systems):

{{8, 2}, {8, 2}, {1, 4}} {{16, 2}, {16, 2}, {7, 4}}
{{16, 4}, {16, 4}, {2, 8}} {{24, 6}, {24, 6}, {3, 12}}
{{32, 8}, {32, 8}, {4, 16}} {{34, 8}, {34, 8}, {7, 16}}
{{46, 8}, {46, 8}, {17, 16}} {{48, 6}, {48, 6}, {21, 12}}
{{48, 12}, {48, 12}, {6, 24}}

3. There exist two groups with two triangles each (by observation of the geometrical pictures it is easy to determine that both have isomorphic dynamical systems):

{ {{2, 24}, {18, 16}, {9, 8}}, {{4, 48}, {36, 32}, {18, 16}} }
{ {{12, 26}, {26, 12}, {3, 14}}, {{26, 12}, {28, 6}, {13, 6}} }

4. Two further groups consist of two triangles each, but the common point is contained twice in one triangle (the dynamical systems are isomorphic, but different from the two groups above):

{ {{24, 22}, {32, 6}, {3, 16}}, {{32, 6}, {32, 6}, {11, 12}} }
{ {{8, 38}, {32, 22}, {11, 16}}, {{38, 8}, {38, 8}, {11, 16}} }

5. As we can see by inspecting their geometrical structures, further seven groups are not isomorphic to any group found above:

{ {{6, 12}, {12, 6}, {3, 6}}, {{12, 24}, {24, 12}, {6, 12}},
{{24, 48}, {48, 24}, {12, 24}} }

{ {{2, 16}, {14, 8}, {1, 8}}, {{4, 32}, {28, 16}, {2, 16}},
{{32, 4}, {32, 4}, {14, 8}} }

{ {{2, 4}, {4, 2}, {1, 2}}, {{4, 8}, {8, 4}, {2, 4}},
{{8, 16}, {16, 8}, {4, 8}}, {{16, 32}, {32, 16}, {8, 16}} }

{ {{4, 22}, {10, 20}, {11, 2}}, {{8, 44}, {20, 40}, {22, 4}},
{{10, 20}, {20, 10}, {5, 10}}, {{20, 40}, {40, 20}, {10, 20}} }

{ {{10, 40}, {26, 32}, {19, 8}}, {{26, 32}, {38, 16}, {13, 16}},
{{32, 26}, {40, 10}, {13, 16}}, {{40, 10}, {40, 10}, {5, 20}} }

{ {{4, 18}, {14, 12}, {7, 6}}, {{8, 36}, {28, 24}, {14, 12}},
{{12, 14}, {14, 12}, {9, 2}}, {{24, 28}, {28, 24}, {18, 4}},
{{36, 42}, {42, 36}, {27, 6}}, {{42, 36}, {21, 18}, {4, 18}} }

{ {{2, 36}, {20, 30}, {17, 6}}, {{4, 6}, {6, 4}, {3, 2}},
{{8, 12}, {12, 8}, {6, 4}}, {{12, 18}, {18, 12}, {9, 6}},
{{16, 24}, {24, 16}, {12, 8}}, {{18, 12}, {9, 6}, {4, 6}},
{{20, 30}, {30, 20}, {15, 10}}, {{20, 30}, {34, 12}, {1, 18}},
{{24, 36}, {36, 24}, {18, 12}}, {{30, 20}, {36, 2}, {1, 18}},
{{32, 48}, {48, 32}, {24, 16}}, {{34, 12}, {36, 2}, {15, 10}},
{{36, 24}, {18, 12}, {8, 12}}, {{45, 30}, {34, 12}, {12, 18}} }

Geometrical interpretation of all topological elements is given below. In cases when there exist more than one element with given structure, wavenumbers are written at the picture corresponding to the element chosen for presentation.

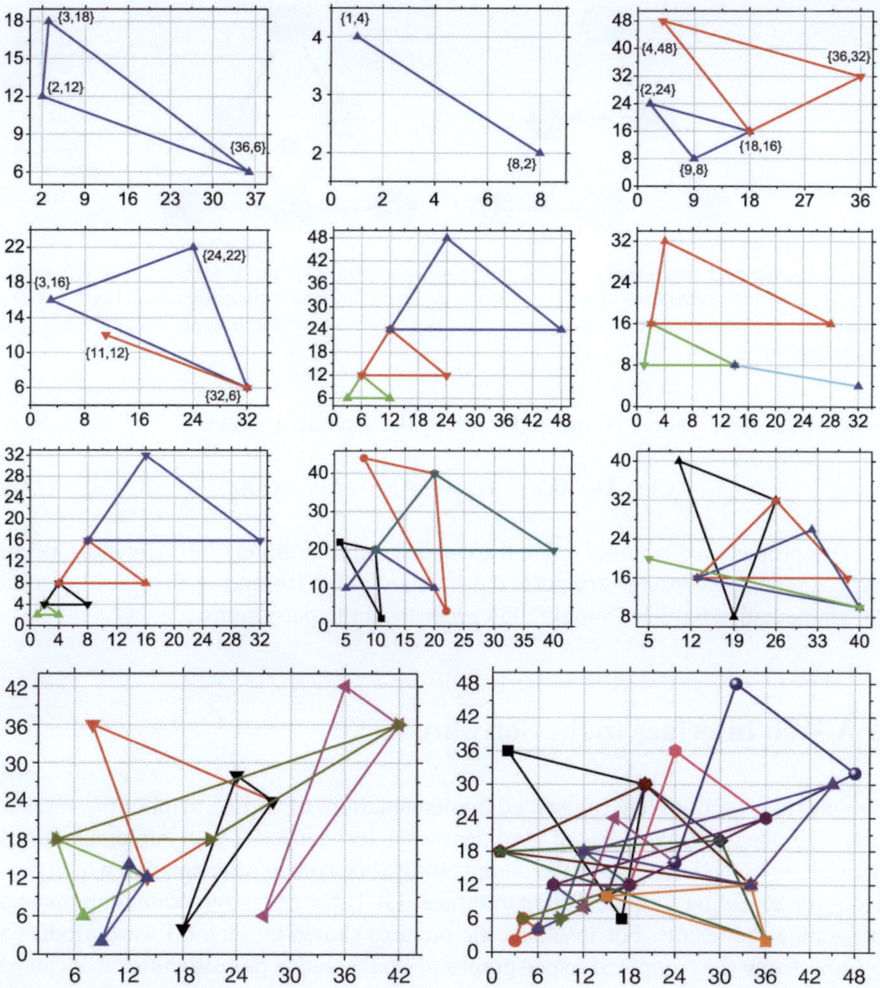

5.4 Important Remark

Computing all non-isomorphic sub-graphs algorithmically is a nontrivial problem. Indeed, all isomorphic graphs presented in previous section are described by similar dynamical systems, only magnitudes of interaction coefficients α_i vary. However, in the general case graph structure thus defined does not present the dynamical system unambiguously. Consider Fig. 2 below where two objects are isomorphic *as graphs*. However, the first object represents four connected triads with dynamical system

$$(A_1, A_2, A_3),\ (A_1, A_2, A_5),\ (A_1, A_3, A_4),\ (A_2, A_3, A_6) \tag{9}$$

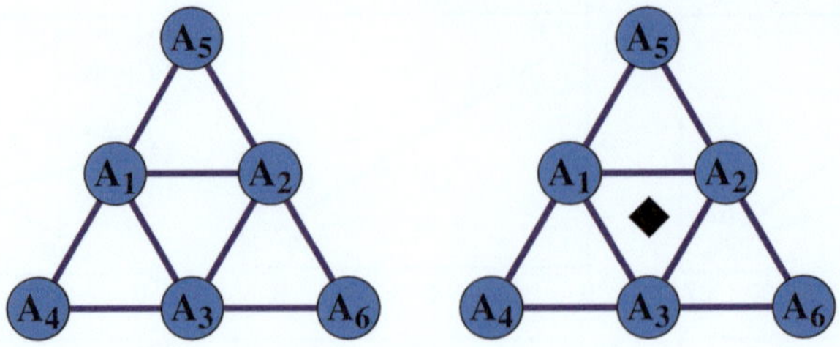

Fig. 2 Example of isomorphic graphs and non-isomorphic dynamical systems. The *left* graph corresponds to the dynamical system (9) and the graph on the *right*—to the dynamical system (10). To discern between these two cases, we set a placeholder inside the triangle not representing a resonance

while the second—three connected triads with dynamical system

$$(A_1, A_2, A_5), \ (A_1, A_3, A_4), \ (A_2, A_3, A_6). \tag{10}$$

This problem has been solved in Kartashova and Mayrhofer (2007) by introducing hyper-graphs of a special structure; the standard graph isomorphism algorithm used by Mathematica has been modified in order to suit hyper-graphs.

6 A Web Interface to the Software

The previous sections have presented implementations of various symbolic computation methods for the analysis of nonlinear wave resonances. These implementations are written in the language of the computer algebra system Mathematica which provides an appealing graphical user interface (GUI) for executing computations and presenting the results. For instance, the pictures shown in Sect. 4.3 were produced by converting the computed hyper-graphs to Mathematica plot structures that can be displayed by the GUI of the system.

However, to run these methods, the user needs an installation of Mathematica on the local computer with the previously described methods installed in a local directory. These requirements make access to the software difficult and hamper its wide-spread usage. In order to overcome this problem, we have implemented a Web interface such that the software can be executed from any computer connected to the Internet via a Web browser without the need for a local installation of mathematical software.

This implementation follows a general trend in computer science which turns away from stand alone software (that is installed on local computers and can be only executed on these computers via a graphical user interface) and proceeds towards *service-oriented software* (Gold et al. 2004) (that is installed on remove server computers and wraps each method into a service that can be invoked over the Internet via standardized Web interfaces). Various projects in computer mathematics have pursued middleware for *mathematical web services*, see for instance MathBroker (2007), MONET (2004), Baraka and Schreiner (2006). On the long term, it is thus envisioned that mathematical methods generally become remote services that can be invoked by humans (or other software) without requiring local software installations.

However, even without sophisticated middleware, it is nowadays relatively simple to provide (for restricted application scenarios) web interfaces to mathematical software by generally available technologies. The web interface presented in the following sections is deliberately kept as simple as possible and makes only use of such technologies; thus it should be easy to take this solution as a blueprint for other mathematical software with similar features. In particular, the web interface is quite independent of Mathematica as the system underlying the implementation of the mathematical methods; the same strategy can be applied to other mathematical software systems such as Maple, MATLAB, etc.

6.1 The Interface

Figure 3 shows the web interface to some of the methods presented in the previous sections. Its functionality is as follows:

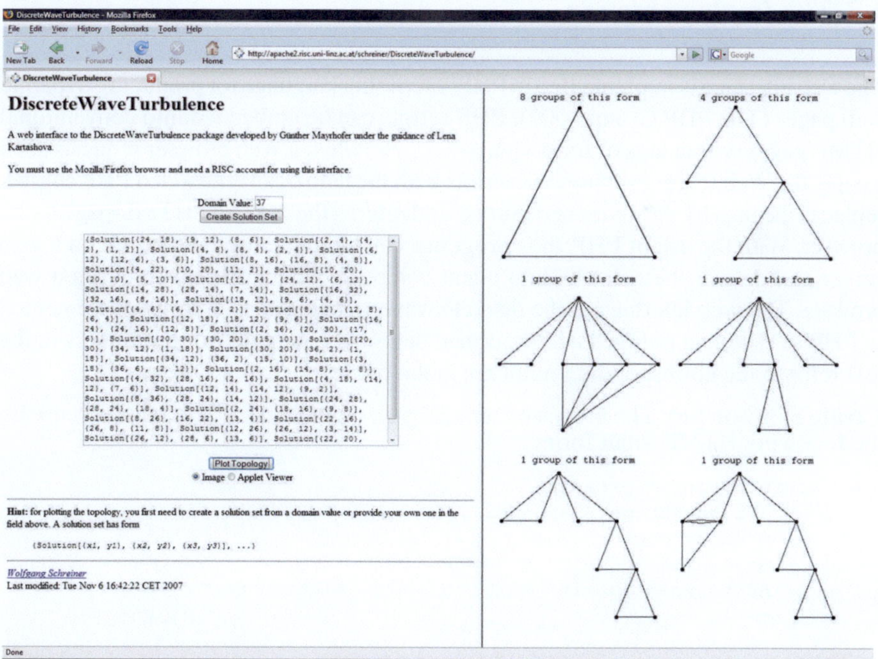

Fig. 3 Web interface to the implementation

Create Solution Set: The user may enter a parameter D in the first (small) text field and then press the button "Create Solution Set." This invokes the method `CreateSolutionSet` which computes the set of all solutions whose values are smaller than or equal to D. This set is written into the second (large) text field in the form

`{Solution[`x_1,y_1,z_1`],...,Solution[`x_n,y_n,z_n`]}`.

Plot Topology: The user may enter into the second (large) text field a specific solution set (or, as show above, compute one), and then press the button "Plot Topology." This first invokes the method `Topology` which computes the topological structure of the solution set as a list of hyper-graphs and then calls the method `PlotTopology` which computes a plot of each hyper-graph. The results are displayed in the right frame of the browser window.

The web interface is available at the URL

`http://www.risc.uni-linz.ac.at/projects/alisa`
(Button "Discrete Wave Turbulence")

To run the computations, an account and a password are needed.

6.2 The Implementation

The web interface is implemented in PHP, a scripting language for producing dynamic web pages (The PHP Group 2007). PHP scripts can be embedded into conventional HTML pages within tags of form `<php?...?>`; when a Web browser requests such a page, the Web server executes the scripts with the help of an embedded PHP engine, replaces the tags by the generated output, and returns the resulting HTML page to the browser. With the use of PHP, thus programs can be implemented that run on a web server and deliver their results to a client computer which displays them in a web browser. The web interface to the discrete wave turbulence package is implemented in PHP as sketched in Fig. 4 and described below (the parenthesized numbers in the text refer to the corresponding numbers in the figure).

Create Solution Set: The browser frame `input` on the left side contains essentially the following HTML input form:

```
<form target="textarea"
     action="https://apache2.../CreateSolutionSet.php"
     method="post">
   <input name="domain" size="3">
   <input type="submit" value="Create Solution Set">
</form>
```

This form consists of an input field `domain` to receive a domain value and a button to trigger the creation of the solution set. When the button is pressed, (1) a request is sent to the web server which carries the value of `domain`; this request asks the

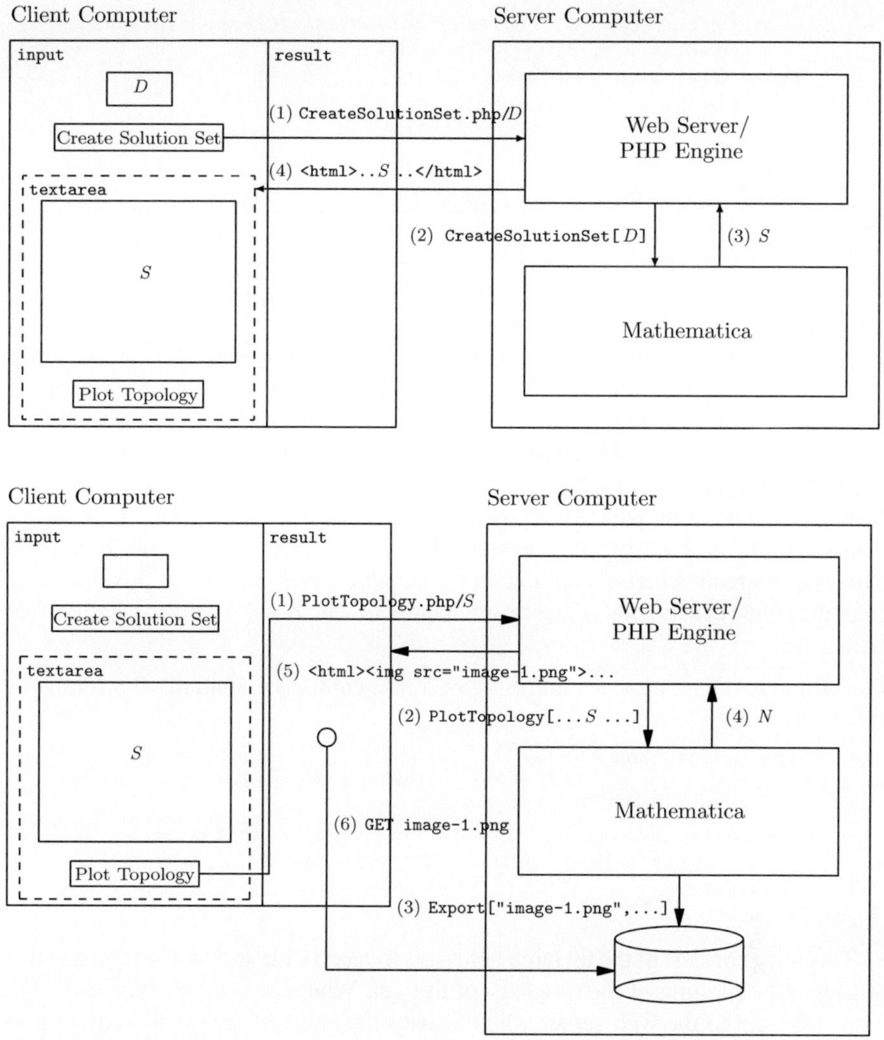

Fig. 4 Implementation of the web interface

server to deliver the PHP-enhanced web page `CreateSolutionSet.php` into the target frame `textarea` which is displayed internally to `input`.

The file `CreateSolutionSet.php` has essentially the content

```
<?php
  $math="/.../math";
  $cwd="/.../DiscreteWaveTurbulence";
  $domain = $_POST['domain'];
  $mcmd =
    "SetDirectory[\"". $cwd ."\"];" .
    "Needs[\"DiscreteWaveTurbulence`SolutionSet`\"];" .
```

```
      "sol=DiscreteWaveTurbulence`SolutionSet`CreateSolutionSet[" .
         $domain . "];";
    $command="$math -noprompt -run'" . $mcmd .
       "Print[StandardForm[sol]]; Quit[];'";
    $result = shell_exec("$command");
    echo
      ...
      "<textarea  name=\"sol\"cols=\"60\"rows=\"20\">" .
      htmlspecialchars($result) .
      "</textarea>" .
     ...;
?>
```

After setting the paths `$math` of the Mathematica binary and `$cwd` of the directory where the `DiscreteWaveTurbulence` package is installed, the script sets the local variable `$domain` to the value of the input field `domain`. Then the Mathematica command `$mcmd` is constructed in order to load the file `SolutionSet.m` and execute the command `CreateSolutionSet` to compute the solution set. Now the system command `$command` is constructed to (2) invoke Mathematica which calls the previously constructed command and (3) prints its result to the standard output stream which is captured in the variable `$result`. From this, the script constructs the HTML code of the result document which is (4) delivered to the Web browser.

Plot Topology: The browser frame `textarea` contains essentially the following HTML input form:

```
<form target="result"
    action="https://apache2..../PlotTopology.php"
    method="post">
  <textarea name="sol" cols="60" rows="20">...</textarea>
  <input type="submit" value="Plot Topology">
  </center>
</form>
```

This form consists of the textarea field `sol` to receive the solution set and a button to trigger the plotting of the topology of this set. When the button is pressed, (1) a request is sent to the web server which carries the value of `sol`; this request asks the server to deliver the PHP-enhanced web page `PlotTopology.php` into the target frame `result` on the right side of the browser.

The file `CreateSolutionSet.php` has essentially the content

```
<?php
   $math="/.../math";
   $basedir ="/.../DiscreteWaveTurbulence";
   $baseurl ="http://apache2/.../DiscreteWaveTurbulence";
   $sol = $_POST['sol'];
   ... // create under $basedir a unique subdirectory $dir
   $mcmd =
     "SetDirectory[\"$basedir/$dir\"];" .
     "Needs[\"DiscreteWaveTurbulence`Topology`\"];" .
     "Needs[\"DiscreteWaveTurbulence`SolutionSet`\"];" .
     "top=DiscreteWaveTurbulence`Topology`Topology[$sol];" .
```

```
    "plots=DiscreteWaveTurbulence`Topology`PlotTopology1[top];";
  $command="/usr/bin/Xvnc :20 & export DISPLAY=:20;" .
    "export MATHEMATICA_USERBASE=$basedir/.Mathematica;" .
    "$math -run'" . $mcmd .
    "Print[ExportList[plots,\"$image\"]]; Quit[];'";
  $result = shell_exec("$command | tail -n 1");
  for ($i=0;$i<$result;$i++)
    echo "<img src=\"$baseurl/$dir/image-$i.png\"/>";
?>
```

For holding the images to be generated later, the script creates a unique directory `$basedir/$dir` which is served by the web server under the url `$baseurl/$dir`. The script extracts the solution set `$sol` from the request and sets up the Mathematica command to compute its topological structure and generate the plots from which ultimately the image files will be produced.

For this purpose, however, Mathematica needs an X11 display server running; since a Web server has not access to an X11 server, we start the virtual X11 server Xvnc (RealVNC Remote Control Software 2007) as a replacement and set the environment variable `DISPLAY` to the display number on which the number listens; Mathematica will subsequently send X11 requests to that display which will be handled by the virtual server. Likewise, Mathematica needs access to a `.Mathematica` configuration directory; the script sets the environment variable `MATHEMATICA_USERBASE` correspondingly.

With these provisions, we can (2) invoke first the command to compute the plots and then the (self-defined) command `ExportList` to generate for every plot an image in the previously created directory. For this purpose, the command uses (3) the Mathematica command `EXPORT[`*`file`*`,`*`plot`*`,"PNG"]` which converts *plot* to an image in PNG format and writes the image to *file*. `ExportList` returns the number of images generated which is (4) written to the standard output stream which in turn is captured in the variable `$result`. From this information, the script generates an HTML document which contains a sequence of `img` elements referencing these images. After this document has been (5) returned to the client browser, the browser (6) requests the referenced images with `GET` messages from the web server.

6.3 Extensions

As an alternative to the display of static images, the Web interface also provides an option "Applet Viewer" with somewhat more flexibility. If this option is selected, Mathematica is instructed to save all generated plots as files in the standard representation. The generated HTML document then embeds (rather than `img` elements) a sequence of `applet` elements that load instances of the "JavaView" applet (The JavaView Project 2007). These applets run in the Java Virtual Machine of the Web browser on the client computer, load the plot files from the web, and visualize them in the browser. Rather than just displaying static images, the viewer allows to perform certain manipulations and transformations of the plots such as scaling, rotating, etc. While this additional flexibility is not of particular importance for the presented methods, they may in the future become useful for others.

To limit access to the software respectively to the computing power of the server computer, it may be protected by authentication mechanisms. For example, on the Apache Web server, it suffices to provide in the installation directory of the software a file `.htaccess` with content

```
<Files"*.php">
  SSLRequireSSL
  AuthName "your account"
  AuthType Basic
  Require valid-user
</Files>
```

With this configuration, the user is asked for the data of a valid account on the computer running the Web server; other authentication mechanisms based e.g. on password files may be provided in a similar fashion.

7 Discussion

Summing up all the results obtained, we would like to make some concluding remarks.

- In general, coefficients α_i can be computed symbolically by hand and only numerically by Mathematica (see Sect. 3.3); at present we are not aware of the possibility to overcome this problem.
- For the known case of spherical barotropic vorticity equation, values of coefficients α_i coincide with known form the literature for all triads but three. These thee triads, though satisfying resonant conditions, are known to be special from the physical point of view in the following sense (see Kartashova and L'vov 2007 for details). Although resonance conditions are fulfilled for the waves of these triads, they, so to say, do not have a place in the physical space to interact and their influence (if any) on the dynamics of the wave system has to be studied separately from all other waves. Our results might indicate that also the coefficients α_i of these triads have to be defined in some other way compare to other resonant triads. For instance, another way of space-averaging has to be chosen.
- The results of Sect. 3.4.2 show that analytical formulae given in Kartashova and Reznik (1992) for α_j *are not correct.*
- The results of Sect. 4.3 show a crucial dependence of the number of solutions on the form of the boundary conditions. In particular, some boundary conditions (for example, $(L_x, L_y) = (11, 29)$) yield *no solutions* which is of most importance for physical applications. From the mathematical point of view, an interesting result has been observed: in all our computations (i.e. for $m, n \leq 300$) indexes corresponding to non-empty classes turned out to be *odd.* It would be interesting to prove this fact analytically because if it keeps true, we can reduce the computational time.

- The algorithm presented in Sect. 4 has been implemented before numerically in Visual Basic, and our purpose here was to show that it works fast enough also in Mathematica. The algorithms presented in Sects. 3 and 5 *have never been implemented before*, the whole work is usually done by hand and some mistakes as in Kartashova and Reznik (1992) are almost unavoidable: it takes sometimes a few weeks of skillful researchers to compute interaction coefficients of dynamical systems for one specific wave system.
- All the algorithms presented above can easily be modified for the case of a four-term mesoscopic system. The only problem left is a procedure to establish all non-isomorphic topological elements for a quadruple graphs, similar to the procedure given in Kartashova and Mayrhofer (2007) for a triangle graphs. The structure of quadruple graphs is much more complicated while some mechanisms of energy transfer in the spectral space do exist (Kartashova 2007) that are absent in three-term mesoscopic systems. A complete classification of quadruple graphs is still an open question but in a given spectral domain it can be done directly (a very time-consuming operation).
- We have developed a Web interface for the presented methods, which turns the implementations from only locally available software to Web-based services that can be accessed from any computer in the Internet that is equipped with a Web browser. The presented implementation strategy is simple and based on generally available technologies; it can be applied as a blueprint for a large variety of mathematical softwares. In particular, the results are not bound to the current Mathematica implementation but can be adapted to any other computer algebra system (e.g. Maple) or numerical software system (e.g. MATLAB) of similar expressiveness.
- At present, an explicit form of eigen-modes (6), (7) is used as one of the input parameters for our program package. Theoretically, at least for some classes of linear partial differential operators and boundary conditions, computing eigen-modes can also be performed symbolically basing on the results in Rosenkranz (2005). If this were done, not an eigen-mode but boundary conditions would play role of input parameter.

Acknowledgments Authors acknowledge the support of the Austrian Science Foundation (FWF) under projects SFB F013/F1301 "Numerical and symbolical scientific computing," P20164-N18 "Discrete resonances in nonlinear wave systems," and P17643-NO4 "MathBroker II: Brokering Distributed Mathematical Services."

References

Baraka R, Schreiner W (2006) Semantic querying of mathematical web service descriptions. In: Bravetti M et al. (eds) Third international workshop on web services and formal methods (WS-FM 2006), Vienna, Austria, September 8–9, 2006. (Lecture Notes in Computer Science 4184, pp. 73–87. Springer)

Berman GP, Israilev FM (2005) The Fermi-Pasta-Ulam problem: fifty years of progress. Chaos 15(1):015104–015104-18

Cheney M (1989) Tesla man out of time. Dorset Press, UK

Gold N, Mohan A, Knight C, Munro M (2004) Understanding service-oriented software. IEEE Softw 21(2):71–77 March-April 2004

Kartashova EA (1998) Wave resonances in systems with discrete spectra. In: Zakharov VE (ed) Nonlinear Waves and Weak Turbulence. pp 95–129 (Series: Advances in the Mathematical Sciences, AMS, 1998)

Kartashova E (2006a) A model of laminated turbulence. JETP Lett 83(7):341–345

Kartashova E (2006b) Fast computation algorithm for discrete resonances among gravity waves. Low Temp Phys 145(1–4):286–295

Kartashova E (2007) Exact and quasi-resonances in discrete water-wave turbulence. Phys Rev Lett 98(21):214502

Kartashova E, Kartashov A (2006) Laminated wave turbulence: generic algorithms I. Int J Mod Phys C 17(11):1579–1596

Kartashova E, Kartashova A (2007a) Laminated wave turbulence: generic algorithms II. Commun Comput Phys 2(4):783–794

Kartashova E, Kartashova A (2007b) Laminated wave turbulence: generic algorithms III. Physics A: Stat Mech Appl 380:66–74

Kartashova E, L'vov VS (2007) A model of intra-seasonal oscillations in the Earth atmosphere. Phys Rev Lett 98(19):198501, (featured in Nature Physics 3(6):368

Kartashova E, Mayrhofer G (2007) Cluster formation in mesoscopic systems. Physica A: Stat Mech Appl 385:527–542

Kartashova EA, Reznik GM (1992) Interactions between Rossby waves in bounded regions. Oceanology 31:385–389

MathBroker II (2007): Brokering distributed mathematical services reseach institute for symbolic computation (RISC). http://www.risc.uni-linz.ac.at/projects/mathbroker2

MONET (2004) Mathematics on the web. The MONET Consortium, (April 2004). http://monet.nag.co.uk

Nayfeh AN (1981) Introduction to perturbation techniques. Wiley-Interscience, NY

Pedlosky J (1987) Geophysical fluid dynamics. Springer, New York

Pushkarev AN, Zakharov VE (2000) Turbulence of capillary waves—theory and numerical simulations. Physica D 135:98–116

Real VNC remote control software (2007). VNC free edition 4.1. http://www.realvnc.com/products/free/4.1

Rosenkranz M (2005) A new symbolic method for solving linear two-point boundary value problems on the level of operators. J Symb Comput 39:171–199

Tanaka M (2007) On the role of resonant interactions in the short-term evolution of deep-water ocean spectra. J Phys Oceanogr 37:1022–1036

The Java view project (2007). Java view—interactive 3D geometry and visualization. http://www.javaview.de

The PHP group (2007). PHP: hypertext preprocessor. http://www.php.net

Zakharov VE, L'vov VS, Falkovich G (1992) Kolmogorov spectra of turbulence. Series in nonlinear dynamics, Springer, New York

Zakharov V, Dias F, Pushkarev A (2004) One-dimensional wave turbulence. Phys Rep 398:1–65

Zakharov VE, Korotkevich AO, Pushkarev AN, Dyachenko AI (2005) Mesoscopic wave turbulence. JETP Lett 82(8):491

Zakharov VE, Filonenko NN (1967) Weak turbulence of capillary waves. J Appl Mech Tech Phys 4:500–515

Index

E. Pelinovsky and C. Kharif (eds.), *Extreme Ocean Waves*,
DOI 10.1007/978-3-319-21575-4

MIX
Papier aus verantwortungsvollen Quellen
Paper from responsible sources
FSC® C105338
FSC
www.fsc.org

If you have any concerns about our products,
you can contact us on
ProductSafety@springernature.com

In case Publisher is established outside the EU,
the EU authorized representative is:
Springer Nature Customer Service Center GmbH
Europaplatz 3, 69115 Heidelberg, Germany

Printed by Libri Plureos GmbH
in Hamburg, Germany